Adaptive Detection of Multichannel Signals Exploiting Persymmetry

This book offers a systematic presentation of persymmetric adaptive detection, including detector derivations and the definitions of key concepts, followed by detailed discussions relating to theoretical underpinnings, design methodology, design considerations, and techniques enabling its practical implementation.

The received data for modern radar systems are usually multichannel, namely, vector-valued, or even matrix-valued. Multichannel signal detection in Gaussian backgrounds is a fundamental problem for radar applications. With an overarching focus on persymmetric adaptive detectors, this book presents the mathematical models and design principles necessary for analyzing the behavior of each kind of persymmetric adaptive detector. Building upon that, it also introduces new design approaches and techniques that will guide engineering students as well as radar engineers toward efficient detector solutions, especially in challenging sample-starved environments where training data are limited.

This book will be of interest to students, scholars, and engineers in the field of signal processing. It will be especially useful for those who have a solid background in statistical signal processing, multivariate statistical analysis, matrix theory, and mathematical analysis.

Jun Liu is an Associate Professor with the Department of Electronic Engineering and Information Science, University of Science and Technology of China. Dr. Liu is a member of the Sensor Array and Multichannel (SAM) Technical Committee, IEEE Signal Processing Society.

Danilo Orlando is an Associate Professor at Università degli Studi "Niccolò Cusano". His research interests focus on signal processing for radar and sonar systems. He has co-authored more than 150 publications in international journals, conferences, and books.

Chengpeng Hao is a Professor at the Institute of Acoustics, Chinese Academy of Sciences. His research interests are in the fields of statistical signal processing, array signal processing, radar, and sonar engineering. He has authored and co-authored more than 100 scientific publications in international journals and conferences.

Weijian Liu is an Associate Professor with the Wuhan Electronic Information Institute, China. His research interests include multichannel signal detection and statistical and array signal processing.

Adaptive Detection of Multichannel Signals Exploiting Persymmetry

Jun Liu
Danilo Orlando
Chengpeng Hao
Weijian Liu

CRC Press
Taylor & Francis Group
Boca Raton London New York

CRC Press is an imprint of the
Taylor & Francis Group, an **informa** business

This book is published with financial support from the National Natural Science Foundation of China under Grants 61871469, 61971412, 62071482, the Youth Innovation Promotion Association CAS (2019447), and the Anhui Provincial Natural Science Foundation under Grant 2208085J17.

First edition published 2023
by CRC Press
6000 Broken Sound Parkway NW, Suite 300, Boca Raton, FL 33487-2742

and by CRC Press
4 Park Square, Milton Park, Abingdon, Oxon, OX14 4RN

CRC Press is an imprint of Taylor & Francis Group, LLC

ISBN: 978-1-032-37424-6 (hbk)
ISBN: 978-1-032-37427-7 (pbk)
ISBN: 978-1-003-34023-2 (ebk)

DOI: 10.1201/9781003340232

Typeset in CMR10 font
by KnowledgeWorks Global Ltd.

Publisher's note: This book has been prepared from camera-ready copy provided by the authors.

In loving memory of my father, Houcheng Liu. To his greatness and uniqueness.
– Jun Liu

To my lovely family.
– Danilo Orlando

To my beloved parents.
– Chengpeng Hao

To my wife and daughter.
– Weijian Liu

Contents

List of Abbreviations

ACE: adaptive coherence estimator

AMF: adaptive matched filter

ASD: adaptive subspace detector

BIC: Bayesian information criterion

CCDF: complementary cumulative distribution function

CDF: cumulative distribution function

CFAR: constant false alarm rate

CNR: clutter-to-noise ratio

CPI: coherent processing interval

CUT: cell under test

DOA: direction of arrival

EM: expectation maximization

FIM: Fisher information matrix

GLRT: generalized likelihood ratio test

HE: homogeneous environment

ICM: interference covariance matrix

IID: independent and identically distributed

INR: interference-to-noise ratio

KA: knowledge-aided

LMPID: locally most powerful invariant detector

LRT: likelihood ratio test

MAP: maximum a posteriori

MC: Monte Carlo

MIMO: multiple-input multiple-output

MIS: maximal invariant statistic

MIT: Massachusetts Institute of Technology

MLE: maximum likelihood estimate

MOS: model order selection

MPID: most powerful invariant detector

PDF: probability density function

PD: probability of detection

PFA: probability of false alarm

PHE: partially homogeneous environment

PRF: pulse repetition frequency

PRI: pulse repetition interval

RCS: radar cross section

RE: relative error

RMB: Reed, Mallett, and Brenann

ROC: receiver operating characteristic

SCM: sample covariance matrix

SCR: signal-to-clutter ratio

SINR: signal-to-interference-plus-noise ratio

SMI: sample matrix inversion

SNR: signal-to-noise ratio

SNT: spectral norm test

ULA: uniform linear array

List of Symbols

x: boldface lowercase letters denote vectors

X: boldface uppercase letters denote matrices

$(\cdot)^*$: the complex conjugate operator

$(\cdot)^T$: the transpose operator

$(\cdot)^\dagger$: the complex conjugate transpose operator

∇: the gradient operator

Σ: the cumulative operator

$\prod$: the product operator

$(\cdot)^M$: the Moore-Penrose pseudoinverse of a square matrix

$\otimes$: the Kronecker product

$\mathbb{N}$: the set of natural numbers

$\mathbb{R}$: the set of real numbers

$\mathbb{R}^+$: the set of positive real numbers

$\mathbb{R}^{N \times M}$: the Euclidean space of ($N \times M$)-dimensional real matrices (or vectors if $M = 1$)

$\mathbb{C}$: the set of complex numbers

$\mathbb{C}^{N \times M}$: the Euclidean space of ($N \times M$)-dimensional complex matrices (or vectors if $M = 1$)

$\mathfrak{Re}\{\cdot\}$: the real part of a complex number

$\mathfrak{Im}\{\cdot\}$: the imaginary part of a complex number

$|\cdot|$: the absolute value of a real number or modulus a complex number

$\|\cdot\|$: the Euclidean norm of a vector

$\jmath$: the imaginary unit, i.e., $\jmath = \sqrt{-1}$

$\mathbf{C}_n^m$: the binomial coefficient, i.e., $\mathrm{C}_n^m = \frac{n!}{m!(n-m)!}$

$X \propto Y$: X is proportional to Y

$\mathbf{A} > \mathbf{B}$: $\mathbf{A} - \mathbf{B}$ is positive definite

$\mathbf{A} \geq \mathbf{B}$: $\mathbf{A} - \mathbf{B}$ is positive semidefinite

$\mathbf{I}_N$ or $\mathbf{I}$: the identity matrix of dimension $N \times N$ or of proper dimensions

$\mathbf{0}$: the null vector or matrix of proper dimensions

$\mathbf{S}_{++}$: the set of positive definite symmetric matrices

$\mathbf{A}(k, l)$: the (k, l)-entry of a generic matrix $\mathbf{A}$

$\mathbf{a}(l)$: the l-entry of a generic vector $\mathbf{a}$

$\overset{\mathrm{d}}{=}$: the random quantities on both sides of the equation have the same distribution

$\triangleq$: is defined as

$\sim$: be distributed as

$\mathrm{vec}(\mathbf{A})$: the column vector formed by stacking the columns of $\mathbf{A}$

$\mathbf{x} \sim \mathcal{CN}_N(\mathbf{m}, \mathbf{R})$: $\mathbf{x} \in \mathbb{C}^{N \times 1}$ is a circular symmetric, N-dimensional complex Gaussian vector with mean $\mathbf{m} \in \mathbb{C}^{N \times 1}$ and covariance matrix $\mathbf{R} \in \mathbb{C}^{N \times N} > \mathbf{0}$

$\mathbf{x} \sim \mathcal{N}_N(\mathbf{m}, \mathbf{R})$: $\mathbf{x} \in \mathbb{R}^{N \times 1}$ is an N-dimensional real Gaussian vector with mean $\mathbf{m} \in \mathbb{R}^{N \times 1}$ and covariance matrix $\mathbf{R} \in \mathbb{R}^{N \times N} > \mathbf{0}$

$\mathbf{X} \sim \mathcal{N}_{n \times m}(\mathbf{M}, \mathbf{R}_n \otimes \mathbf{R}_m)$: the random matrix $\mathbf{X} \in \mathbb{R}^{n \times m}$ has a real matrix variate Gaussian distribution with mean matrix $\mathbf{M} \in \mathbb{R}^{n \times m}$ and covariance matrix $\mathbf{R}_n \otimes \mathbf{R}_m$ where $\mathbf{R}_n \in \mathbb{R}^{n \times n}$ and $\mathbf{R}_m \in \mathbb{R}^{m \times m}$, if $\mathrm{vec}(\mathbf{X}^T) \sim \mathcal{N}_{nm}(\mathrm{vec}(\mathbf{M})^T, \mathbf{R}_n \otimes \mathbf{R}_m)$.

$\mathbf{R} \sim \mathcal{CW}^{-1}(\mathbf{R}_0, v)$: $\mathbf{R}$ has an inverse complex Wishart distribution with v degrees of freedom and scale matrix $\mathbf{R}_0$

χ_n^2: real central Chi-squared distribution with n degrees of freedom

$\chi_n'^2(\zeta)$: real non-central Chi-squared distribution with n degrees of freedom and non-centrality parameter ζ

$F_{n,m}$: real central F-distribution with n and m degrees of freedom

$F_{n,m}(\delta)$: real non-central F-distribution with n and m degrees of freedom and noncentrality parameter δ

$\beta(a, b)$: real Beta distribution of parameters a and b

$U_{n,m,j}$: central real Wilks' distribution with n, m, and i being the number of variates, the error degrees of freedom, and the hypothesis degrees of freedom, respectively.

$\mathbf{M} \sim \mathcal{W}_n(v, \mathbf{M}_0)$: $\mathbf{M}$ has an n-dimensional real Wishart distribution with v degrees of freedom and scale matrix $\mathbf{M}_0$

$\ln(\cdot)$: the natural logarithm function

$\exp(\cdot)$: the exponential function

$\max(\cdot)$: the maximum function

$\min(\cdot)$: the minimum function

$\Gamma(\cdot)$: the Gamma function

$B(a, b)$: the Beta function

$_1F_1(\cdot; \cdot; \cdot)$, $_2F_1(\cdot, \cdot; \cdot; \cdot)$: the hypergeometric function

$F_1(\cdot, \cdot, \cdot, \cdot; \cdot, \cdot)$: the Appell hypergeometric function

$\mathcal{O}(n)$: the implementation requires a number of flops proportional to n

$\lfloor \cdot \rfloor$: the largest integer less than or equal to the argument

$\lceil \cdot \rceil$: the smallest integer greater than or equal to a given number

$E[\cdot]$: the statistical expectation

$\mathcal{P}(\cdot)$: probability of a random event

$\mathcal{P}_e(\cdot)$: the principal eigenvector

$\frac{dh(x)}{dx}$: the gradient of $h(\cdot) \in \mathbb{R}$ with respect to x

$\frac{\partial h(x,y)}{\partial x}$ or $\frac{\partial h(x,y)}{\partial y}$: the partial derivatives of $h(\cdot) \in \mathbb{R}$ with respect to x or y

$\frac{\partial f(\mathbf{x})}{\partial \mathbf{x}}$: the gradient of $f(\cdot) \in \mathbb{R}$ with respect to $\mathbf{x}$ arranged in a column vector

$\frac{\partial f(\mathbf{x})}{\partial \mathbf{x}^T}$: the gradient of $f(\cdot) \in \mathbb{R}$ with respect to $\mathbf{x}$ arranged in a row vector

$\frac{\partial f(\hat{\mathbf{x}})}{\partial \mathbf{x}}$: the gradient of $f(\cdot) \in \mathbb{R}$ with respect to $\mathbf{x}$ evaluated at $\hat{\mathbf{x}}$

$\lim_{x \to \pm\infty} h(x)$: the limit values of $h(x)$ with respect to $\pm\infty$

$\det(\cdot)$: the determinant of a square matrix

$\mathrm{tr}(\cdot)$: the trace of a square matrix

$\mathrm{diag}(\mathbf{A})$: the diagonal matrix with the diagonal entries of $\mathbf{A}$

$\mathbf{P_B}$: the orthogonal projection matrix onto the subspace spanned by the columns of the matrix $\mathbf{B}$, i.e., $\mathbf{P_B} = \mathbf{B}(\mathbf{B}^{\dagger}\mathbf{B})^{-1}\mathbf{B}^{\dagger}$

$\mathbf{P_B^{\perp}} = \mathbf{I} - \mathbf{P_B}$: the orthogonal projection matrix onto the subspace orthogonal to that spanned by the columns of the matrix $\mathbf{B}$

$\lambda_i(\mathbf{A})$: the ith largest eigenvalue of the matrix $\mathbf{A}$

$\lambda_{\max}(\mathbf{A})$: the maximum eigenvalue of the matrix $\mathbf{A}$

$\arg\max_x f(x)$: the argument x which maximizes $f(x)$

$\arg\min_x f(x)$: the argument x which minimizes $f(x)$

H_1: the target-present hypothesis

H_0: the target-absent hypothesis

1

Basic Concept

1.1   Multichannel Radar

As an electromagnetic-based system, "radar" was stemmed from an acronym of RAdio Detection And Ranging (RADAR). It finds wide applications in various fields so that radar has been accepted as a standard English word. A radar system is usually equipped with a transmitter and a receiver. The transmitter sends electromagnetic waves to cover a region of interest, and then the receiver collects echoes reflected from this region. The collected echoes are usually composed of target and unwanted signals (such as clutter, interference, and thermal noise). The primary function of radar is to detect the target of interest based on the collected data.

Early radar systems use a single channel (e.g., one pulse or one antenna), and as a result the collected data are scalar-valued. In contrast, a modern radar system is usually equipped with multiple coherent pulses and/or multiple antenna elements [1, 2], due to the applications of pulsed Doppler techniques and/or multiple transmit/receive (T/R) modules, along with the increase in computation power and advances in hardware design. This kind of system is often referred to as multichannel radar, where data are collected from multiple temporal, spatial, or spatial-temporal channels. As a result, the received data are vector-valued or even matrix-valued.

1.2   Adaptive Detection of Multichannel Signal

Multichannel signal detection in radar has received significant interest [3, 4] due to two main reasons. On the one hand, more information is involved in multichannel data compared to single-channel data, and hence more degrees of freedom are provided to develop advanced detection architectures. On the other hand, the correlation between different channels can be employed to devise a filter to suppress unwanted components in the received data, which is very helpful to enhance SNR and improve radar detection performance.

Since the location of a target to be detected is generally unknown in practice, a grid search is often performed, which divides the desired radar

DOI: 10.1201/9781003340232-1

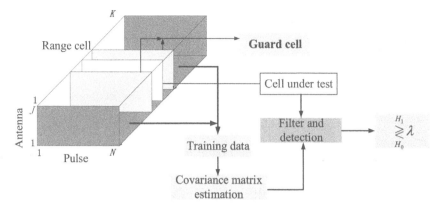

FIGURE 1.1
Detection architecture.

surveillance area into many (range) cells or bins. We need to test each cell one by one to decide whether an interested target is present or not. The data in the CUT are referred to as the test (or primary) data. In practice, noise is ubiquitous, which, in a general sense, possibly includes interference, clutter, and thermal noise.

If all parameters are known *a priori*, the Neyman-Pearson criterion can be used to design the best detector which exhibits the highest PD for a given PFA. However, the noise covariance matrix for multichannel data in the CUT is in practice unknown and has to be estimated. The Neyman-Pearson criterion cannot be used to design detectors in such a case.

A typical diagram for multichannel signal detection in noise with unknown covariance matrix is shown in Fig. 1. A standard approach for the covariance matrix estimation is to employ a set of IID training (or secondary) data samples which are assumed to contain noise only. In practice, these training data samples are often collected from range cells adjacent to the CUT. Usually, guard cells are set between the test and training data to avoid target signal contamination in the noise covariance matrix estimate. When the test data and training data have the same covariance matrix, it is commonly referred to as HE. In contrast, when the primary and secondary data share the same covariance matrix structure but with unknown power mismatch, it is often referred to as PHE [5,6].

The test and training data are jointly used to design various detectors for adaptive detection of multichannel signal. A target is declared to be present (or absent) if the value of test statistic exceeds (or is lower than) a threshold. It is desired that a detection strategy is adaptive to unknown noise spectral characteristic which usually changes in operational environments. It means that a desirable strategy should possess a CFAR property with respect to nuisance parameters (e.g., noise power and covariance matrix), namely, the PFA is irrelevant to noise parameters [7]. For a preassigned PFA, the CFARness of a test gives the possibility to set the detection threshold independent of unknown noise parameters [2].

Adaptive processing uses training data collected from the neighborhood of the CUT to estimate the noise covariance matrix in the CUT. It is known that the estimation error is non-negligible when the number of training data is limited. A fundamental question in adaptive processing is that how many training data are required to ensure the performance loss below a preassigned level. It is proved by RMB in [8] that the adaptive processing loss caused by this estimation error is more than 3 dB when the number of training data is less than twice the dimension of the received data. This is the famous RMB's rule, which can be served as a thumb of rule on the requirement of training data. For adaptive detection, to ensure the SNR loss (defined as the required additional SNR for an adaptive detector achieving the same PD with the optimum detector) within 3 dB, the required number of training data is larger than that for adaptive filtering [4, 9].

There are three widely used criteria for adaptive CFAR detector design: GLRT, Wald test, and Rao test, as presented in Appendix 1.A. In [9], Kelly designed a GLRT detector for the multichannel signal detection problem in the HE. The test statistic of the GLRT is obtained by replacing all unknown parameters with their MLEs in the likelihood ratio. In [10], Robey *et al.* devised the AMF in the HE according to the design criterion of two-step GLRT. Specifically, the test statistic of the AMF is firstly obtained by assuming noise covariance matrix is known in the likelihood ratio, then the unknown covariance matrix is replaced by the estimated covariance matrix with the secondary data. Although the performance of the detector designed by the two-step method may be worse than that of the detector designed by the one-step method, the AMF is more computationally efficient than the GLRT. Besides, the performance loss is bearable; hence the AMF is widely used in practice. De Maio derived two adaptive detectors in the HE according to the criteria of the Wald test and the Rao test in [11] and [12], respectively, and the Wald test is found to be the same as the AMF. Remarkably, these detectors bear CFAR properties with respect to the noise covariance matrix. In order to flexibly adjust between selectivity and robustness, researchers have proposed many detection algorithms in the past. For example, two-stage detectors combining different classical adaptive detectors were proposed to achieve flexible selectivity through choosing an appropriate threshold pair [13].

In practice, there may exist a mismatch in the target steering vector, e.g., due to beam-pointing errors and multipath [13]. To account for the uncertainty in the target steering vector, a subspace model is widely used in open literature [6, 14–20]. More specifically, in the subspace model, the target signal is expressed as the product of a known full-column-rank matrix and an unknown column vector. It means that the target signal lies in a known subspace spanned by the columns of a matrix, but its exact location is unknown since the coordinate vector is unknown. More detailed explanations about the subspace model can be found in [15–17]. In [17], several matched subspace detectors were developed in Gaussian clutter, where the clutter covariance matrix is assumed known. The authors in [21] designed a one-step GLRT detector for detecting subspace signals when the clutter covariance matrix is unknown.

In some situations of practical interest, the environments are not always homogeneous, due to environmental and instrumental factors. A class of widely used nonhomogeneity is partial homogeneity, for which the primary and secondary data share the same structure of the covariance matrix up to an unknown scaling factor. The introduction of the scaling factor is to improve robustness to the variations of the noise power levels between the test and training data. In [22, 23], the ACE was proposed to address the target detection problem in the PHE. This detector was also independently obtained for the detection problem in compound-Gaussian clutter [22]. It is proved in [24] that in the PHE the ACE coincides with the detectors derived according to the design criteria of Rao test and Wald test. The detection problem with subspace interference in partially homogeneous environment is investigated in [15, 25, 26].

It is required in the above studies that the number of training data is sufficient for acceptable accuracy in noise covariance matrix estimation. In practice, however, this requirement may be prohibitive due to two reasons. First, the dimension of the received data may be very high, e.g., in the space-time adaptive processing where hundreds of antennas and pulses are employed [1]. It leads to a challenge that the homogeneous training data available are much less than what are required. Second, the number of homogeneous training data available is small due to outliers and environmental heterogeneity [27]. It may be less than the requirement, even though the dimension of the received data is not high. So, this sample-starved environment is a more interesting and also practically motivated case.

Dimension reduction and *a priori* information-based method are two main approaches to alleviate the requirement of sufficient IID training data. In dimension reduction, a transformation is applied to the test and training data prior to adaptive processing, which has the effect of projecting the noise covariance matrix onto a low-dimension subspace. As a result, the required number of IID training data can be reduced considerably, and computational complexity is also reduced. Various reduced-dimension approaches have been proposed, e.g., the space-time multiple-beam method [28], alternating low-rank decomposition [29], and modified GLRT [30].

Prior information-based methods include Bayesian methods, parametric methods, and structure-based methods. For Bayesian methods, the noise covariance matrix is ruled by a certain statistical distribution, e.g., Wishart distribution [31] and inverse Wishart distribution [32, 33]. Bayesian one-step GLRT and Bayesian two-step GLRT were proposed in [33] where numerical examples based on simulated and experimental data show that these Bayesian detectors could provide better detection performance than the conventional ones with low sample support. The parametric (or model-based) method

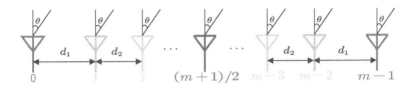

FIGURE 1.2
Symmetrically spaced linear array.

approximates the interference spectrum with a low-order multichannel autoregressive model [34]. In other words, the noise covariance matrix can be well characterized using only a few parameters. Thus, this method largely reduces the required training data. The parametric GLRT, AMF, and Rao tests were proposed in [34–36], respectively. The parametric AMF was demonstrated to be equivalent to the parametric Rao test in [36], where the asymptotic (in the case of sufficiently large samples) statistical distribution was also derived.

The structure-based methods exploit different types of noise covariance matrix structures which result from different antenna configurations or different radar operating environments. The special structures for noise covariance matrix include low-rank structure [1], Toeplitz [37], Kronecker [38], spectral symmetry [39], and persymmetry [40]. The focus of this book is on the exploitation of persymmetry for multichannel signal detection in sample-starved environments where the number of training data is limited. This is a more interesting and also practically motivated case.

1.3 Persymmetric Structure of Covariance Matrix

In practical applications, the target steering vector and the noise covariance matrix are Hermitian persymmetric (also called centrohermitian), when radar system uses a symmetrically spaced linear array with its center for spatial processing [41, Chap. 7] (see Fig. 1.2) or symmetrically spaced pulse trains for temporal processing [40] (see Fig. 1.3). It has to be emphasized that the

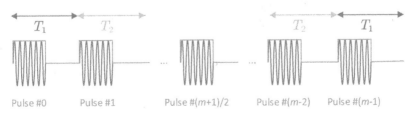

FIGURE 1.3
Symmetrically spaced pulse trains.

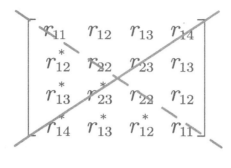

FIGURE 1.4
Persymmetric covariance matrix.

widely used ULA is a special case of the symmetrically spaced linear array, and the widely used pulse trains with uniform spacing are a special case of the symmetrically spaced pulse trains. Hermitian persymmetry of covariance matrix has a property of double symmetry, i.e., Hermitian about its principal diagonal and symmetric about its cross diagonal. Fig. 1.4 gives an example of a 4×4 Hermitian persymmetric covariance matrix. Hermitian persymmetry of a vector means that it is Hermitian about its center. The above properties can be expressed in mathematical forms as follows. If an $N \times 1$ column vector $\mathbf{s}$ and an $N \times N$ covariance matrix $\mathbf{R}$ possess Hermitian persymmetric structures, we have

$$\begin{cases} \mathbf{R} = \mathbf{J}\mathbf{R}^*\mathbf{J}, \\ \mathbf{s} = \mathbf{J}\mathbf{s}^*, \end{cases} \tag{1.1}$$

where $\mathbf{J}$ is a permutation matrix with unit anti-diagonal elements and zero elsewhere, i.e.,

$$\mathbf{J} = \begin{bmatrix} 0 & 0 & \cdots & 0 & 1 \\ 0 & 0 & \cdots & 1 & 0 \\ \vdots & \vdots & \vdots & \vdots & \vdots \\ 0 & 1 & \cdots & 0 & 0 \\ 1 & 0 & \cdots & 0 & 0 \end{bmatrix}_{N \times N}. \tag{1.2}$$

Unless otherwise stated, "persymmetric" always denotes "Hermitian persymmetric" for brevity in the following parts of this book. Note that the Toeplitz structure exists in a uniformly spaced array or pulse train, and hence the persymmetry includes the Toeplitz structure as a special case.

In [42], a statistical test was developed for ascertaining whether multichannel radar data have persymmetry. It is verified that the measured data collected by the MIT Lincoln Laboratory Phase One radar in May 1985 at the Katahdin Hill site, MIT-LL, have covariance persymmetry. Fig. 1.5 shows the intensity of the measured data. Many challenging realistic factors are involved in these measured data, such as heterogeneous terrain and array errors. The

TABLE 1.1
Specifications of the analyzed land clutter dataset

	Dataset H067037 2.iq
Date	May 3, 1985
Number of Pulses N_t	30720
Number of Cells N_s	76
Polarizations	HH, VV
RF Frequency	1.23 Ghz
Pulse Length	100 ns
Pulse Repetition Frequency	100 Hz
Sampling Frequency	10MHz
Radar Scan Mode	Fixed Azimuth
Radar Azimuth Angle	235 Deg
Grazing angle	0.65 Deg
Range	2001–3125 m
Radar Beam Width	3.4 Deg
Range Resolution	15 m
Quantization Bit	13 bit
Mean/Max Wind Speed	20/20 mph

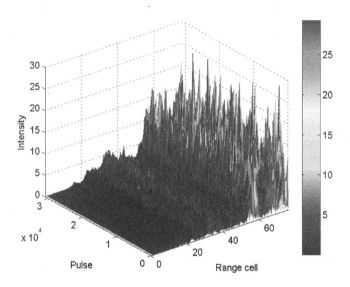

FIGURE 1.5
Spectrum of range-Doppler of the data received by the Phase One radar.

observed area contains wind-blown vegetation consisting of mixed deciduous trees (without leaves), occasional pine, and cedar. The specifications of the L-Band clutter dataset used in the rest are given in Table 1.1. More details on the Phase One radar and the experiment can be found in [43].

1.4 Organization and Outline of the Book

The exploitation of persymmetry can significantly improve detection performance of adaptive detectors in sample-starved environments [40, 44–58]. Following the growing demands on target detection in sample-starved environments, this book provides a comprehensive review and thorough discussions of persymmetric adaptive detectors, including convergence analysis, detector design, and analytical performance. Educating and imparting a holistic understanding of persymmetric adaptive detection lays a strong foundation for postgraduate students, research scholars, and practicing engineers in generating and innovating solutions and products for a broad range of applications. Readers are required to have a quite solid background in statistical signal processing, multivariate statistical analysis, matrix theory, and mathematical analysis.

The remainder of this book is organized as follows:

Chapter 2 conducts comprehensive analysis on statistical properties of the normalized output SINR of the persymmetric SMI beamformer, including both the mismatched case and non-homogeneous case. Then, an answer is given to the question on how many training data are required to achieve a desired level of performance loss when persymmetry is exploited.

Chapter 3 discusses invariance issues on the multichannel signal detection problem under persymmetry and shows how the invariance principle allows us to design decision rules exhibiting some natural symmetries that are of practical interests for radar systems.

Chapter 4 exploits persymmetry to address the detection problem of detecting subspace signal which is expressed as the product of a known full-column rank matrix and an unknown column vector. An adaptive detector is designed according to the criterion of one-step GLRT, and an analytical expression for the PFA is derived.

Chapter 5 takes into account beampointing errors for multichannel signal detection. With exploitation of persymmetry, two selective receivers are devised according to the criteria of Rao test and GLRT, which exhibit capabilities of rejecting with high probability signals whose signatures are unlikely to correspond to that of interest.

Chapter 6 addresses the problem of detecting a distributed target in HEs by using persymmetry. Two adaptive detectors with persymmetry are developed resorting to the principles of one-step and two-step GLRTs. Analytical performance is provided for both detectors.

Chapter 7 examines the detection problem similar to that in Chapter 6. The difference is that Chapter 7 considers the mismatched case where the nominal steering vector is not aligned with the true one. A subspace model is employed to describe the steering vector uncertainty. Several adaptive detectors are designed by using persymmetry, which exhibits strong robustness to the steering vector uncertainty.

Chapter 8 further deals with the case where the steering vector is totally unknown. Two adaptive detectors are derived by using persymmetry. Then, analytical expressions are obtained for the probabilities of false alarm of the proposed detectors, which indicate their CFAR properties against the noise covariance matrix.

Chapter 9 considers the distributed target detection problem in the presence of interference. Two persymmetric adaptive detectors are designed with the principles of one-step and two-step GLRTs. The statistical properties of the one-step detector are discussed in detail.

Chapter 10 (or Chapter 11) addresses the similar problems to those in Chapter 6 (or Chapter 7), respectively. The difference is that the Chapter 10 (or Chapter 11) deals with PHEs, while Chapter 6 (or Chapter 7) considers HEs. Several persymmetric detectors are given in Chapter 10 (or Chapter 11).

Chapter 12 jointly exploits persymmetry and symmetric spectrum to come up with adaptive detectors in HE and PHE. The joint exploitation of persymmetry and symmetric spectrum helps to improve the accuracy in noise covariance matrix estimate, which leads to detection performance gains.

Chapter 13 presents a detection architecture composed of noise covariance matrix structure classier before a bank of adaptive radar detectors. The classifier accounts for six classes of covariance matrix structures: Hermitian, symmetric, persymmetric, persymmetric and symmetric, Toeplitz, and Toeplitz and symmetric. This architecture exhibits an improved estimation of covariance matrix, and better detection performance, especially in the case of limited training samples.

Chapter 14 focuses on MIMO radar detection problems by using persymmetry. Two adaptive detectors are developed for both colocated and distributed MIMO radars. Analytical performance of the persymmetric detector in colocated MIMO radar is provided, and an expression for the PFA of the persymmetric detector in distributed MIMO radar is derived.

Acknowledgments This work was supported in part by the National Natural Science Foundation of China under Grants 61871469, 61971412, 62071482, the Youth Innovation Promotion Association CAS (2019447), and the Anhui Provincial Natural Science Foundation under Grant 2208085J17. Phase One radar data are courtesy of Mr. J. B. Billingsley.

1.A   Detector Design Criteria

Assume that a complex-valued data $\mathbf{x}$ under H_0 or H_1 has the PDF $p(\mathbf{x}; \boldsymbol{\theta}_0, H_0)$ or $p(\mathbf{x}; \boldsymbol{\theta}_1, H_1)$, where $\boldsymbol{\theta}_1$ is the parameter of the PDF under H_1, and $\boldsymbol{\theta}_0$ is the parameter under H_0.

The total complex-valued parameter vector is defined as $\boldsymbol{\theta} = [\boldsymbol{\theta}_r^T \ \boldsymbol{\theta}_s^T]^T$, where $\boldsymbol{\theta}_r$ is the so-called relevant parameter, and $\boldsymbol{\theta}_s$ is the so-called nuisance parameter. In general, $\boldsymbol{\theta}_0$ is the same as $\boldsymbol{\theta}_s$. Radar target detection is to discriminate between the target-absent hypothesis H_0 that $\boldsymbol{\theta}_r = \boldsymbol{\theta}_{r0}$ and the target-present hypothesis H_1 that $\boldsymbol{\theta}_r \neq \boldsymbol{\theta}_{r0}$, where $\boldsymbol{\theta}_{r0}$ is the value of $\boldsymbol{\theta}_r$ under H_0. That is why $\boldsymbol{\theta}_s$ is called the nuisance parameter.

1.A.1   Nuisance Parameter

First, the case of nuisance parameter is considered. Assume that the parameter $\boldsymbol{\theta}_0$ is unknown, namely, there exists an unknown nuisance parameter $\boldsymbol{\theta}_s$.

1.A.1.1   Rao Test

Denote the relevant, nuisance, and the total complex parameter sets by $\boldsymbol{\vartheta}_r = [\boldsymbol{\theta}_r^T \ \boldsymbol{\theta}_r^{\dagger T}]^T$, $\boldsymbol{\vartheta}_s = [\boldsymbol{\theta}_s^T \ \boldsymbol{\theta}_s^{\dagger T}]^T$, and $\boldsymbol{\vartheta} = [\boldsymbol{\vartheta}_r^T \ \boldsymbol{\vartheta}_s^T]^T$, respectively. The general complex parameter Rao test with the nuisance parameter is given as [59]

$$T_{\text{Rao}} = \left. \frac{\partial \ln p_{\mathbf{x}}(\mathbf{x}; \boldsymbol{\vartheta})}{\partial \boldsymbol{\vartheta}_r^T} \right|_{\boldsymbol{\vartheta} = \hat{\boldsymbol{\vartheta}}_0} \left[\mathbf{F}^{-1}(\hat{\boldsymbol{\vartheta}}_0) \right]_{\boldsymbol{\vartheta}_r, \boldsymbol{\vartheta}_r} \left. \frac{\partial \ln p_{\mathbf{x}}(\mathbf{x}; \boldsymbol{\vartheta})}{\partial \boldsymbol{\vartheta}_r^*} \right|_{\boldsymbol{\vartheta} = \hat{\boldsymbol{\vartheta}}_0}, \qquad (1.A.1)$$

where $p_{\mathbf{x}}(\mathbf{x}; \boldsymbol{\vartheta})^1$ is the PDF of $\mathbf{x}$ under H_1, $\hat{\boldsymbol{\vartheta}}_0$ is the MLE of $\boldsymbol{\vartheta}$ under H_0, the FIM of $\boldsymbol{\vartheta}$ can be partitioned as

$$\mathbf{F}(\boldsymbol{\vartheta}) = \mathrm{E}\left[\begin{bmatrix} \frac{\partial \ln p_{\mathbf{x}}(\mathbf{x};\boldsymbol{\vartheta})}{\partial \boldsymbol{\vartheta}_r} \\ \frac{\partial \ln p_{\mathbf{x}}(\mathbf{x};\boldsymbol{\vartheta})}{\partial \boldsymbol{\vartheta}_s} \end{bmatrix}^* \begin{bmatrix} \frac{\partial \ln p_{\mathbf{x}}(\mathbf{x};\boldsymbol{\vartheta})}{\partial \boldsymbol{\vartheta}_r} \\ \frac{\partial \ln p_{\mathbf{x}}(\mathbf{x};\boldsymbol{\vartheta})}{\partial \boldsymbol{\vartheta}_s} \end{bmatrix}^T \right]$$

$$= \begin{bmatrix} \mathbf{F}_{\boldsymbol{\vartheta}_r, \boldsymbol{\vartheta}_r}(\boldsymbol{\vartheta}) & \mathbf{F}_{\boldsymbol{\vartheta}_r, \boldsymbol{\vartheta}_s}(\boldsymbol{\vartheta}) \\ \mathbf{F}_{\boldsymbol{\vartheta}_s, \boldsymbol{\vartheta}_r}(\boldsymbol{\vartheta}) & \mathbf{F}_{\boldsymbol{\vartheta}_s, \boldsymbol{\vartheta}_s}(\boldsymbol{\vartheta}) \end{bmatrix}, \qquad (1.A.2)$$

and

$$\left\{ \left[\mathbf{F}^{-1}(\boldsymbol{\vartheta}) \right]_{\boldsymbol{\vartheta}_r, \boldsymbol{\vartheta}_r} \right\}^{-1} = \mathbf{F}_{\boldsymbol{\vartheta}_r, \boldsymbol{\vartheta}_r}(\boldsymbol{\vartheta}) - \mathbf{F}_{\boldsymbol{\vartheta}_r, \boldsymbol{\vartheta}_s}(\boldsymbol{\vartheta}) \mathbf{F}_{\boldsymbol{\vartheta}_s, \boldsymbol{\vartheta}_s}^{-1}(\boldsymbol{\vartheta}) \mathbf{F}_{\boldsymbol{\vartheta}_s, \boldsymbol{\vartheta}_r}(\boldsymbol{\vartheta}). \tag{1.A.3}$$

If the FIM of $\boldsymbol{\vartheta}$ has a special form as

$$\mathbf{F}(\boldsymbol{\vartheta}) = \begin{bmatrix} \mathbf{F}_{\boldsymbol{\theta}_r, \boldsymbol{\theta}_r}(\boldsymbol{\vartheta}) & 0 & \mathbf{F}_{\boldsymbol{\theta}_r, \boldsymbol{\theta}_s}(\boldsymbol{\vartheta}) & 0 \\ 0 & \mathbf{F}_{\boldsymbol{\theta}_r, \boldsymbol{\theta}_r}^*(\boldsymbol{\vartheta}) & 0 & \mathbf{F}_{\boldsymbol{\theta}_r, \boldsymbol{\theta}_s}^*(\boldsymbol{\vartheta}) \\ \mathbf{F}_{\boldsymbol{\theta}_s, \boldsymbol{\theta}_r}(\boldsymbol{\vartheta}) & 0 & \mathbf{F}_{\boldsymbol{\theta}_s, \boldsymbol{\theta}_s}(\boldsymbol{\vartheta}) & 0 \\ 0 & \mathbf{F}_{\boldsymbol{\theta}_s, \boldsymbol{\theta}_r}^*(\boldsymbol{\vartheta}) & 0 & \mathbf{F}_{\boldsymbol{\theta}_s, \boldsymbol{\theta}_s}^*(\boldsymbol{\vartheta}) \end{bmatrix}, \qquad (1.A.4)$$

where

$$\begin{bmatrix} \mathbf{F}_{\boldsymbol{\theta}_r, \boldsymbol{\theta}_r}(\boldsymbol{\vartheta}) & \mathbf{F}_{\boldsymbol{\theta}_r, \boldsymbol{\theta}_s}(\boldsymbol{\vartheta}) \\ \mathbf{F}_{\boldsymbol{\theta}_s, \boldsymbol{\theta}_r}(\boldsymbol{\vartheta}) & \mathbf{F}_{\boldsymbol{\theta}_s, \boldsymbol{\theta}_s}(\boldsymbol{\vartheta}) \end{bmatrix} = \mathrm{E}\left[\frac{\partial \ln p_{\mathbf{x}}(\mathbf{x}; \boldsymbol{\vartheta})}{\partial \boldsymbol{\theta}^*} \frac{\partial \ln p_{\mathbf{x}}(\mathbf{x}; \boldsymbol{\vartheta})}{\partial \boldsymbol{\theta}^T} \right] \stackrel{\mathrm{d}}{=} \mathbf{F}(\boldsymbol{\theta}),$$

$$\tag{1.A.5}$$

---

[1]Note that $p_{\mathbf{x}}(\mathbf{x}; \boldsymbol{\vartheta})$ is used for complex criteria of detector design, which is the same as $p(\mathbf{x}; \boldsymbol{\theta}_1, H_1)$.

then the Rao test with the nuisance parameter in this case can be simplified as

$$T_{\text{Rao}} = 2 \left. \frac{\partial \ln p_{\mathbf{x}}(\mathbf{x}; \boldsymbol{\vartheta})}{\partial \boldsymbol{\theta}_r^T} \right|_{\boldsymbol{\theta} = \hat{\boldsymbol{\theta}}_0} \left[\mathbf{F}^{-1}(\hat{\boldsymbol{\theta}}_0) \right]_{\boldsymbol{\theta}_r, \boldsymbol{\theta}_r} \left. \frac{\partial \ln p_{\mathbf{x}}(\mathbf{x}; \boldsymbol{\vartheta})}{\partial \boldsymbol{\theta}_r^*} \right|_{\boldsymbol{\theta} = \hat{\boldsymbol{\theta}}_0}, \qquad (1.A.6)$$

where $\hat{\boldsymbol{\theta}}_0$ being the MLE of $\boldsymbol{\theta}$ under H_0.

1.A.1.2 Wald Test

The general complex parameter Wald test with the nuisance parameter is given as [59]

$$T_{\text{Wald}} = (\hat{\boldsymbol{\vartheta}}_{r1} - \boldsymbol{\vartheta}_{r0})^\dagger \left\{ \left[\mathbf{F}^{-1}(\hat{\boldsymbol{\vartheta}}_1) \right]_{\boldsymbol{\vartheta}_r, \boldsymbol{\vartheta}_r} \right\}^{-1} (\hat{\boldsymbol{\vartheta}}_{r1} - \boldsymbol{\vartheta}_{r0}), \qquad (1.A.7)$$

where $\hat{\boldsymbol{\vartheta}}_{r1}$ and $\hat{\boldsymbol{\vartheta}}_1$ are the MLEs of $\boldsymbol{\vartheta}_r$ and $\boldsymbol{\vartheta}$ under H_1, respectively, and $\boldsymbol{\vartheta}_{r0}$ is the value of $\boldsymbol{\vartheta}_r$ under H_0.

If the FIM of $\boldsymbol{\vartheta}$ has a special form as (1.A.4), then the complex parameter Wald test can be simplified as

$$T_{\text{Wald}} = 2 (\hat{\boldsymbol{\theta}}_{r1} - \boldsymbol{\theta}_{r0})^\dagger \left\{ \left[\mathbf{F}^{-1}(\hat{\boldsymbol{\theta}}_1) \right]_{\boldsymbol{\theta}_r, \boldsymbol{\theta}_r} \right\}^{-1} (\hat{\boldsymbol{\theta}}_{r1} - \boldsymbol{\theta}_{r0}), \qquad (1.A.8)$$

where $\hat{\boldsymbol{\theta}}_{r1}$ and $\hat{\boldsymbol{\theta}}_1$ are the MLEs of $\boldsymbol{\theta}_r$ and $\boldsymbol{\theta}$ under H_1, respectively, and $\boldsymbol{\theta}_{r0}$ is the value of $\boldsymbol{\theta}_r$ under H_0.

1.A.1.3 GLRT

The GLRT with the nuisance parameter is given by

$$T_{\text{GLRT}} = \frac{\max_{\boldsymbol{\theta}_1} \, p(\mathbf{x}; \boldsymbol{\theta}_1, H_1)}{\max_{\boldsymbol{\theta}_0} \, p(\mathbf{x}; \boldsymbol{\theta}_0, H_0)}. \qquad (1.A.9)$$

1.A.2 No Nuisance Parameter

Now we consider the case where there is no nuisance parameter, namely, $\boldsymbol{\theta}_s$ does not exist and $\boldsymbol{\theta}_0$ is known. In such a case, the total complex-valued parameter vector $\boldsymbol{\theta}$ is reduced to be $\boldsymbol{\theta} = \boldsymbol{\theta}_r$. Define $\boldsymbol{\vartheta} \triangleq [\boldsymbol{\theta}^T \; \boldsymbol{\theta}^\dagger]^T$.

1.A.2.1 Rao Test

The general complex parameter Rao test with no nuisance parameter is given as [60]

$$T_{\text{Rao}} = \left. \frac{\partial \ln p_{\mathbf{x}}(\mathbf{x}; \boldsymbol{\vartheta})}{\partial \boldsymbol{\vartheta}^T} \right|_{\boldsymbol{\vartheta} = \boldsymbol{\vartheta}_0} \mathbf{F}^{-1}(\boldsymbol{\vartheta}_0) \left. \frac{\partial \ln p_{\mathbf{x}}(\mathbf{x}; \boldsymbol{\vartheta})}{\partial \boldsymbol{\vartheta}^*} \right|_{\boldsymbol{\vartheta} = \boldsymbol{\vartheta}_0}, \qquad (1.A.10)$$

where $\boldsymbol{\vartheta}_0$ is the value of $\boldsymbol{\vartheta}$ under H_0, and $\mathbf{F}(\boldsymbol{\vartheta})$ is the FIM of $\boldsymbol{\vartheta}$, i.e.,

$$\mathbf{F}(\boldsymbol{\vartheta}) = \begin{bmatrix} \mathbf{F}(\boldsymbol{\theta}) & \mathbf{G}(\boldsymbol{\theta}) \\ \mathbf{G}^*(\boldsymbol{\theta}) & \mathbf{F}^*(\boldsymbol{\theta}) \end{bmatrix}, \tag{1.A.11}$$

with

$$\mathbf{F}(\boldsymbol{\theta}) = \mathrm{E}\left[\frac{\partial \ln p_{\mathbf{x}}(\mathbf{x};\boldsymbol{\vartheta})}{\partial \boldsymbol{\theta}^*} \frac{\partial \ln p_{\mathbf{x}}(\mathbf{x};\boldsymbol{\vartheta})}{\partial \boldsymbol{\theta}^T} \right], \tag{1.A.12}$$

$$\mathbf{G}(\boldsymbol{\theta}) = \mathrm{E}\left[\frac{\partial \ln p_{\mathbf{x}}(\mathbf{x};\boldsymbol{\vartheta})}{\partial \boldsymbol{\theta}^*} \frac{\partial \ln p_{\mathbf{x}}(\mathbf{x};\boldsymbol{\vartheta})}{\partial \boldsymbol{\theta}^\dagger} \right]. \tag{1.A.13}$$

If the FIM of $\boldsymbol{\vartheta}$ has a special form as

$$\mathbf{F}(\boldsymbol{\vartheta}) = \begin{bmatrix} \mathbf{F}(\boldsymbol{\theta}) & \mathbf{0} \\ \mathbf{0} & \mathbf{F}^*(\boldsymbol{\theta}) \end{bmatrix}, \tag{1.A.14}$$

then the Rao test with no nuisance parameter in this case can be simplified as

$$T_{\mathrm{Rao}} = 2 \left. \frac{\partial \ln p_{\mathbf{x}}(\mathbf{x};\boldsymbol{\vartheta})}{\partial \boldsymbol{\theta}^T} \right|_{\boldsymbol{\theta}=\boldsymbol{\theta}_{r0}} \mathbf{F}^{-1}(\boldsymbol{\theta}_{r0}) \left. \frac{\partial \ln p_{\mathbf{x}}(\mathbf{x};\boldsymbol{\vartheta})}{\partial \boldsymbol{\theta}^*} \right|_{\boldsymbol{\theta}=\boldsymbol{\theta}_{r0}}, \tag{1.A.15}$$

where $\boldsymbol{\theta}_{r0}$ is the value of $\boldsymbol{\theta}$ under H_0.

1.A.2.2 Wald Test

The general complex parameter Wald test with no nuisance parameter is given as

$$T_{\mathrm{Wald}} = (\hat{\boldsymbol{\vartheta}}_1 - \boldsymbol{\vartheta}_0)^\dagger \, \mathbf{F}(\hat{\boldsymbol{\vartheta}}_1) \, (\hat{\boldsymbol{\vartheta}}_1 - \boldsymbol{\vartheta}_0), \tag{1.A.16}$$

where $\hat{\boldsymbol{\vartheta}}_1$ is the MLE of $\boldsymbol{\vartheta}$ under H_1.

If the FIM of $\boldsymbol{\vartheta}$ has a special form as (1.A.14), then the complex parameter Wald test can be simplified as

$$T_{\mathrm{Wald}} = 2 \, (\hat{\boldsymbol{\theta}}_1 - \boldsymbol{\theta}_{r0})^\dagger \mathbf{F}(\hat{\boldsymbol{\theta}}_1)(\hat{\boldsymbol{\theta}}_1 - \boldsymbol{\theta}_{r0}), \tag{1.A.17}$$

where $\hat{\boldsymbol{\theta}}_1$ is the MLE of $\boldsymbol{\theta}$ under H_1.

1.A.2.3 GLRT

The GLRT without the nuisance parameter is given by

$$T_{\mathrm{GLRT}} = \frac{\max_{\boldsymbol{\theta}_1} \; p(\mathbf{x}; \boldsymbol{\theta}_1, H_1)}{p(\mathbf{x}; \boldsymbol{\theta}_0, H_0)}. \tag{1.A.18}$$

Bibliography

[1] J. Ward, "Space-time adaptive processing for airborne radar," Lincoln Laboratory, MIT, Technical Report 1015, December 1994.

[2] M. A. Richards, *Fundamentals of Radar Signal Processing*. New York McGraw-Hill, 2005.

[3] C. Hao, D. Orlando, J. Liu, and C. Yin, *Advances in Adaptive Radar Detection and Range Estimation*. Singapore: Springer, 2022.

[4] W. Liu, J. Liu, C. Hao, Y. Gao, and Y.-L. Wang, "Multichannel adaptive signal detection: Basic theory and literature review," *SCIENCE CHINA Information Sciences*, vol. 65, 121301:1–121301:40, 2022.

[5] S. Kraut and L. L. Scharf, "The CFAR adaptive subspace detector is a scale-invariant GLRT," *IEEE Transactions on Signal Processing*, vol. 47, no. 9, pp. 2538–2541, September 1999.

[6] J. Liu, Z.-J. Zhang, Y. Yang, and H. Liu, "A CFAR adaptive subspace detector for first-order or second-order Gaussian signals based on a single observation," *IEEE Transactions on Signal Processing*, vol. 59, no. 11, pp. 5126–5140, November 2011.

[7] A. De Maio and M. S. Greco, Eds., *Modern Radar Detection Theory*. ser. Radar, Sonar and Navigation. Stevenage: U.K.: Inst. Eng. Technol., 2015.

[8] I. S. Reed, J. D. Mallett, and L. E. Brennan, "Rapid convergence rate in adaptive arrays," *IEEE Transactions on Aerospace and Electronic Systems*, vol. 10, no. 6, pp. 853–863, 1974.

[9] E. J. Kelly, "An adaptive detection algorithm," *IEEE Transactions on Aerospace and Electronic Systems*, vol. 22, no. 1, pp. 115–127, March 1986.

[10] F. C. Robey, D. R. Fuhrmann, E. J. Kelly, and R. Nitzberg, "A CFAR adaptive matched filter detector," *IEEE Transactions on Aerospace and Electronic Systems*, vol. 28, no. 1, pp. 208–216, January 1992.

[11] A. De Maio, "A new derivation of the adaptive matched filter," *IEEE Signal Processing Letters*, vol. 11, no. 10, pp. 792–793, October 2004.

[12] ——, "Rao test for adaptive detection in Gaussian interference with unknown covariance matrix," *IEEE Transactions on Signal Processing*, vol. 55, no. 7, pp. 3577–3584, July 2007.

[13] F. Bandiera, D. Orlando, and G. Ricci, *Advanced Radar Detection Schemes under Mismatched Signal Models in Synthesis Lectures on Signal Processing*. San Rafael, CA, USA. Morgan & Claypool, 2009.

[14] O. Besson, L. L. Scharf, and F. Vincent, "Matched direction detectors and estimators for array processing with subspace steering vector uncertainties," *IEEE Transactions on Signal Processing*, vol. 53, no. 12, pp. 4453–4463, December 2005.

[15] D. Ciuonzo, A. De Maio, and D. Orlando, "On the statistical invariance for adaptive radar detection in partially homogeneous disturbance plus structured interference," *IEEE Transaction on Signal Processing*, vol. 5, no. 65, pp. 1222–1234, March 1 2017.

[16] F. Gini and A. Farina, "Vector subspace detection in compound-Gaussian clutter. part I: survey and new results," *IEEE Transactions on Aerospace and Electronic Systems*, vol. 38, no. 4, pp. 1295–1311, October 2002.

[17] L. L. Scharf and B. Friedlander, "Matched subspace detectors," *IEEE Transactions on Signal Processing*, vol. 42, no. 8, pp. 2146–2157, August 1994.

[18] S. Kraut, L. L. Scharf, and L. T. McWhorter, "Adaptive subspace detectors," *IEEE Transactions on Signal Processing*, vol. 49, no. 1, pp. 1–16, January 2001.

[19] W. Liu, W. Xie, J. Liu, and Y. Wang, "Adaptive double subspace signal detection in Gaussian background–Part I: Homogeneous environments," *IEEE Transactions on Signal Processing*, vol. 62, no. 9, pp. 2345–2357, May 2014.

[20] ——, "Adaptive double subspace signal detection in Gaussian background–Part II: Partially homogeneous environments," *IEEE Transactions on Signal Processing*, vol. 62, no. 9, pp. 2358–2369, May 2014.

[21] R. S. Raghavan, N. Pulsone, and D. J. McLaughlin, "Performance of the GLRT for adaptive vector subspace detection," *IEEE Transactions on Aerospace and Electronic Systems*, vol. 32, no. 4, pp. 1473–1487, October 1996.

[22] E. Conte, M. Lops, and G. Ricci, "Asymptotically optimum radar detection in compound-Gaussian clutter," *IEEE Transactions on Aerospace and Electronic Systems*, vol. 31, no. 2, pp. 617–625, April 1995.

[23] S. Kraut, L. L. Scharf, and R. W. Butler, "The adaptive coherence estimator: A uniformly-most-powerful-invariant adaptive detection statistic," *IEEE Transactions on Signal Processing*, vol. 53, no. 2, pp. 417–438, Febuary 2005.

[24] A. De Maio and S. Iommelli, "Coincidence of the Rao test, Wald test, and GLRT in partially homogeneous environment," *IEEE Signal Processing Letters*, vol. 15, no. 4, pp. 385–388, April 2008.

[25] F. Bandiera, A. De Maio, A. S. Greco, and G. Ricci, "Adaptive radar detection of distributed targets in homogeneous and partially homogeneous noise plus subspace interference," *IEEE Transactions on Signal Processing*, vol. 55, no. 4, pp. 1223–1237, April 2007.

[26] D. Ciuonzo, D. Orlando, and L. Pallotta, "On the maximal invariant statistic for adaptive radar detection in partially homogeneous disturbance with persymmetric covariance," *IEEE Signal Processing Letters*, vol. 23, no. 12, pp. 1830–1834, December 2016.

[27] J. Guerci and E. J. Baranoski, "Knowledge-aided adaptive radar at DARPA: an overview," *IEEE Signal Processing Magazine*, vol. 23, no. 1, pp. 41–50, January 2006.

[28] Y. Wang, J. Chen, Z. Bao, and Y. Peng, "Robust space-time adaptive processing for airborne radar in nonhornogeneous clutter environments," *IEEE Transactions on Aerospace and Electronic Systems*, vol. 39, no. 1, pp. 70–81, January 2003.

[29] Y. Cai, X. Wu, M. Zhao, R. C. de Lamare, and B. Champagne, "Low-complexity reduced-dimension space-time adaptive processing for navigation receivers," *IEEE Transactions on Aerospace and Electronic Systems*, vol. 54, no. 6, pp. 3160–3168, December 2018.

[30] T. E. Ayoub and A. M. Haimovich, "Modified GLRT signal detection algorithm," *IEEE Transactions on Aerospace and Electronic Systems*, vol. 36, no. 3, pp. 810–818, July 2000.

[31] S. Bidon, O. Besson, and J.-Y. Tourneret, "A Bayesian approach to adaptive detection in non-homogeneous environments," *IEEE Transactions on Signal Processing*, vol. 56, no. 1, pp. 205–217, January 2008.

[32] O. Besson, J. Y. Tourneret, and S. Bidon, "Knowledge-aided Bayesian detection in heterogeneous environments," *IEEE Signal Processing Letters*, vol. 14, no. 5, pp. 355–358, May 2007.

[33] A. De Maio, A. Farina, and G. Foglia, "Knowledge-aided bayesian radar detectors & their application to live data," *IEEE Transactions on Aerospace and Electronic Systems*, vol. 46, no. 1, pp. 170–183, 2010.

[34] J. R. Román, M. Rangaswamy, D. W. Davis, Q. Zhang, B. Himed, and J. H. Michels, "Parametric adaptive matched filter for airborne radar applications," *IEEE Transactions on Aerospace and Electronic Systems*, vol. 36, no. 2, pp. 677–692, April 2000.

[35] K. J. Sohn, H. Li, and B. Himed, "Parametric GLRT for multichannel adaptive signal detection," *IEEE Transactions on Signal Processing*, vol. 55, no. 11, pp. 5351–5360, November 2007.

[36] ——, "Parametric Rao test for multichannel adaptive signal detection," *IEEE Transactions on Aerospace and Electronic Systems*, vol. 43, no. 3, pp. 920–933, July 2007.

[37] D. R. Fuhrmann, "Application of Toeplitz covariance estimation to adaptive beamforming and detection," *IEEE Transactions on Signal Processing*, vol. 39, no. 10, pp. 2194–2198, October 1991.

[38] R. S. Raghavan, "CFAR detection in clutter with a Kronecker covariance structure," *IEEE Transactions on Aerospace and Electronic Systems*, vol. 53, no. 2, pp. 619–629, April 2017.

[39] A. De Maio, D. Orlando, C. Hao, and G. Foglia, "Adaptive detection of point-like targets in spectrally symmetric interference," *IEEE Transactions on Signal Processing*, vol. 64, no. 12, pp. 3207–3220, December 2016.

[40] L. Cai and H. Wang, "A persymmetric multiband GLR algorithm," *IEEE Transactions on Aerospace and Electronic Systems*, vol. 28, no. 3, pp. 806–816, July 1992.

[41] H. L. Van Trees, *Optimum Array Processing, Part IV of Detection, Estimation, and Modulation Theory*. Wiley-Interscience, New York, 2002.

[42] E. Conte, A. De Maio, and A. Farina, "Statistical tests for higher order analysis of radar clutter: their application to L-band measured data," *IEEE Transactions on Aerospace and Electronic Systems*, vol. 41, no. 1, pp. 205–218, January 2005.

[43] J. B. Billingsley, A. Farina, F. Gini, M. V. Greco, and L. Verrazzani, "Statistical analyses of measured radar ground clutter data," *IEEE Transactions on Aerospace and Electronic Systems*, vol. 35, no. 2, pp. 579–593, April 1999.

[44] L. Cai and H. Wang, "A persymmetric modified-SMI algorithm," *Signal Processing*, vol. 23, no. 1, pp. 27–34, January 1991.

[45] Y. Gao, G. Liao, S. Zhu, X. Zhang, and D. Yang, "Persymmetric adaptive detectors in homogeneous and partially homogeneous environments," *IEEE Transactions on Signal Processing*, vol. 62, no. 2, pp. 331–342, February 2014.

[46] Z. Wang, M. Li, H. Chen, Y. Lu, R. Cao, P. Zhang, L. Zuo, and Y. Wu, "Persymmetric detectors of distributed targets in partially homogeneous disturbance," *Signal Processing*, vol. 128, pp. 382–388, 2016.

[47] J. Liu, S. Sun, and W. Liu, "One-step persymmetric GLRT for subspace signals," *IEEE Transaction on Signal Processing*, vol. 14, no. 67, pp. 3639–3648, July 15 2019.

[48] J. Liu, D. Orlando, P. Addabbo, and W. Liu, "SINR distribution for the persymmetric SMI beamformer with steering vector mismatches," *IEEE Transactions on Signal Processing*, vol. 67, no. 5, pp. 1382–1392, March 1 2019.

[49] J. Liu, W. Liu, H. Liu, C. Bo, X.-G. Xia, and F. Dai, "Average SINR calculation of a persymmetric sample matrix inversion beamformer," *IEEE Transactions on Signal Processing*, vol. 64, no. 8, pp. 2135–2145, April 15 2016.

[50] J. Liu and J. Li, "Mismatched signal rejection performance of the persymmetric GLRT detector," *IEEE Transactions on Signal Processing*, vol. 67, no. 6, pp. 1610–1619, March 15 2019.

[51] J. Liu, T. Jian, and W. Liu, "Persymmetric detection of subspace signals based on multiple observations in the presence of subspace interference," *Signal Processing*, vol. 183, 2021.

[52] J. Liu, Z. Gao, Z. Sun, T. Jian, and W. Liu, "Detection architecture with improved classification capabilities for covariance structures," *Digital Signal Processing*, vol. 123, 2022.

[53] J. Liu, W. Liu, B. Tang, J. Zheng, and S. Xu, "Distributed target detection exploiting persymmetry in Gaussian clutter," *IEEE Transactions on Signal Processing*, vol. 67, no. 4, pp. 1022–1033, February 2019.

[54] J. Liu, W. Liu, C. Hao, and D. Orlando, "Persymmetric subspace detectors with multiple observations in homogeneous environments," *IEEE Transactions on Aerospace and Electronic Systems*, vol. 56, no. 4, pp. 3276–3284, August 2020.

[55] J. Liu, J. Chen, J. Li, and W. Liu, "Persymmetric adaptive detection of distributed targets with unknown steering vectors," *IEEE Transactions on Signal Processing*, vol. 68, pp. 4123–4134, 2020.

[56] J. Liu, T. Jian, W. Liu, C. Hao, and D. Orlando, "Persymmetric adaptive detection with improved robustness to steering vector mismatches," *Signal Processing*, vol. 176, p. 107669, 2020.

[57] J. Liu, W. Liu, J. Han, B. Tang, Y. Zhao, and H. Yang, "Persymmetric GLRT detection in MIMO radar," *IEEE Transactions on Vehicular Technology*, vol. 67, no. 12, pp. 11 913–11 923, December 2018.

[58] J. Liu, H. Li, and B. Himed, "Persymmetric adaptive target detection with distributed MIMO radar," *IEEE Transactions on Aerospace and Electronic Systems*, vol. 51, no. 1, pp. 372–382, January 2015.

[59] M. Sun, W. Liu, J. Liu, and C. Hao, "Complex parameter Rao, Wald, gradient, and Durbin tests for multichannel signal detection," *IEEE Transactions on Signal Processing*, vol. 70, pp. 117–131, 2022.

[60] S. K. Kay and Z. Zhu, "The complex parameter Rao test," *IEEE Transactions on Signal Processing*, vol. 64, no. 24, pp. 6580–6588, December 2016.

2

Output SINR Analysis

Chapter 1 introduces the multichannel adaptive detection problem when per-symmetry is exploited for covariance matrix estimation. However, it does not answer the question how many secondary data are required to achieve a desired level of performance loss.

A common metric is the average SINR loss defined as the expectation of the ratio of the output SINR of the SMI beamformer to the optimal SINR [1]. This average SINR loss measures the level of how the performance of the SMI beamformer is close to the optimal one. The average SINR loss plays an important role in the performance evaluation of the SMI beamformer. In [1], RMB derived the distribution of the normalized output SINR of the SMI beamformer, and further obtained its expectation (i.e., the average SINR loss). Based on it, the well-known RMB rule is proposed: the average SINR loss is 3 dB when the amount of training data used to estimate the noise covariance matrix is about twice the dimension of the received signal. Richmond examined the statistical properties of many adaptive beamformers in a series of his papers [2–5]. Notice that the results in [1–5] are derived without taking into account any information on the structure of the covariance matrix. In the following, the SMI beamformer without using any structural knowledge is referred to as the unstructured SMI beamformer.

In many practical scenarios, the actual signal steering vector is not always aligned with the presumed one [4,6,7]. For example, a mismatch in the steering vector often exists in an array system, e.g., due to errors in calibration or look direction, distortions in signal waveform or array geometry. Boroson in [8] derived the distribution of the normalized SINR of the unstructured SMI beamformer in the case where mismatch exists in the steering vector. Thus, the average SINR loss can be easily obtained by computing the expectation of the distribution derived. This theoretical result can be used to evaluate the effect of the steering vector mismatch on the output SINR of the unstructured SMI beamformer.

Another practical scenario required to take into consideration is non-homogeneity between the test and training data, namely, the covariance matrix of the training data may be different from that of the test data [9]. In practice, the non-homogeneity may be due to many factors, such as interfering targets and variations in terrain.

Note that the work in [1, 8] does not consider the structure of the covariance matrix. When persymmetry is exploited in the SMI beamformer, it

DOI: 10.1201/9781003340232-2

is referred to as persymmetric SMI beamformer. This chapter is to conduct comprehensive analysis on the statistical properties of the normalized output SINR of the persymmetric SMI beamformer, including both the mismatched case and non-homogeneous case. Section 2.1 formulates the considered problem. Section 2.2 gives an exact expression for the average SINR loss of the persymmetric SMI beamformer in the matched case. Section 2.3 analyzes the normalized output SINR in mismatched cases, including homogeneous and non-homogeneous environments. Numerical examples are provided in Section 2.4, which indicate that the exploitation of persymmetry is equivalent to doubling the training data size and hence can greatly alleviate the requirement of training data in adaptive processing.

2.1 Problem Formulation

Consider the following model of an N-dimensional received data:

$$\mathbf{x} = a\,\mathbf{s} + \mathbf{n}, \tag{2.1}$$

where $\mathbf{s}$ is a signal steering vector of dimension $N \times 1$; a is a deterministic but unknown complex scalar accounting for the target reflectivity and the channel propagation effects; the term $\mathbf{n} \in \mathbb{C}^{N \times 1}$ including interference (or clutter) and noise is assumed to have a circularly symmetric, complex Gaussian distribution with zero mean and covariance matrix $\mathbf{R} \in \mathbb{C}^{N \times N}$, i.e., $\mathbf{n} \sim \mathcal{CN}_N(\mathbf{0}, \mathbf{R})$.

The received data is filtered through weighting the components of the vector $\mathbf{x}$. The optimal weight vector $\mathbf{w} \in \mathbb{C}^{N \times 1}$ can be designed using different criteria. Here, we employ the criterion that minimizes the interference-plus-noise energy without distorting the signal of interest. Under this criterion, the output SINR of the filter is the highest [10]. It results in the well-known minimum variance distortionless response beamformer. More precisely, the optimal weight vector in this beamformer can be obtained by solving the following optimization problem [11]:

$$\begin{cases} \min_{\mathbf{w}} \mathbf{w}^\dagger \mathbf{R} \mathbf{w} \\ \text{s. t. } \mathbf{w}^\dagger \mathbf{s} = 1. \end{cases} \tag{2.2}$$

The optimal weight vector $\mathbf{w}_{\text{opt}}$ is

$$\mathbf{w}_{\text{opt}} = \frac{\mathbf{R}^{-1}\mathbf{s}}{\mathbf{s}^\dagger \mathbf{R}^{-1}\mathbf{s}}. \tag{2.3}$$

In many practical applications, the actual steering vector $\mathbf{s}$ may not be exactly known [12, 13], and instead the presumed steering vector is used. We denote the presumed steering vector by $\mathbf{q} \in \mathbb{C}^{N \times 1}$. The angle denoted by ϕ

between the actual steering vector $\mathbf{s}$ and the assumed one $\mathbf{q}$ is defined as [13]

$$\cos^2 \phi = \frac{|\mathbf{q}^\dagger \mathbf{R}^{-1} \mathbf{s}|^2}{(\mathbf{q}^\dagger \mathbf{R}^{-1} \mathbf{q})(\mathbf{s}^\dagger \mathbf{R}^{-1} \mathbf{s})}, \tag{2.4}$$

where $0 \leqslant \phi \leqslant \pi/2$. This angle measures the degree of mismatch between $\mathbf{s}$ and $\mathbf{q}$. In a particular case, $\phi = 0$ corresponds to the matched case, i.e., $\mathbf{q} = \mathbf{s}$.

The interference-plus-noise covariance matrix $\mathbf{R}$ is usually unknown in practice and has to be estimated. A standard assumption is often imposed that there exists a set of training data $\{\mathbf{y}_k\}_{k=1}^K$ free of target signal components, i.e., $\{\mathbf{y}_k | \mathbf{y}_k \sim \mathcal{CN}_N(\mathbf{0}, \mathbf{R}_s), k = 1, 2, \ldots, K\}$.

2.1.1 Unstructured SMI Beamformer

Without using the structural information on the covariance matrix, one can obtain the MLE of $\mathbf{R}$ (up to a scaling) to be

$$\hat{\mathbf{R}} = \sum_{k=1}^K \mathbf{y}_k \mathbf{y}_k^\dagger. \tag{2.5}$$

Obviously, $\hat{\mathbf{R}}$ has a complex Wishart distribution [14].

2.1.1.1 Matched Case

Assume that there is no mismatch in the signal steering vector, and the environment is homogeneous, i.e., $\phi = 0$ and $\mathbf{R}_s = \mathbf{R}$. Replacing $\mathbf{R}$ with $\hat{\mathbf{R}}$ in (2.3), we obtain the adaptive weight $\hat{\mathbf{w}}$ as

$$\hat{\mathbf{w}} = \frac{\hat{\mathbf{R}}^{-1} \mathbf{s}}{\mathbf{s}^\dagger \hat{\mathbf{R}}^{-1} \mathbf{s}}, \tag{2.6}$$

which is referred to as the SMI beamformer due to using the inversion of the sample matrix [15, Chap. 7]. After the adaptive array-filter processor, the normalized output SINR denoted by ρ^{match} can be written as

$$\begin{aligned} \rho^{\text{match}} &= \frac{|\hat{\mathbf{w}}^\dagger \mathbf{s}|^2}{(\hat{\mathbf{w}}^\dagger \mathbf{R} \hat{\mathbf{w}})(\mathbf{s}^\dagger \mathbf{R}^{-1} \mathbf{s})} \\ &= \frac{|\mathbf{s}^\dagger \hat{\mathbf{R}}^{-1} \mathbf{s}|^2}{(\mathbf{s}^\dagger \hat{\mathbf{R}}^{-1} \mathbf{R} \hat{\mathbf{R}}^{-1} \mathbf{s})(\mathbf{s}^\dagger \mathbf{R}^{-1} \mathbf{s})}. \end{aligned} \tag{2.7}$$

Note that ρ^{match} is random due to the random quantity $\mathbf{y}_k$ involved in $\hat{\mathbf{R}}$. In [1], RMB derived the expected value of the normalized output SINR as

$$\mathrm{E}(\rho^{\text{match}}) = \frac{K - N + 2}{K + 1}. \tag{2.8}$$

It leads to the well-known RMB rule that at least $K = 2N - 3 \approx 2N$ training data are required if one wishes that the average SINR loss is less than 3 dB.

2.1.1.2 Mismatched Case

Assume that the actual signal steering vector **s** is not aligned with the assumed vector **q**, and the environment is homogeneous, i.e., $\phi \neq 0$ and $\mathbf{R}_s = \mathbf{R}$. Replacing **R** and **s** with $\hat{\mathbf{R}}$ and **q** in (2.3), respectively, we obtain the adaptive weight $\hat{\mathbf{w}}$ to be

$$\hat{\mathbf{w}} = \frac{\hat{\mathbf{R}}^{-1}\mathbf{q}}{\mathbf{q}^\dagger\hat{\mathbf{R}}^{-1}\mathbf{q}}. \tag{2.9}$$

As a consequence, the normalized output SINR denoted by ρ^{mismatch} in the mismatched case can be expressed as

$$\rho^{\text{mismatch}} = \frac{|\mathbf{q}^\dagger\hat{\mathbf{R}}^{-1}\mathbf{s}|^2}{(\mathbf{q}^\dagger\hat{\mathbf{R}}^{-1}\mathbf{R}\hat{\mathbf{R}}^{-1}\mathbf{q})(\mathbf{s}^\dagger\mathbf{R}^{-1}\mathbf{s})}. \tag{2.10}$$

The expected value of the normalized output SINR in the mismatched case is shown to be [8]

$$E(\rho^{\text{mismatch}}) = \frac{1}{K+1}\left[1 + (K - N + 1)\cos^2\phi\right]. \tag{2.11}$$

2.1.2 Persymmetric SMI Beamformer

Note that **s**, **q**, and **R** possess persymmetric structures when a symmetrically spaced linear array with its center at the origin and/or symmetrically spaced pulse trains are used. In such cases, we have

$$\begin{cases} \mathbf{R} = \mathbf{J}\mathbf{R}^*\mathbf{J}, \\ \mathbf{q} = \mathbf{J}\mathbf{q}^*, \\ \mathbf{s} = \mathbf{J}\mathbf{s}^*, \end{cases} \tag{2.12}$$

where **J** is defined in (1.2). Using the training data and the persymmetric structure of **R**, we can obtain the MLE of **R** to be [16]

$$\breve{\mathbf{R}} = \frac{1}{2}\sum_{k=1}^{K}\left[\mathbf{y}_k\mathbf{y}_k^\dagger + \mathbf{J}(\mathbf{y}_k\mathbf{y}_k^\dagger)^*\mathbf{J}\right] \in \mathbb{C}^{N\times N}, \tag{2.13}$$

where the constraint $K \geqslant \lceil\frac{N}{2}\rceil$ has to be met to ensure with probability one the non-singularity of the estimated covariance matrix.

Using $\breve{\mathbf{R}}$ to replace $\hat{\mathbf{R}}$ in (2.7) or (2.10), we can obtain the normalized output SINR of the persymmetric beamformer. In the following two sections, we derive the expectation of the normalized output SINR of the persymmetric beamformer in the matched and mismatched cases.

2.2 Average SINR in Matched Case

In this section, we derive the expected value of the normalized output SINR of the persymmetric SMI beamformer for the matched case where $\phi = 0$ and $\mathbf{R}_s = \mathbf{R}$. Replacing $\mathbf{R}$ with $\breve{\mathbf{R}}$ in (2.3) leads to

$$\breve{\mathbf{w}} = \frac{\breve{\mathbf{R}}^{-1}\mathbf{s}}{\mathbf{s}^\dagger\breve{\mathbf{R}}^{-1}\mathbf{s}} \in \mathbb{C}^{N \times 1}. \tag{2.14}$$

As a result, the normalized output SINR of the persymmetric SMI beamformer in the absence of mismatch in the steering vector is

$$\rho_{\text{per}}^{\text{match}} = \frac{|\mathbf{s}^\dagger\breve{\mathbf{R}}^{-1}\mathbf{s}|^2}{(\mathbf{s}^\dagger\breve{\mathbf{R}}^{-1}\mathbf{R}\breve{\mathbf{R}}^{-1}\mathbf{s})(\mathbf{s}^\dagger\mathbf{R}^{-1}\mathbf{s})}. \tag{2.15}$$

To analyze the statistical property of $\rho_{\text{per}}^{\text{match}}$, we transform all quantities from the complex-valued domain to the real-valued domain. To do so, we define

$$\begin{cases} \mathbf{y}_k^{\text{e}} = \frac{1}{2}(\mathbf{y}_k + \mathbf{J}\mathbf{y}_k^*) \in \mathbb{C}^{N \times 1}, \\ \mathbf{y}_k^{\text{o}} = \frac{1}{2}(\mathbf{y}_k - \mathbf{J}\mathbf{y}_k^*) \in \mathbb{C}^{N \times 1}, \end{cases} \tag{2.16}$$

and then (2.13) can be rewritten as

$$\breve{\mathbf{R}} = \sum_{k=1}^{K} \left[\mathbf{y}_k^{\text{e}}(\mathbf{y}_k^{\text{e}})^\dagger + \mathbf{y}_k^{\text{o}}(\mathbf{y}_k^{\text{o}})^\dagger \right]. \tag{2.17}$$

It is easy to show that

$$\begin{cases} \mathbf{y}_k^{\text{e}} \sim \mathcal{CN}_N\left(\mathbf{0}_{N \times 1}, \mathbf{R}/2\right), \\ \mathbf{y}_k^{\text{o}} \sim \mathcal{CN}_N\left(\mathbf{0}_{N \times 1}, \mathbf{R}/2\right), \end{cases} \tag{2.18}$$

and

$$\mathrm{E}\{\mathbf{y}_k^{\text{e}}(\mathbf{y}_k^{\text{o}})^\dagger\} = \mathbf{0}_{N \times N}. \tag{2.19}$$

This means that $\mathbf{y}_k^{\text{e}}$ and $\mathbf{y}_k^{\text{o}}$ are IID.

Define a unitary transformation

$$\mathbf{D} = \frac{1}{2}\left[(\mathbf{I}_N + \mathbf{J}) + \jmath\,(\mathbf{I}_N - \mathbf{J})\right] \in \mathbb{C}^{N \times N}, \tag{2.20}$$

where $\mathbf{J}$ is the permutation matrix. Define

$$\begin{cases} \mathbf{y}_k^{\text{er}} \triangleq \mathbf{D}\mathbf{y}_k^{\text{e}}, \\ \mathbf{y}_k^{\text{or}} \triangleq -\jmath\,\mathbf{D}\mathbf{y}_k^{\text{o}}, \end{cases} \tag{2.21}$$

and then we can obtain [17]

$$\begin{cases} \mathbf{y}_k^{\text{er}} = \frac{1}{2}[(\mathbf{I}_N + \mathbf{J})\mathfrak{Re}(\mathbf{y}_k) - (\mathbf{I}_N - \mathbf{J})\mathfrak{Im}(\mathbf{y}_k)] \in \mathbb{R}^{N \times 1}, \\ \mathbf{y}_k^{\text{or}} = \frac{1}{2}[(\mathbf{I}_N - \mathbf{J})\mathfrak{Re}(\mathbf{y}_k) + (\mathbf{I}_N + \mathbf{J})\mathfrak{Im}(\mathbf{y}_k)] \in \mathbb{R}^{N \times 1}. \end{cases} \tag{2.22}$$

Clearly, both $\mathbf{y}_k^{\text{er}}$ and $\mathbf{y}_k^{\text{or}}$ are real-valued vectors. Moreover,

$$\begin{aligned} \mathrm{E}[\mathbf{y}_k^{\text{er}}(\mathbf{y}_k^{\text{er}})^\dagger] &= \mathrm{E}[\mathbf{y}_k^{\text{or}}(\mathbf{y}_k^{\text{or}})^\dagger] = \frac{1}{2}\mathbf{D}\mathbf{R}\mathbf{D}^\dagger \\ &= \frac{1}{2}[\mathfrak{Re}(\mathbf{R}) + \mathbf{J}\mathfrak{Im}(\mathbf{R})] \\ &\triangleq \mathbf{R}_r \in \mathbb{R}^{N \times N}, \end{aligned} \tag{2.23}$$

and

$$\mathrm{E}[\mathbf{y}_k^{\text{er}}(\mathbf{y}_k^{\text{or}})^\dagger] = \mathbf{0}_{N \times N}. \tag{2.24}$$

Hence, both $\mathbf{y}_k^{\text{er}}$ and $\mathbf{y}_k^{\text{or}}$ have the same real-valued Gaussian distribution with zero mean and covariance matrix $\mathbf{R}_r$ defined in (2.23), and they are independent.

Similarly, we define

$$\mathbf{s}_r \triangleq \mathbf{D}\mathbf{s} = \mathfrak{Re}(\mathbf{s}) - \mathfrak{Im}(\mathbf{s}) \in \mathbb{R}^{N \times 1}, \tag{2.25}$$

and

$$\begin{aligned} \check{\mathbf{R}}_r &\triangleq \mathbf{D}\check{\mathbf{R}}\mathbf{D}^\dagger = \mathfrak{Re}(\check{\mathbf{R}}) + \mathbf{J}\mathfrak{Im}(\check{\mathbf{R}}) \\ &= \sum_{k=1}^{K} \left[\mathbf{y}_k^{\text{er}}(\mathbf{y}_k^{\text{er}})^\dagger + \mathbf{y}_k^{\text{or}}(\mathbf{y}_k^{\text{or}})^\dagger \right] \\ &\sim \mathcal{W}_N(2K, \mathbf{R}_r), \end{aligned} \tag{2.26}$$

where the last equality is obtained from the fact that $\mathbf{y}_k^{\text{er}}$ and $\mathbf{y}_k^{\text{or}}$ are independent real-valued Gaussian vectors with zero mean and covariance matrix $\mathbf{R}_r$. Obviously, $\mathbf{s}_r$ and $\check{\mathbf{R}}_r$ are real-valued. Using (2.25) and (2.26), we can recast $\rho_{\text{per}}^{\text{match}}$ defined in (2.15) as

$$\rho_{\text{per}}^{\text{match}} = \frac{(\mathbf{s}_r^\dagger \check{\mathbf{R}}_r^{-1} \mathbf{s}_r)^2}{(\mathbf{s}_r^\dagger \check{\mathbf{R}}_r^{-1} \mathbf{R}_r \check{\mathbf{R}}_r^{-1} \mathbf{s}_r)(\mathbf{s}_r^\dagger \mathbf{R}_r^{-1} \mathbf{s}_r)}. \tag{2.27}$$

So far, the quantities $\mathbf{s}_r$, $\check{\mathbf{R}}_r$, and $\mathbf{R}_r$ in (2.27) are all real. Therefore, all the processing below for deriving the distribution of $\rho_{\text{per}}^{\text{match}}$ can be performed in the real domain. Note that the RMB's result cannot be directly extended to the real case, because the real Gaussian distribution is not a special case of the circularly symmetric, complex Gaussian distribution.

As derived in Appendix 2.A, we obtain an exact expression for $\mathrm{E}(\rho_{\text{per}}^{\text{match}})$ to be

$$\mathrm{E}(\rho_{\text{per}}^{\text{match}}) = \frac{2K - N + 2}{2K + 1}. \tag{2.28}$$

Remark 1: Compared to the RMB's rule in (2.8), (2.28) reveals that the exploitation of the persymmetric structures is equivalent to doubling the amount of secondary data. Therefore, the required amount of secondary data in the persymmetric SMI beamformer is reduced by half for yielding the same performance as the unstructured SMI beamformer. Interestingly, the average SINR loss is 3 dB when the amount of secondary data is $K = N - \frac{3}{2} \approx N$ for the persymmetric SMI beamformer, compared to $K = 2N - 3 \approx 2N$ for the unstructured SMI beamformer.

2.3 Average SINR in Mismatched Cases

2.3.1 Homogeneous Case

In this section, we consider the mismatched case in the homogeneous environment, i.e., $\mathbf{q} \neq \mathbf{s}$, and $\mathbf{R}_s = \mathbf{R}$.

Replacing $\mathbf{R}$ and $\mathbf{s}$ with $\breve{\mathbf{R}}$ and $\mathbf{q}$ in (2.3), respectively, the persymmetric SMI beamformer in the mismatched case is

$$\breve{\mathbf{w}} = \frac{\breve{\mathbf{R}}^{-1}\mathbf{q}}{\mathbf{q}^{\dagger}\breve{\mathbf{R}}^{-1}\mathbf{q}} \in \mathbb{C}^{N \times 1}. \tag{2.29}$$

Moreover, the normalized output SINR in this case, denoted by $\rho_{\text{per}}^{\text{mismatch1}}$, can be expressed as

$$\rho_{\text{per}}^{\text{mismatch1}} = \frac{|\mathbf{q}^{\dagger}\breve{\mathbf{R}}^{-1}\mathbf{s}|^2}{(\mathbf{q}^{\dagger}\breve{\mathbf{R}}^{-1}\mathbf{R}\breve{\mathbf{R}}^{-1}\mathbf{q})(\mathbf{s}^{\dagger}\mathbf{R}^{-1}\mathbf{s})}. \tag{2.30}$$

In the following, we derive the distribution of $\rho_{\text{per}}^{\text{mismatch1}}$ for the mismatched case.

Define

$$\mathbf{q}_r \triangleq \mathbf{Dq} = \mathfrak{Re}(\mathbf{q}) - \mathfrak{Im}(\mathbf{q}) \in \mathbb{R}^{N \times 1}. \tag{2.31}$$

Using (2.23), (2.25), (2.26), and (2.31), we can recast $\rho_{\text{per}}^{\text{mismatch1}}$ defined in (2.30) as

$$\rho_{\text{per}}^{\text{mismatch1}} = \frac{(\mathbf{q}_r^T \breve{\mathbf{R}}_r^{-1}\mathbf{s}_r)^2}{(\mathbf{q}_r^T \breve{\mathbf{R}}_r^{-1}\mathbf{R}_r\breve{\mathbf{R}}_r^{-1}\mathbf{q}_r)(\mathbf{s}_r^T \mathbf{R}_r^{-1}\mathbf{s}_r)}, \tag{2.32}$$

where all the quantities are real-valued. Therefore, all the processing below for deriving the distribution of $\rho_{\text{per}}^{\text{mismatch1}}$ can be performed in the real domain.

Theorem 2.3.1. *When $\breve{\mathbf{R}}_r \sim \mathcal{W}_N(2K, \mathbf{R}_r)$ with $K \geqslant \lceil \frac{N}{2} \rceil$, and*

$$\cos^2 \phi = \frac{(\mathbf{q}_r^T \mathbf{R}_r^{-1}\mathbf{s}_r)^2}{(\mathbf{q}_r^T \mathbf{R}_r^{-1}\mathbf{q}_r)(\mathbf{s}_r^T \mathbf{R}_r^{-1}\mathbf{s}_r)} \neq 0, \tag{2.33}$$

the PDF of $\rho_{per}^{mismatch1}$ is

$$\mathcal{P}(\bar{\rho}) = \begin{cases} c\,\bar{h}(\bar{\rho})\bar{g}_1(\bar{\rho}), & \text{for } \sin^2\phi < \bar{\rho} \leqslant 1, \\ \bar{k}, & \text{for } \bar{\rho} = \sin^2\phi, \\ c\,\bar{h}(\bar{\rho})\bar{g}_2(\bar{\rho}), & \text{for } 0 < \bar{\rho} < \sin^2\phi, \end{cases} \tag{2.34}$$

where

$$c = \frac{1}{B\left(\frac{1}{2}, \frac{N-2}{2}\right) B\left(\frac{2K-N+2}{2}, \frac{N-1}{2}\right)}, \tag{2.35}$$

$$\bar{k} = c\,(4\cos^2\phi)^{K-1}(\sin^2\phi)^{\frac{2K-N}{2}} B\left(\frac{N-2}{2}, \frac{4K-N}{2}\right), \tag{2.36}$$

$$\bar{h}(\bar{\rho}) = \frac{|\bar{\rho} - \sin^2\phi|^{\frac{4K-N}{2}}(1-\bar{\rho})^{\frac{N-3}{2}}}{\bar{\rho}^K\,(\cos^2\phi)^{\frac{2K-1}{2}}}, \tag{2.37}$$

$$\bar{g}_1(\bar{\rho}) = B\left(\frac{N-2}{2}, 1\right)\left[F_1\left(1, 2K-1, -\frac{N-4}{2}, \frac{N}{2}; \sqrt{\bar{w}}, -1\right)\right.$$
$$\left. + F_1\left(1, 2K-1, -\frac{N-4}{2}, \frac{N}{2}; -\sqrt{\bar{w}}, -1\right)\right], \tag{2.38}$$

and

$$\bar{g}_2(\bar{\rho}) = w^{-\frac{2K-1}{2}} B\left(\frac{N-2}{2}, 2K-N+2\right)$$
$$\times\left[F_1\left(2K-N+2, 2K-1, -\frac{N-4}{2}, L; \frac{1}{\sqrt{\bar{w}}}, -1\right)\right.$$
$$\left. + F_1\left(2K-N+2, 2K-1, -\frac{N-4}{2}, L; \frac{-1}{\sqrt{\bar{w}}}, -1\right)\right] \tag{2.39}$$

with

$$L = \frac{4K-N+2}{2}, \tag{2.40}$$

and

$$\bar{w} = \frac{(1-\bar{\rho})\sin^2\phi}{\bar{\rho}\cos^2\phi}. \tag{2.41}$$

Proof. The proof is given in Appendix 2.B. □

Remark 2: In the matched case where $\cos^2\phi = 1$ and $\sin^2\phi = 0$, the PDF of $\rho_{per}^{mismatch1}$ in (2.34) reduces to be

$$\mathcal{P}(\bar{\rho}) = \begin{cases} c\,\alpha\bar{\rho}^{\frac{2K-N}{2}}(1-\bar{\rho})^{\frac{N-3}{2}}, & \text{for } 0 < \bar{\rho} \leqslant 1, \\ 0, & \text{for } \bar{\rho} = 0, \end{cases} \tag{2.42}$$

where

$$\alpha \triangleq 2 \int_0^1 (1 - \gamma^2)^{\frac{N-4}{2}} d\gamma$$
$$= B\left(\frac{1}{2}, \frac{N-2}{2}\right). \tag{2.43}$$

The second equality in (2.43) is obtained using [20, p. 324, eq. (3.249.5)]. Substituting (2.35) and (2.43) into (2.42) yields

$$\mathcal{P}(\bar\rho) = \frac{1}{B\left(\frac{2K-N+2}{2}, \frac{N-1}{2}\right)} \bar\rho^{\frac{2K-N}{2}} (1 - \bar\rho)^{\frac{N-3}{2}}, \quad 0 \leqslant \bar\rho \leqslant 1, \tag{2.44}$$

which is exactly the result in [19, eq. (A.25)]. The expectation can be obtained by numerical calculations using the theoretical expression PDF in (2.44).

2.3.2   Non-Homogeneous Case

Now, we consider the mismatced case in the non-homogeneous environment, i.e., $\phi \neq 0$ and $\mathbf{R}_s \neq \mathbf{R}$. Replacing $\mathbf{R}$ and $\mathbf{s}$ with $\check{\mathbf{R}}$ and $\mathbf{q}$ in (2.3), respectively, leads to the persymmetric SMI beamformer in the mismatched case as

$$\check{\mathbf{w}} = \frac{\check{\mathbf{R}}^{-1}\mathbf{q}}{\mathbf{q}^\dagger \check{\mathbf{R}}^{-1}\mathbf{q}}. \tag{2.45}$$

Moreover, the normalized output SINR in this case, denoted by $\rho_{\text{per}}^{\text{mismatch2}}$, can be expressed as

$$\rho_{\text{per}}^{\text{mismatch2}} = \frac{|\mathbf{q}^\dagger \check{\mathbf{R}}^{-1}\mathbf{s}|^2}{(\mathbf{q}^\dagger \check{\mathbf{R}}^{-1}\mathbf{R}\check{\mathbf{R}}^{-1}\mathbf{q})(\mathbf{s}^\dagger \mathbf{R}^{-1}\mathbf{s})}. \tag{2.46}$$

It seems difficult, if not impossible, to derive an exact expression for the average SINR loss in such a case. In the following, we turn to seek an approximate expression for it.

According to [21, eq. (30)], we can obtain

$$\mathrm{E}(\rho_{\text{per}}^{\text{mismatch2}}) = \delta_{\text{appro}}\Phi + \Delta, \tag{2.47}$$

where

$$\delta_{\text{appro}} = \frac{\mathbf{s}^\dagger \mathrm{E}\{\check{\mathbf{w}}\check{\mathbf{w}}^\dagger\}\mathbf{s}}{\mathrm{tr}\left[\mathbf{R}\,\mathrm{E}\{\check{\mathbf{w}}\check{\mathbf{w}}^\dagger\}\right](\mathbf{s}^\dagger \mathbf{R}^{-1}\mathbf{s})}, \tag{2.48}$$

$$\Phi = \mathrm{E}\left\{\frac{\mathrm{E}(\check{\mathbf{w}}^\dagger \mathbf{R}\check{\mathbf{w}})}{\check{\mathbf{w}}^\dagger \mathbf{R}\check{\mathbf{w}}}\right\}, \tag{2.49}$$

and

$$\Delta = \mathrm{E}\left\{\frac{\mathbf{s}^\dagger \check{\mathbf{w}}\check{\mathbf{w}}^\dagger \mathbf{s} - \mathrm{E}(\mathbf{s}^\dagger \check{\mathbf{w}}\check{\mathbf{w}}^\dagger \mathbf{s})}{\check{\mathbf{w}}^\dagger \mathbf{R}\check{\mathbf{w}}\,\mathbf{s}^\dagger \mathbf{R}^{-1}\mathbf{s}}\right\} \tag{2.50}$$

with $\breve{\mathbf{w}}$ defined in (2.45). When $\Phi \approx 1$ and $\Delta \approx 0$, we have

$$E(\rho_{\text{per}}^{\text{mismatch2}}) \approx \delta_{\text{appro}}$$

$$= \frac{\mathbf{s}^\dagger \, E\{\breve{\mathbf{w}}\breve{\mathbf{w}}^\dagger\} \, \mathbf{s}}{\text{tr}\left[\mathbf{R} \, E\{\breve{\mathbf{w}}\breve{\mathbf{w}}^\dagger\}\right] (\mathbf{s}^\dagger \mathbf{R}^{-1} \mathbf{s})}. \tag{2.51}$$

Remark 3: Here we discuss the situation where $\Phi \approx 1$ and $\Delta \approx 0$. In fact, for $K \to \infty$, $\breve{\mathbf{R}}$ tends to be the true covariance matrix $\mathbf{R}_s$. Then, $\breve{\mathbf{w}}$ can be approximately seen as a constant. As a result, $\Phi \approx 1$ and $\Delta \approx 0$. This theoretical prediction will be verified in Section 2.4. In particular, simulation results in Section 2.4 show that the approximation is very good even for moderate K.

Until now, the problem of deriving the average SINR loss turns to calculate $E\{\breve{\mathbf{w}}\breve{\mathbf{w}}^\dagger\}$. As derived in Appendix 2.F, $E\{\breve{\mathbf{w}}\breve{\mathbf{w}}^\dagger\}$ is given as

$$E(\breve{\mathbf{w}}\breve{\mathbf{w}}^\dagger) = \frac{1}{(2K-N)(\mathbf{q}^\dagger \mathbf{R}_s^{-1} \mathbf{q})} \left[\mathbf{R}_s^{-1} + (2K-N-1) \frac{\mathbf{R}_s^{-1} \mathbf{q}\mathbf{q}^\dagger \mathbf{R}_s^{-1}}{\mathbf{q}^\dagger \mathbf{R}_s^{-1} \mathbf{q}}\right]. \tag{2.52}$$

Inserting (2.52) into (2.47) results in

$$E(\rho_{\text{per}}^{\text{mismatch2}}) \approx \frac{\frac{\mathbf{s}^\dagger \mathbf{R}_s^{-1} \mathbf{s}}{\mathbf{s}^\dagger \mathbf{R}^{-1} \mathbf{s}} + \frac{(2K-N-1)|\mathbf{s}^\dagger \mathbf{R}_s^{-1} \mathbf{q}|^2}{(\mathbf{q}^\dagger \mathbf{R}_s^{-1} \mathbf{q})(\mathbf{s}^\dagger \mathbf{R}^{-1} \mathbf{s})}}{\text{tr}\left[\mathbf{R}\mathbf{R}_s^{-1} + \frac{(2K-N-1)\mathbf{R}\mathbf{R}_s^{-1}\mathbf{q}\mathbf{q}^\dagger \mathbf{R}_s^{-1}}{\mathbf{q}^\dagger \mathbf{R}_s^{-1} \mathbf{q}}\right]}. \tag{2.53}$$

We consider a special case where the environment is homogeneous (i.e., $\mathbf{R}_s = \mathbf{R}$), but mismatch exists in the signal steering vector (i.e., $\mathbf{s} \neq \mathbf{q}$). In such a case, the average SINR loss has the simple form as

$$E(\rho_{\text{per}}^{\text{mismatch2}}) \approx \frac{1}{2K-1}\left[1 + (2K-N-1)\cos^2 \phi\right], \tag{2.54}$$

where $\cos^2 \phi$ is defined in (2.4).

Remark 4: Comparison between (2.11) and (2.54) reveals that the amount of training data is approximately doubled when the persymmetric structure is exploited.

2.4   Simulation Results

In this section, numerical simulations are conducted to attest to the validity of the above theoretical results. The average SINR loss is defined as

$$\text{Average SINR Loss} = -10\log_{10}\{E(\rho)\}, \tag{2.55}$$

where ρ is the normalized output SINR in the corresponding case. For example, $\rho = \rho_{\text{per}}^{\text{mismatch1}}$ is the normalized output SINR of the persymmetric SMI beamformer in the mismatched case in HEs.

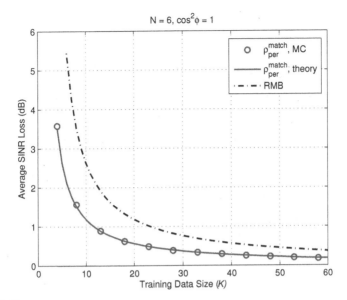

FIGURE 2.1
Average SINR loss versus K in matched case in homogeneous environment.

Assume that a ULA with N sensors spaced a half wavelength separation is used. The steering vector for a signal coming from the DOA θ can be written as

$$\mathbf{t}(\theta) = \frac{1}{\sqrt{N}} \left[e^{-j\frac{N}{2}\pi \sin\theta}, \ldots, e^{-j\pi \sin\theta}, \ e^{j\pi \sin\theta}, \ldots, e^{j\frac{N}{2}\pi \sin\theta} \right]^T \quad (2.56)$$

for even N, and

$$\mathbf{t}(\theta) = \frac{1}{\sqrt{N}} \left[e^{-j\frac{N-1}{2}\pi \sin\theta}, \ldots, e^{-j\pi \sin\theta}, 1, e^{j\pi \sin\theta}, \ldots, e^{j\frac{N-1}{2}\pi \sin\theta} \right]^T \quad (2.57)$$

for odd N. The DOA of a signal of interest is assumed to be $\theta_0 = 40°$. Suppose that there are two interference signals with the DOAs $\theta_1 = -30°$ and $\theta_2 = 20°$, respectively. Therefore, the covariance matrix $\mathbf{R}$ in the test data has the form

$$\mathbf{R} = \sum_{k=1}^{2} \sigma_k^2 \mathbf{t}(\theta_k)\mathbf{t}(\theta_k)^\dagger + \sigma^2 \mathbf{I}, \quad (2.58)$$

where σ_k^2 is the kth interference power for $k = 1, 2$, and σ^2 denotes the white noise power. The INR is set to be 20 dB.

First, we consider the matched case where $\mathbf{s} = \mathbf{q}$ and $\mathbf{R}_s = \mathbf{R}$. Fig. 2.1 plots the average SINR loss as a function of K for $N = 6$. The solid and dotted-dashed lines represent the results obtained by using (2.28) and (2.8), respectively. The symbol "∘" denotes the results obtained by MC simulation techniques. The number of independent trials used for each case is 1 000.

It can be observed from Fig. 2.1 that the theoretical results obtained by (2.28) are in good agreement with the MC simulation results. It is also demonstrated that the persymmetric SMI beamformer outperforms the unstructured counterpart in terms of the output SINR, when using the same number of training data. It is worth noticing that within the same SINR loss, the amount of secondary data required by the unstructured SMI beamformer is twice as that required by the persymmetric SMI beamformer. This observation is consistent with the theoretical analysis in Section 2.2. It should be pointed out that K can lie in the range $N/2 \leqslant K < N$ in the persymmetric SMI beamformer, compared to the unstructured SMI beamformer where K has to be no less than N. This is to say, the persymmetric SMI beamformer can work in a more restrictive case of limited training data (i.e., $N/2 \leqslant K < N$).

We plot the average SINR loss with respect to K in Fig. 2.2 in the mismatched case in HEs, where the symbol "∘" denotes the MC results, the solid line represents the results obtained by numerical calculations using the theoretical expression in (2.34), the dotted-dashed line denotes the Boroson's result in [8], and the symbol "+" denotes the results calculated with the approximate expression in (2.54). We can observe that the requirement of homogeneous training data can be greatly alleviated by using persymmetry to achieve the same performance. As expected, the number of training data required in the persymmetric SMI beamformer is approximately half of that required in the unstructured SMI beamformer, when the average SINR loss is 3 dB.

The above simulations are conducted in the HE (i.e., $\mathbf{R}_s = \mathbf{R}$). In the following, we consider the non-homogeneous environment where $\mathbf{R}_s \neq \mathbf{R}$. The unwanted signal $\mathbf{n}$ is modeled as an exponentially correlated complex Gaussian vector with one-lag correlation coefficient [22]. For simplicity, we select the (n, m)th elements of the covariance matrices in the test and training data as, respectively,

$$\mathbf{R}(n, m) = \dot{\sigma}^2 0.9^{|n-m|} \exp(\jmath\pi(n - m)/2) \tag{2.59}$$

and

$$\mathbf{R}_s(n, m) = 2\dot{\sigma}^2 0.98^{|n-m|} \exp(\jmath\pi(n - m)/2), \tag{2.60}$$

where $\dot{\sigma}^2$ is the power of $\mathbf{n}$ in the test data. It means that both the power and one-lag correlation coefficient of $\mathbf{n}$ are different in the test and training data.

In the non-homogeneous case where $\mathbf{R}_s \neq \mathbf{R}$, we derived the approximate expression (2.53) for the average SINR loss of the persymmetric SMI beamformer. The average SINR loss curves of the persymmetric SMI beamformer versus K are plotted in Fig. 2.3. We can observe that in the non-homogeneous environment, the results obtained by the approximate expression (2.53) are very close to those obtained by the MC simulations, especially for the case of large K.

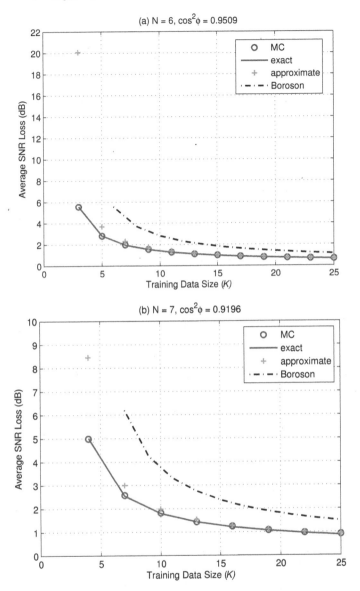

FIGURE 2.2
Average SINR loss versus K in mismatched case in homogeneous environment.
(a) $N = 6$; (b) $N = 7$.

2.A   Derivation of $\mathbf{E}(\rho_{\mathbf{per}}^{\mathbf{match}})$

Define by $\mathbf{e}_1$ the first elementary vector in $\mathbb{R}^{N \times 1}$, namely,

$$\mathbf{e}_1 \triangleq [1, 0, \dots, 0]^T \in \mathbb{R}^{N \times 1}. \tag{2.A.1}$$

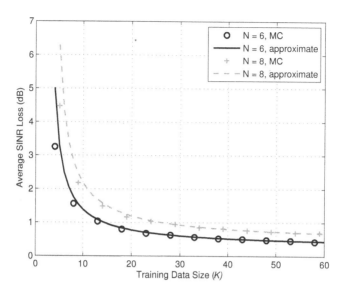

FIGURE 2.3

Average SINR loss versus K in the presence of steering vector mismatch in the non-homogeneous environment.

There always exists a real orthogonal matrix $\bar{\mathbf{U}} \in \mathbb{R}^{N \times N}$ such that

$$\mathbf{e}_1 = \frac{\bar{\mathbf{U}} \mathbf{R}_r^{-1/2} \mathbf{s}_r}{(\mathbf{s}_r^\dagger \mathbf{R}_r^{-1} \mathbf{s}_r)^{1/2}}. \tag{2.A.2}$$

As a result, ρ_{per} defined in (2.27) can be rewritten as

$$\rho_{\text{per}}^{\text{match}} = \frac{(\mathbf{e}_1^T \mathbf{V}^{-1} \mathbf{e}_1)^2}{\mathbf{e}_1^T \mathbf{V}^{-2} \mathbf{e}_1}, \tag{2.A.3}$$

where

$$\mathbf{V} = \bar{\mathbf{U}} \mathbf{R}_r^{-1/2} \breve{\mathbf{R}}_r \mathbf{R}_r^{-1/2} \bar{\mathbf{U}}^T \in \mathbb{R}^{N \times N}. \tag{2.A.4}$$

Since $\mathbf{y}_k^{\text{er}}$ and $\mathbf{y}_k^{\text{or}}$ are independent real Gaussian vectors with zero mean and covariance matrix $\mathbf{R}_r$, we have

$$\breve{\mathbf{R}}_r \sim \mathcal{W}_N(2K, \mathbf{R}_r). \tag{2.A.5}$$

Hence, $\mathbf{V}$ has a real Wishart distribution, i.e.,

$$\mathbf{V} \sim \mathcal{W}_N(2K, \mathbf{I}). \tag{2.A.6}$$

It is easy to verify that

$$\mathbf{J} \mathbf{V} \mathbf{J} \sim \mathcal{W}_N(2K, \mathbf{I}). \tag{2.A.7}$$

This is to say, $\mathbf{V}$ and $\mathbf{JVJ}$ have the same distribution $\mathcal{W}_N(K, \mathbf{I})$. Hence, the distribution of ρ_{per} in (2.A.3) remains the same when $\mathbf{V}$ is replaced with $\mathbf{JVJ}$, namely,

$$\rho_{\text{per}}^{\text{match}} \overset{\text{d}}{=} \frac{(\mathbf{e}_1^T \mathbf{J V}^{-1} \mathbf{J e}_1)^2}{\mathbf{e}_1^T \mathbf{J V}^{-2} \mathbf{J e}_1}. \tag{2.A.8}$$

According to the Bartlett's decomposition of a real Wishart distribution [23, p. 253, Corollary 7.2.1], we have $\mathbf{V} = \tilde{\mathbf{V}} \tilde{\mathbf{V}}^T$ where $\tilde{\mathbf{V}}$ is a lower triangular matrix with

$$\tilde{v}_{i,i}^2 \sim \chi_{2K-i+1}^2, \tag{2.A.9}$$

and

$$\tilde{v}_{i,j} \sim \mathcal{N}_1(0, 1), \quad 1 \leqslant j < i \leqslant N. \tag{2.A.10}$$

In addition, these entries in $\tilde{\mathbf{V}}$ are independent of each other. Define

$$\mathbf{Q} = \mathbf{J} \tilde{\mathbf{V}} \mathbf{J} \in \mathbb{R}^{N \times N}. \tag{2.A.11}$$

It follows that $\mathbf{Q}$ is upper triangular, and

$$q_{i,j} = \tilde{v}_{N-j+1,N-i+1}, \quad 1 \leqslant i < j \leqslant N. \tag{2.A.12}$$

Therefore,

$$q_{i,i}^2 = \tilde{v}_{N-i+1,N-i+1}^2 \sim \chi_{2K-N+i}^2, \tag{2.A.13}$$

and

$$q_{i,j} = \tilde{v}_{N-j+1,N-i+1} \sim \mathcal{N}_1(0, 1), \quad 1 \leqslant i < j \leqslant N. \tag{2.A.14}$$

Moreover, the elements in the random matrix $\mathbf{Q}$ are independent of each other. Let $\mathbf{G} = \mathbf{Q}^{-1} \in \mathbb{R}^{N \times N}$. Obviously, $\mathbf{G}$ is also upper triangular, because $\mathbf{Q}$ is upper triangular. Thus,

$$\mathbf{G} \mathbf{e}_1 = g_{1,1} \mathbf{e}_1. \tag{2.A.15}$$

Due to $\mathbf{J} = \mathbf{J}^T$ and $\mathbf{J}^{-1} = \mathbf{J}$, we have

$$\begin{aligned}
\mathbf{V}^{-1} &= (\tilde{\mathbf{V}}^T)^{-1} (\tilde{\mathbf{V}})^{-1} \\
&= (\mathbf{J} \mathbf{Q}^T \mathbf{J})^{-1} (\mathbf{J} \mathbf{Q} \mathbf{J})^{-1} \\
&= \mathbf{J} \mathbf{Q}^{-T} \mathbf{Q}^{-1} \mathbf{J}.
\end{aligned} \tag{2.A.16}$$

Taking (2.A.16) into (2.A.8) produces

$$\begin{aligned}
\rho_{\text{per}}^{\text{match}} &\overset{\text{d}}{=} \frac{(\mathbf{e}_1^T \mathbf{G}^T \mathbf{G} \mathbf{e}_1)^2}{\mathbf{e}_1^T \mathbf{G}^T \mathbf{G} \mathbf{G}^T \mathbf{G} \mathbf{e}_1} \\
&= \frac{g_{1,1}^2}{\mathbf{e}_1^T \mathbf{G} \mathbf{G}^T \mathbf{e}_1} \\
&= \frac{g_{1,1}^2}{\sum_{m=1}^N g_{1,m}^2}.
\end{aligned} \tag{2.A.17}$$

Since $\mathbf{GQ} = \mathbf{I}$ and $\mathbf{G}$, $\mathbf{Q}$ are both upper triangular, we have

$$\sum_{i=1}^{n+1} g_{1,i} q_{i,n+1} = 0, \quad 1 \leqslant n < N. \tag{2.A.18}$$

Consequently,

$$\begin{aligned} g_{1,n+1}^2 &= \frac{\left(\sum_{i=1}^n g_{1,i} q_{i,n+1}\right)^2}{q_{n+1,n+1}^2} \\ &= \alpha_n \frac{\sum_{i=1}^n g_{1,i}^2}{q_{n+1,n+1}^2}, \end{aligned} \tag{2.A.19}$$

where

$$\begin{aligned} \alpha_n &= \frac{\left(\sum_{i=1}^n g_{1,i} q_{i,n+1}\right)^2}{\sum_{i=1}^n g_{1,i}^2} \\ &\sim \chi_1^2 \end{aligned} \tag{2.A.20}$$

with the last equation obtained from

$$\frac{\sum_{i=1}^n g_{1,i}\, q_{i,n+1}}{\left(\sum_{i=1}^n g_{1,i}^2\right)^{1/2}} \sim \mathcal{N}_1(0,1). \tag{2.A.21}$$

In addition, α_n is independent of $q_{n+1,n+1}$, since α_n is the square of a linear combination of $q_{i,n+1}, i = 1, 2, \ldots, n$. Using (2.A.19), we can rewrite the denominator of (2.A.17) as

$$\begin{aligned} \sum_{m=1}^N g_{1,m}^2 &= g_{1,N}^2 + \sum_{m=1}^{N-1} g_{1,m}^2 \\ &= \left(1 + \frac{\alpha_{N-1}}{q_{N,N}^2}\right) \sum_{m=1}^{N-1} g_{1,m}^2 \\ &= \left(1 + \frac{\alpha_{N-1}}{q_{N,N}^2}\right)\left(1 + \frac{\alpha_{N-2}}{q_{N-1,N-1}^2}\right) \sum_{m=1}^{N-2} g_{1,m}^2 \\ &= g_{1,1}^2 \prod_{j=2}^N \left(1 + \frac{\alpha_{j-1}}{q_{j,j}^2}\right), \end{aligned} \tag{2.A.22}$$

where the last equality is obtained by repeatedly using (2.A.19). Inserting (2.A.22) into (2.A.17) yields

$$\rho_{\text{per}}^{\text{match}} = \prod_{j=2}^N \left(1 + \frac{\alpha_{j-1}}{q_{j,j}^2}\right)^{-1}. \tag{2.A.23}$$

Using (2.A.13) and (2.A.20), we can obtain that

$$\left(1 + \frac{\alpha_{j-1}}{q_{j,j}^2}\right)^{-1} \sim \beta\left(\frac{2K - N + j}{2}, \frac{1}{2}\right),\qquad(2.A.24)$$

where $j = 2, 3, \ldots, N$. According to Theorem 2 of [24], we have

$$\rho_{\text{per}}^{\text{match}} \sim \beta\left(\frac{2K - N + 2}{2}, \frac{N - 1}{2}\right).\qquad(2.A.25)$$

Accordingly, the expectation of ρ_{per} can be easily calculated using the above distribution, which corresponds to the result in (2.28).

2.B Proof of Theorem 2.3.1

Let us start by defining

$$\begin{aligned}
\rho &= \frac{\rho_{\text{per}}^{\text{mismatch1}}}{\cos^2 \phi} \\
&= \frac{(\mathbf{q}_r^T \check{\mathbf{R}}_r^{-1} \mathbf{s}_r)^2 (\mathbf{q}_r^T \mathbf{R}_r^{-1} \mathbf{q}_r)}{(\mathbf{q}_r^T \check{\mathbf{R}}_r^{-1} \mathbf{R}_r \check{\mathbf{R}}_r^{-1} \mathbf{q}_r)(\mathbf{q}_r^T \mathbf{R}_r^{-1} \mathbf{s}_r)^2}.
\end{aligned}\qquad(2.B.1)$$

Our focus now is to derive the distribution of ρ. Then, we can obtain the distribution of $\rho_{\text{per}}^{\text{mismatch1}}$ by using the transformation $\rho = \frac{\rho_{\text{per}}^{\text{mismatch1}}}{\cos^2 \phi}$.

There always exists a real-valued orthogonal matrix $\mathbf{U}$ such that

$$\mathbf{e}_1 = \frac{\mathbf{U}\mathbf{R}_r^{-1/2}\mathbf{q}_r}{(\mathbf{q}_r^T \mathbf{R}_r^{-1} \mathbf{q}_r)^{1/2}},\qquad(2.B.2)$$

where $\mathbf{e}_1$ is defined in (2.A.1). Let

$$\begin{aligned}
\tilde{\mathbf{s}} &= \mathbf{U}\mathbf{R}_r^{-1/2}\mathbf{s}_r \\
&\triangleq \left[\tilde{s}_1, \tilde{\mathbf{s}}_2^T\right]^T \in \mathbb{R}^{N \times 1},
\end{aligned}\qquad(2.B.3)$$

where $\tilde{s}_1$ is a real-valued scalar, and $\tilde{\mathbf{s}}_2 \in \mathbb{R}^{(N-1) \times 1}$. Then, we have

$$\begin{aligned}
(\mathbf{q}_r^T \mathbf{R}_r^{-1} \mathbf{s}_r)^2 &= (\mathbf{q}_r^T \mathbf{R}_r^{-1} \mathbf{q}_r)\mathbf{e}_1^T \mathbf{U}\mathbf{R}_r^{-1/2} \mathbf{s}_r \mathbf{s}_r^T \mathbf{R}_r^{-1/2} \mathbf{U}^T \mathbf{e}_1 \\
&= (\mathbf{q}_r^T \mathbf{R}_r^{-1} \mathbf{q}_r)\mathbf{e}_1^T \tilde{\mathbf{s}}\tilde{\mathbf{s}}^T \mathbf{e}_1 \\
&= (\mathbf{q}_r^T \mathbf{R}_r^{-1} \mathbf{q}_r)\tilde{s}_1^2,
\end{aligned}\qquad(2.B.4)$$

and

$$\begin{aligned}
\mathbf{q}_r^T \check{\mathbf{R}}_r^{-1} \mathbf{R}_r \check{\mathbf{R}}_r^{-1} \mathbf{q}_r &= (\mathbf{q}_r^T \mathbf{R}_r^{-1} \mathbf{q}_r)\mathbf{e}_1^T \mathbf{U}\mathbf{R}_r^{1/2}\check{\mathbf{R}}_r^{-1} \mathbf{R}_r \check{\mathbf{R}}_r^{-1}\mathbf{R}_r^{1/2}\mathbf{U}^T \mathbf{e}_1 \\
&= (\mathbf{q}_r^T \mathbf{R}_r^{-1} \mathbf{q}_r)\mathbf{e}_1^T \mathbf{W}^{-2}\mathbf{e}_1,
\end{aligned}\qquad(2.B.5)$$

where

$$\mathbf{W} = \mathbf{U}\mathbf{R}_r^{-1/2}\breve{\mathbf{R}}_r\mathbf{R}_r^{-1/2}\mathbf{U}^T$$

$$\triangleq \begin{bmatrix} w_{11} & \mathbf{w}_{21}^T \\ \mathbf{w}_{21} & \mathbf{W}_{22} \end{bmatrix} \in \mathbb{R}^{N \times N}. \tag{2.B.6}$$

Clearly,

$$\mathbf{W} \sim \mathcal{W}_N(2K, \mathbf{I}_N). \tag{2.B.7}$$

In addition,

$$(\mathbf{q}_r^T\breve{\mathbf{R}}_r^{-1}\mathbf{s}_r)^2 = (\mathbf{q}_r^T\mathbf{R}_r^{-1}\mathbf{q}_r)\mathbf{e}_1^T\mathbf{U}\mathbf{R}_r^{1/2}\breve{\mathbf{R}}_r^{-1}\mathbf{s}_r\mathbf{s}_r^T\breve{\mathbf{R}}_r^{-1}\mathbf{R}_r^{1/2}\mathbf{U}^T\mathbf{e}_1$$

$$= (\mathbf{q}_r^T\mathbf{R}_r^{-1}\mathbf{q}_r)\mathbf{e}_1^T\mathbf{W}^{-1}\tilde{\mathbf{s}}\tilde{\mathbf{s}}^T\mathbf{W}^{-1}\mathbf{e}_1. \tag{2.B.8}$$

Substituting (2.B.4), (2.B.5), and (2.B.8) into (2.B.1) leads to

$$\rho = \frac{\mathbf{e}_1^T\mathbf{W}^{-1}\tilde{\mathbf{s}}\tilde{\mathbf{s}}^T\mathbf{W}^{-1}\mathbf{e}_1}{\mathbf{e}_1^T\mathbf{W}^{-2}\mathbf{e}_1\tilde{s}_1^2}. \tag{2.B.9}$$

Define

$$\mathbf{M} = \mathbf{W}^{-1} \triangleq \begin{bmatrix} m_{11} & \mathbf{m}_{21}^T \\ \mathbf{m}_{21} & \mathbf{M}_{22} \end{bmatrix} \in \mathbb{R}^{N \times N}, \tag{2.B.10}$$

where m_{11} is a real-valued scalar, $\mathbf{m}_{21} \in \mathbb{R}^{(N-1) \times 1}$, and $\mathbf{M}_{22} \in \mathbb{R}^{(N-1) \times (N-1)}$. According to the partitioned matrix inversion theorem [25], we have

$$\mathbf{m}_{21} = -m_{11}\mathbf{W}_{22}^{-1}\mathbf{w}_{21}. \tag{2.B.11}$$

Further,

$$\mathbf{e}_1^T\mathbf{W}^{-1}\tilde{\mathbf{s}}\tilde{\mathbf{s}}^T\mathbf{W}^{-1}\mathbf{e}_1 = \left(\tilde{s}_1 m_{11} + \mathbf{m}_{21}^T\tilde{\mathbf{s}}_2\right)^2, \tag{2.B.12}$$

and

$$\mathbf{e}_1^T\mathbf{W}^{-2}\mathbf{e}_1 = m_{11}^2 + \mathbf{m}_{21}^T\mathbf{m}_{21}. \tag{2.B.13}$$

It is known that a real-valued orthogonal matrix $\mathbf{T} \in \mathbb{R}^{(N-1) \times (N-1)}$ always exists such that

$$\mathbf{e}_2 = \frac{\mathbf{T}\tilde{\mathbf{s}}_2}{\left(\tilde{\mathbf{s}}_2^T\tilde{\mathbf{s}}_2\right)^{1/2}} = [1, 0, \ldots, 0]^T \in \mathbb{R}^{(N-1) \times 1}. \tag{2.B.14}$$

Define

$$\mathbf{y} = \frac{1}{m_{11}}\mathbf{T}\mathbf{m}_{21} \triangleq [y_1, \mathbf{y}_2^T]^T \in \mathbb{R}^{(N-1) \times 1}, \tag{2.B.15}$$

where y_1 is a real-valued scalar, and

$$\mathbf{y}_2 \triangleq [y_2, y_3, \ldots, y_{N-1}]^T \in \mathbb{R}^{(N-2) \times 1}. \tag{2.B.16}$$

Then, we can obtain

$$\tilde{s}_1 m_{11} + \mathbf{m}_{21}^T\tilde{\mathbf{s}}_2 = \tilde{s}_1 m_{11} + \left(\tilde{\mathbf{s}}_2^T\tilde{\mathbf{s}}_2\right)^{1/2}\mathbf{m}_{21}^T\mathbf{T}^T\mathbf{e}_2$$

$$= m_{11}\left[\tilde{s}_1 + \left(\tilde{\mathbf{s}}_2^T\tilde{\mathbf{s}}_2\right)^{1/2}\mathbf{y}^T\mathbf{e}_2\right] \tag{2.B.17}$$

$$= m_{11}\left[\tilde{s}_1 + \left(\tilde{\mathbf{s}}_2^T\tilde{\mathbf{s}}_2\right)^{1/2}y_1\right].$$

Using (2.B.12), (2.B.13), and (2.B.17), we can rewrite (2.B.9) as

$$\rho = \frac{\left[1 + \left(\frac{\tilde{\mathbf{s}}_2^T \tilde{\mathbf{s}}_2}{\tilde{s}_1^2}\right)^{1/2} y_1\right]^2}{1 + \frac{\mathbf{m}_{21}^T \mathbf{m}_{21}}{m_{11}^2}} \tag{2.B.18}$$

$$= \frac{(1 + a y_1)^2}{1 + \mathbf{y}^T \mathbf{y}},$$

where

$$a = \left(\frac{\tilde{\mathbf{s}}_2^T \tilde{\mathbf{s}}_2}{\tilde{s}_1^2}\right)^{1/2} \geqslant 0. \tag{2.B.19}$$

It is obvious that

$$a^2 = \frac{\tilde{\mathbf{s}}^T \tilde{\mathbf{s}} - \tilde{s}_1^2}{\tilde{s}_1^2} \tag{2.B.20}$$

$$= \frac{\sin^2 \phi}{\cos^2 \phi},$$

where the second equality is obtained using (2.B.3) and (2.B.4). From (2.B.11) and (2.B.15), we can obtain that

$$\mathbf{y}^T \mathbf{y} = \mathbf{w}_{21}^T \mathbf{W}_{22}^{-2} \mathbf{w}_{21}. \tag{2.B.21}$$

As derived in Appendix 2.C, the PDF of $\mathbf{y}^T \mathbf{y}$ is

$$\mathcal{P}(\mathbf{y}^T \mathbf{y}) = \frac{c_3}{(1 + \mathbf{y}^T \mathbf{y})^{\frac{2K+1}{2}}}, \tag{2.B.22}$$

where c_3 is defined in (2.C.18). In the following, we use a series of transformations to derive the distribution of ρ from the distribution of $\mathbf{y}^T \mathbf{y}$ in (2.B.22), based on the relationship in (2.B.18).

First, we express the PDF (2.B.22) in polar coordinates by using the transformation:

$$\mathfrak{T}1 : \begin{cases} y_1 = r_1, \\ y_2 = r_2 \sin \psi_2, \\ y_3 = r_2 \cos \psi_2 \sin \psi_3, \\ \quad \vdots \\ y_{N-1} = r_2 \cos \psi_2 \cos \psi_3 \ldots \cos \psi_{N-2}, \end{cases} \tag{2.B.23}$$

where $-\pi/2 < \psi_i \leqslant \pi/2$ for $i = 2, 3, \ldots, N-3$, and $-\pi < \psi_{N-2} \leqslant \pi$. The Jacobian of the transformation in (2.B.23) is

$$J_3 = |r_2^{N-3} f(\psi)|, \tag{2.B.24}$$

where

$$f(\psi) = \cos^{N-4} \psi_2 \cos^{N-5} \psi_3 \ldots \cos \psi_{N-3} \tag{2.B.25}$$

with $\boldsymbol{\psi} = [\psi_2, \psi_3, \ldots, \psi_{N-3}]^T$. Based on the results in [23, p. 48], we have

$$\int_{-\pi}^{\pi} \int_{-\frac{\pi}{2}}^{\frac{\pi}{2}} \cdots \int_{-\frac{\pi}{2}}^{\frac{\pi}{2}} f(\boldsymbol{\psi}) \mathrm{d}\psi_2 \ldots \mathrm{d}\psi_{N-3} \mathrm{d}\psi_{N-2} = \frac{2\pi^{\frac{N-2}{2}}}{\Gamma\left(\frac{N-2}{2}\right)}. \tag{2.B.26}$$

Therefore, the marginal PDF of r_1 and r_2 is

$$\mathcal{P}(r_1, r_2) = \frac{2c|r_2|^{N-3}}{\left(1 + r_1^2 + r_2^2\right)^{\frac{2K+1}{2}}}, \tag{2.B.27}$$

where

$$c = \frac{c_3 \pi^{\frac{N-2}{2}}}{\Gamma\left(\frac{N-2}{2}\right)}$$

$$= \frac{1}{B\left(\frac{1}{2}, \frac{N-2}{2}\right) B\left(\frac{2K-N+2}{2}, \frac{N-1}{2}\right)}. \tag{2.B.28}$$

The derivations of the second equality in (2.B.28) are given in Appendix 2.D.
Define

$$\mathfrak{T}2 : \begin{cases} z = r_1^2 + r_2^2, \\ x = r_1, \end{cases} \tag{2.B.29}$$

and then the Jacobian of this transformation is

$$J_4 = \frac{1}{2(z - x^2)^{1/2}}. \tag{2.B.30}$$

Note that $z > x^2$. Then the joint PDF of z and x is

$$\mathcal{P}(z, x) = c(1 + z)^{-\frac{2K+1}{2}} \left(z - x^2\right)^{\frac{N-4}{2}}, \quad 0 \leqslant x^2 \leqslant z. \tag{2.B.31}$$

Further, we define

$$\mathfrak{T}3 : \begin{cases} x = x, \\ \rho = \frac{(1+ax)^2}{1+z}, \end{cases} \tag{2.B.32}$$

whose Jacobian is

$$J_5 = \frac{(1 + ax)^2}{\rho^2}. \tag{2.B.33}$$

Then the joint PDF of x and ρ is

$$\mathcal{P}(\rho, x) = \frac{c \left[\frac{(1+ax)^2}{\rho} - (1 + x^2)\right]^{\frac{N-4}{2}}}{\rho \left[\frac{(1+ax)^2}{\rho}\right]^{\frac{2K-1}{2}}}, \tag{2.B.34}$$

where $\rho > 0$, and

$$\frac{(1 + ax)^2}{\rho} - (1 + x^2) \geqslant 0. \tag{2.B.35}$$

It follows that $0 < \rho \leqslant 1 + a^2$.

Here, remarks about some special cases are now in order. It can be observed from (2.B.34) that the PDF of ρ cannot be defined for $\rho = 0$. So we have to remove this point from the domain of the PDF of ρ. On the other hand, in Appendix 2.E we provide the values of $\mathcal{P}(\rho)$ for the special cases $\rho = 1 + a^2$ and $\rho = a^2$, namely,

$$\mathcal{P}(\rho) = \begin{cases} 0, & \text{for } \rho = 1 + a^2, \\ k, & \text{for } \rho = a^2, \end{cases} \tag{2.B.36}$$

where

$$k = c\, 4^{K-1} (\cos^2 \phi)^K (\sin^2 \phi)^{\frac{2K-N}{2}} B\left(\frac{N-2}{2}, \frac{4K-N}{2}\right) \tag{2.B.37}$$

with c defined in (2.B.28).

Let us proceed by considering the case where $0 < \rho < 1 + a^2$, and $\rho \neq a^2$. Now we can define

$$\mathcal{T}4 : \begin{cases} r = x + \frac{a}{a^2 - \rho}, \\ \rho = \rho, \end{cases} \tag{2.B.38}$$

and the Jacobian of this transformation is $J_6 = 1$. Moreover, we have

$$\begin{cases} \frac{(1+ax)^2}{\rho} - (1 + x^2) = \frac{1+b}{b}\left[\frac{r^2 b^2}{(1+b)\rho} - 1\right], \\ \frac{(1+ax)^2}{\rho} = \frac{\rho}{b^2}\left(1 - \frac{abr}{\rho}\right)^2, \end{cases} \tag{2.B.39}$$

where

$$b = a^2 - \rho > -1. \tag{2.B.40}$$

The joint PDF of r and ρ can be expressed as

$$\mathcal{P}(\rho, r) = \frac{c|b|^{\frac{4K-N+2}{2}}(1+b)^{\frac{N-4}{2}}}{\rho^{\frac{2K+1}{2}}\left|1 - \frac{abr}{\rho}\right|^{2K-1}} \left|\frac{r^2 b^2}{(1+b)\rho} - 1\right|^{\frac{N-4}{2}}, \tag{2.B.41}$$

where

$$\begin{cases} r^2 > \frac{\rho(1+b)}{b^2}, & \text{for } 0 < \rho < a^2, \\ r^2 < \frac{\rho(1+b)}{b^2}, & \text{for } a^2 < \rho < 1 + a^2. \end{cases} \tag{2.B.42}$$

Define

$$\mathcal{T}5 : \begin{cases} y = \frac{rb}{\sqrt{\rho(1+b)}}, \\ \rho = \rho, \end{cases} \tag{2.B.43}$$

and the Jacobian of this transformation is

$$J_7 = \frac{\sqrt{\rho(1+b)}}{|b|}. \tag{2.B.44}$$

The joint PDF of y and ρ is given as

$$\mathcal{P}(\rho, y) = \frac{c|b|^{\frac{4K-N}{2}}(1+b)^{\frac{N-3}{2}}|y^2-1|^{\frac{N-4}{2}}}{\rho^K|1-\sqrt{w}y|^{2K-1}}, \tag{2.B.45}$$

where

$$w = \frac{a^2}{\rho}(1+b) > 0. \tag{2.B.46}$$

It follows from (2.B.42) and (2.B.43) that

$$\begin{cases} y \in (-1, 1), & \text{for } 0 < \rho < a^2, \\ y \in (1, \infty) \cup (-\infty, -1), & \text{for } a^2 < \rho < 1 + a^2. \end{cases} \tag{2.B.47}$$

Define

$$\mathfrak{T}6 : \begin{cases} t = y^2, \\ \rho = \rho, \end{cases} \tag{2.B.48}$$

and the Jacobian of this transformation is

$$J_8 = \frac{1}{2\sqrt{t}}. \tag{2.B.49}$$

The joint PDF of t and ρ can be written as

$$\mathcal{P}(\rho, t) = \frac{c|b|^{\frac{4K-N}{2}}(1+b)^{\frac{N-3}{2}}|t-1|^{\frac{N-4}{2}}}{2\rho^K\sqrt{t}} \\ \times \left(\frac{1}{|1-\sqrt{wt}|^{2K-1}} + \frac{1}{|1+\sqrt{wt}|^{2K-1}} \right), \tag{2.B.50}$$

where

$$\begin{cases} t > 1, & \text{for } 0 < \rho < a^2, \\ 0 < t < 1, & \text{for } a^2 < \rho < 1 + a^2. \end{cases} \tag{2.B.51}$$

When $a^2 < \rho < 1 + a^2$, we have $0 < w < 1$ and $0 < t < 1$. Thus, $0 < \sqrt{wt} < 1$. The marginal PDF of ρ can be obtained as

$$\mathcal{P}(\rho) = \frac{c(\rho - a^2)^{\frac{4K-N}{2}}(1 + a^2 - \rho)^{\frac{N-3}{2}}}{2\rho^K} \\ \times \int_0^1 \left[\frac{(1-t)^{\frac{N-4}{2}}}{\sqrt{t}\left(1 - \sqrt{wt}\right)^{2K-1}} + \frac{(1-t)^{\frac{N-4}{2}}}{\sqrt{t}\left(1 + \sqrt{wt}\right)^{2K-1}} \right] dt \\ = \frac{c(\rho - a^2)^{\frac{4K-N}{2}}(1 + a^2 - \rho)^{\frac{N-3}{2}}}{\rho^K} \\ \times \int_0^1 \left[\frac{(1-\gamma^2)^{\frac{N-4}{2}}}{(1 - \sqrt{w}\gamma)^{2K-1}} + \frac{(1-\gamma^2)^{\frac{N-4}{2}}}{(1 + \sqrt{w}\gamma)^{2K-1}} \right] d\gamma, \tag{2.B.52}$$

where the last equality is obtained with $\gamma = \sqrt{t}$. Note that the PDF of ρ in (2.B.52) is zero when $\rho = 1 + a^2$, which is consistent with that in (2.B.36). So the case of $\rho = 1 + a^2$ can be included in the domain of the expression (2.B.52).

When $0 < \rho < a^2$, we have $t > 1$ and $w > 1$. Hence, $\sqrt{wt} > 1$. The marginal PDF of ρ can be obtained as

$$
\mathcal{P}(\rho) = \frac{c(a^2 - \rho)^{\frac{4K-N}{2}}(1 + a^2 - \rho)^{\frac{N-3}{2}}}{2\rho^K}
$$

$$
\times \int_1^\infty \left[\frac{(t-1)^{\frac{N-4}{2}}}{\sqrt{t}\left(\sqrt{wt} - 1\right)^{2K-1}} + \frac{(t-1)^{\frac{N-4}{2}}}{\sqrt{t}\left(\sqrt{wt} + 1\right)^{2K-1}} \right] dt
$$

(2.B.53)

$$
= \frac{c(a^2 - \rho)^{\frac{4K-N}{2}}(1 + a^2 - \rho)^{\frac{N-3}{2}}}{\rho^K}
$$

$$
\times \int_0^1 \left[\frac{\gamma^{2K-N+1}(1 - \gamma^2)^{\frac{N-4}{2}}}{(\sqrt{w} - \gamma)^{(2K-1)}} + \frac{\gamma^{2K-N+1}(1 - \gamma^2)^{\frac{N-4}{2}}}{(\sqrt{w} + \gamma)^{(2K-1)}} \right] d\gamma,
$$

where the last equality is obtained with $\gamma = \frac{1}{\sqrt{t}}$.

In summary, the PDF of ρ can be expressed as

$$
\mathcal{P}(\rho) = \begin{cases} c\,h(\rho)g_1(\rho), & \text{for } a^2 < \rho \leqslant 1 + a^2, \\ k, & \text{for } \rho = a^2, \\ c\,h(\rho)g_2(\rho), & \text{for } 0 < \rho < a^2, \end{cases}
$$

(2.B.54)

where c and k are defined in (2.B.28) and (2.B.37), respectively,

$$
h(\rho) = \frac{|\rho - a^2|^{\frac{4K-N}{2}}(1 + a^2 - \rho)^{\frac{N-3}{2}}}{\rho^K},
$$

(2.B.55)

$$
g_1(\rho) = \int_0^1 \left[\frac{(1 - \gamma^2)^{\frac{N-4}{2}}}{(1 - \sqrt{w}\gamma)^{2K-1}} + \frac{(1 - \gamma^2)^{\frac{N-4}{2}}}{(1 + \sqrt{w}\gamma)^{2K-1}} \right] d\gamma,
$$

(2.B.56)

and

$$
g_2(\rho) = \int_0^1 \left[\frac{\gamma^{2K-N+1}(1 - \gamma^2)^{\frac{N-4}{2}}}{(\sqrt{w} - \gamma)^{(2K-1)}} + \frac{\gamma^{2K-N+1}(1 - \gamma^2)^{\frac{N-4}{2}}}{(\sqrt{w} + \gamma)^{(2K-1)}} \right] d\gamma
$$

(2.B.57)

with w defined in (2.B.46). Using (2.B.1) and [20, p. 318, eq. (3.211)], we can derive the PDF of $\rho_{\text{per}}^{\text{mismatch1}}$ to be that in (2.34). The proof is completed.

2.C Derivations of (2.B.22)

According to (2.B.7), the PDF of $\mathbf{W}$ can be written as [26, p. 62]

$$\mathcal{P}(\mathbf{W}) = \frac{\det(\mathbf{W})^{\frac{2K-N-1}{2}}}{2^{KN}\Gamma_N(K)} \exp\left[-\frac{1}{2}\operatorname{tr}(\mathbf{W})\right]. \tag{2.C.1}$$

Using the partitioned form in (2.B.6), we can obtain [23, p. 637]

$$\det(\mathbf{W}) = \det(\mathbf{W}_{22})(w_{11} - \mathbf{w}_{21}^T\mathbf{W}_{22}^{-1}\mathbf{w}_{21}), \tag{2.C.2}$$

and

$$\operatorname{tr}(\mathbf{W}) = w_{11} + \operatorname{tr}(\mathbf{W}_{22}). \tag{2.C.3}$$

Taking (2.C.2) and (2.C.3) back into (2.C.1) produces

$$\begin{aligned}
\mathcal{P}(\mathbf{W}) =& \frac{\det(\mathbf{W}_{22})^{\frac{2K-N-1}{2}}}{2^{KN}\Gamma_N(K)}\left(w_{11} - \mathbf{w}_{21}^T\mathbf{W}_{22}^{-1}\mathbf{w}_{21}\right)^{\frac{2K-N-1}{2}} \\
&\times \exp\left[-\frac{1}{2}\left(w_{11} - \mathbf{w}_{21}^T\mathbf{W}_{22}^{-1}\mathbf{w}_{21}\right)\right] \\
&\times \exp\left[-\frac{1}{2}\left(\mathbf{w}_{21}^T\mathbf{W}_{22}^{-1}\mathbf{w}_{21}\right) - \frac{1}{2}\operatorname{tr}(\mathbf{W}_{22})\right].
\end{aligned} \tag{2.C.4}$$

Define

$$\begin{cases}
v_{11} = w_{11} - \mathbf{w}_{21}^T\mathbf{W}_{22}^{-1}\mathbf{w}_{21} \in \mathbb{R}^{1\times 1}, \\
\mathbf{v}_{21} = \mathbf{w}_{21} \in \mathbb{R}^{(N-1)\times 1}, \\
\mathbf{V}_{22} = \mathbf{W}_{22} \in \mathbb{R}^{(N-1)\times(N-1)},
\end{cases} \tag{2.C.5}$$

and the Jacobian of this transformation is $J_1 = 1$. So the joint PDF of v_{11}, $\mathbf{v}_{21}$ and $\mathbf{V}_{22}$ is

$$\begin{aligned}
\mathcal{P}(v_{11}, \mathbf{v}_{21}, \mathbf{V}_{22}) =& \frac{[v_{11}\det(\mathbf{V}_{22})]^{\frac{2K-N-1}{2}}}{2^{KN}\Gamma_N(K)} \exp\left[-\frac{1}{2}\operatorname{tr}(\mathbf{V}_{22})\right] \\
&\times \exp\left[-\frac{1}{2}\left(v_{11} + \mathbf{v}_{21}^T\mathbf{V}_{22}^{-1}\mathbf{v}_{21}\right)\right] \\
=& \mathcal{P}(v_{11})\mathcal{P}(\mathbf{v}_{21}, \mathbf{V}_{22}),
\end{aligned} \tag{2.C.6}$$

where

$$\mathcal{P}(v_{11}) = c_1 v_{11}^{\frac{2K-N-1}{2}} \exp\left(-\frac{v_{11}}{2}\right), \tag{2.C.7}$$

and

$$\mathcal{P}(\mathbf{v}_{21}, \mathbf{V}_{22}) = c_2 \det(\mathbf{V}_{22})^{\frac{2K-N-1}{2}} \exp\left\{-\frac{1}{2}\left[\mathbf{v}_{21}^T\mathbf{V}_{22}^{-1}\mathbf{v}_{21} + \operatorname{tr}(\mathbf{V}_{22})\right]\right\} \tag{2.C.8}$$

with

$$c_1 c_2 = \frac{1}{2^{KN}\Gamma_N(K)}. \tag{2.C.9}$$

Since $\mathcal{P}(v_{11})$ is the PDF of v_{11} subject to the central Chi-squared distribution with $2K - N + 1$ degrees of freedom, its normalizing constant c_1 is given by

$$c_1 = \frac{1}{2^{\frac{2K-N+1}{2}}\Gamma\left(\frac{2K-N+1}{2}\right)}. \tag{2.C.10}$$

Further, we have

$$c_2 = \frac{2^{\frac{2K-N+1}{2}}\Gamma\left(\frac{2K-N+1}{2}\right)}{2^{KN}\Gamma_N(K)}. \tag{2.C.11}$$

Define

$$\begin{cases} \mathbf{t}_{21} = \mathbf{V}_{22}^{-1}\mathbf{v}_{21} \in \mathbb{R}^{N-1 \times 1}, \\ \mathbf{T}_{22} = \mathbf{V}_{22} \in \mathbb{R}^{N-1 \times N-1}, \end{cases} \tag{2.C.12}$$

and the Jacobian of this transformation is

$$J_2 = \det(\mathbf{T}_{22}). \tag{2.C.13}$$

Then, the joint PDF of $\mathbf{t}_{21}$ and $\mathbf{T}_{22}$ is

$$\begin{aligned}
\mathcal{P}(\mathbf{t}_{21}, \mathbf{T}_{22}) &= c_2 \det(\mathbf{T}_{22})^{\frac{2K-N+1}{2}} \exp\left\{-\frac{1}{2}\mathrm{tr}\left[\left(\mathbf{I}_{N-1} + \mathbf{t}_{21}\mathbf{t}_{21}^T\right)\mathbf{T}_{22}\right]\right\} \\
&= \mathcal{P}(\mathbf{T}_{22}|\mathbf{t}_{21})\mathcal{P}(\mathbf{t}_{21}),
\end{aligned} \tag{2.C.14}$$

where

$$\mathcal{P}(\mathbf{t}_{21}) = \frac{c_2 2^{\frac{(2K+1)(N-1)}{2}}\Gamma_{N-1}\left(\frac{2K+1}{2}\right)}{\det\left(\mathbf{I}_{N-1} + \mathbf{t}_{21}\mathbf{t}_{21}^T\right)^{\frac{2K+1}{2}}}, \tag{2.C.15}$$

and

$$\begin{aligned}
\mathcal{P}(\mathbf{T}_{22}|\mathbf{t}_{21}) &= \frac{\det\left(\mathbf{I}_{N-1} + \mathbf{t}_{21}\mathbf{t}_{21}^T\right)^{\frac{2K+1}{2}}}{2^{\frac{(2K+1)(N-1)}{2}}\Gamma_{N-1}\left(\frac{2K+1}{2}\right)} \det(\mathbf{T}_{22})^{\frac{2K-N+1}{2}} \\
&\quad \times \exp\left\{-\frac{1}{2}\mathrm{tr}\left[\left(\mathbf{I}_{N-1} + \mathbf{t}_{21}\mathbf{t}_{21}^T\right)\mathbf{T}_{22}\right]\right\}.
\end{aligned} \tag{2.C.16}$$

We temporarily fix the random vector $\mathbf{t}_{21}$. It can be observed that $\mathcal{P}(\mathbf{T}_{22}|\mathbf{t}_{21})$ is the PDF of $\mathbf{T}_{22} \sim \mathcal{W}_{N-1}(2K+1, (\mathbf{I}_{N-1} + \mathbf{t}_{21}\mathbf{t}_{21}^T)^{-1})$. As a result, the PDF of $\mathbf{t}_{21}$ is $\mathcal{P}(\mathbf{t}_{21})$, which can be rewritten as

$$\mathcal{P}(\mathbf{t}_{21}) = \frac{c_3}{\left(1 + \mathbf{t}_{21}^T\mathbf{t}_{21}\right)^{\frac{2K+1}{2}}}, \tag{2.C.17}$$

where

$$c_3 = c_2 2^{\frac{(2K+1)(N-1)}{2}}\Gamma_{N-1}\left(\frac{2K+1}{2}\right) \tag{2.C.18}$$

with c_2 defined in (2.C.11). According to (2.B.21), (2.C.5), and (2.C.12), we obtain

$$\mathbf{t}_{21}^T \mathbf{t}_{21} = \mathbf{y}^T \mathbf{y}. \tag{2.C.19}$$

So the PDF of $\mathbf{y}^T \mathbf{y}$ can be derived as (2.B.22).

2.D   Derivations of (2.B.28)

Using (2.C.11) and (2.C.18), we can recast the constant number c in (2.B.28) as

$$c = \frac{\pi^{\frac{N-2}{2}} \Gamma\left(\frac{2K-N+1}{2}\right) \Gamma_{N-1}\left(\frac{2K+1}{2}\right)}{\Gamma\left(\frac{N-2}{2}\right) \Gamma_N(K)}. \tag{2.D.1}$$

According to the definition of multivariate Gamma function [26, p. 62], we have

$$\Gamma_N(K) = \pi^{\frac{N(N-1)}{4}} \prod_{j=1}^{N} \Gamma\left(K + \frac{1-j}{2}\right), \tag{2.D.2}$$

and

$$\Gamma_{N-1}\left(\frac{2K+1}{2}\right) = \pi^{\frac{(N-1)(N-2)}{4}} \prod_{j=1}^{N-1} \Gamma\left(K + \frac{1}{2} + \frac{1-j}{2}\right). \tag{2.D.3}$$

As a result,

$$\frac{\Gamma_{N-1}\left(\frac{2K+1}{2}\right)}{\Gamma_N(K)} = \frac{\pi^{\frac{-N+1}{2}} \Gamma\left(\frac{2K+1}{2}\right)}{\Gamma\left(\frac{2K-N+2}{2}\right) \Gamma\left(\frac{2K-N+1}{2}\right)}. \tag{2.D.4}$$

Substituting (2.D.4) into (2.D.1) leads to

$$\begin{aligned}
c &= \frac{\Gamma\left(\frac{2K+1}{2}\right)}{\Gamma\left(\frac{1}{2}\right) \Gamma\left(\frac{N-2}{2}\right) \Gamma\left(\frac{2K-N+2}{2}\right)} \\
&= \frac{1}{B\left(\frac{1}{2}, \frac{N-2}{2}\right) B\left(\frac{2K-N+2}{2}, \frac{N-1}{2}\right)}.
\end{aligned} \tag{2.D.5}$$

2.E   Derivations of (2.B.36)

According to (2.B.34), the joint PDF of x and ρ can be written as

$$\mathcal{P}(\rho, x) = \frac{c\rho^{\frac{2K-N+1}{2}} \left[(a^2 - \rho)x^2 + 2ax + 1 - \rho^2\right]^{\frac{N-4}{2}}}{\left[(1+ax)^2\right]^{\frac{2K-1}{2}}}, \tag{2.E.1}$$

where $\rho > 0$, and $(a^2 - \rho)x^2 + 2ax + 1 - \rho^2 \geqslant 0$. When $\rho = 1 + a^2$, we have

$$-x^2 + 2ax - a^2 \geqslant 0, \tag{2.E.2}$$

which means that $x = a$. Then,

$$\mathcal{P}(\rho = 1 + a^2) = \lim_{\rho \to 1 + a^2} \mathcal{P}(\rho, a) = 0. \tag{2.E.3}$$

When $\rho = a^2$, we have

$$x \geqslant \frac{a^2 - 1}{2a}. \tag{2.E.4}$$

So the marginal PDF of ρ is

$$\begin{aligned}
\mathcal{P}(\rho = a^2) &= ca^{2K-N+1} \int_{\frac{a^2-1}{2a}}^{\infty} \frac{(1 - a^2 + 2ax)^{\frac{N-4}{2}}}{(1 + ax)^{2K-1}} dx \\
&= \frac{c2^{\frac{4K+N-6}{2}} a^{\frac{4K-N-2}{2}}}{(1 + a^2)^{2K-1}} \int_0^{\infty} \frac{\zeta^{\frac{N-4}{2}}}{\left(1 + \frac{2a}{a^2+1}\zeta\right)^{2K-1}} d\zeta,
\end{aligned} \tag{2.E.5}$$

where the second equality is obtained using $\zeta = x - \frac{a^2-1}{2a}$. According to [20, p. 315, eq. (3.194.3)], we have

$$\int_0^{\infty} \frac{\zeta^{\frac{N-4}{2}}}{\left(1 + \frac{2a}{a^2+1}\zeta\right)^{2K-1}} d\zeta = \frac{B\left(\frac{N-2}{2}, \frac{4K-N}{2}\right)}{\left(\frac{2a}{a^2+1}\right)^{\frac{N-1}{2}}}. \tag{2.E.6}$$

Substituting (2.E.6) into (2.E.5) yields

$$\mathcal{P}(\rho = a^2) = c4^{K-1}(\cos^2 \phi)^K (\sin^2 \phi)^{\frac{2K-N}{2}} B\left(\frac{N-2}{2}, \frac{4K-N}{2}\right), \tag{2.E.7}$$

where we use $a^2 = \frac{\sin^2 \phi}{\cos^2 \phi}$ in (2.B.20).

2.F   Derivation of E($\check{\mathbf{w}}\check{\mathbf{w}}^\dagger$) in the Mismatched Case

It is shown in [17] that the distribution of $\check{\mathbf{R}}$ is

$$\check{\mathbf{R}} \sim \mathcal{W}_N(2K, \mathbf{R}_s/2). \tag{2.F.1}$$

Define $\tilde{\mathbf{R}} = \sqrt{2}\check{\mathbf{R}} \in \mathbb{R}^{N \times N}$. Then, we have

$$\tilde{\mathbf{R}} \sim \mathcal{W}_N(2K, \mathbf{R}_s). \tag{2.F.2}$$

The adaptive weight in (2.45) can be rewritten as

$$\breve{w} = \frac{\tilde{R}^{-1}q}{q^\dagger\tilde{R}^{-1}q}. \tag{2.F.3}$$

There always exists a unitary matrix $\tilde{U}$ such that

$$\tilde{U}R_s^{-1/2}q = (qR_s^{-1}q)^{1/2}e_1, \tag{2.F.4}$$

where e_1 is defined in (2.A.1). Furthermore, we have

$$\tilde{R} \stackrel{\mathrm{d}}{=} R_s^{1/2}\tilde{U}^\dagger V\tilde{U}R_s^{1/2}, \tag{2.F.5}$$

where V defined in (2.A.6) is distributed as $\mathcal{W}_N(2K, I)$. Accordingly,

$$\breve{w} \stackrel{\mathrm{d}}{=} \frac{R_s^{-1/2}\tilde{U}^\dagger V^{-1}\tilde{U}R_s^{-1/2}q}{q^\dagger R_s^{-1/2}\tilde{U}^\dagger V^{-1}\tilde{U}R_s^{-1/2}q}. \tag{2.F.6}$$

Applying (2.F.4) to (2.F.6) leads to

$$\breve{w} \stackrel{\mathrm{d}}{=} \frac{R_s^{-1/2}\tilde{U}^\dagger V^{-1}e_1}{(e_1^\dagger V^{-1}e_1)(q^\dagger R_s^{-1}q)^{1/2}}. \tag{2.F.7}$$

Therefore,

$$\breve{w}\breve{w}^\dagger \stackrel{\mathrm{d}}{=} \frac{R_s^{-1/2}\tilde{U}^\dagger V^{-1}e_1 e_1^\dagger V^{-1}\tilde{U}R_s^{-1/2}}{(e_1^\dagger W^{-1}e_1)^2(q^\dagger R_s^{-1}q)}$$
$$\stackrel{\mathrm{d}}{=} \frac{R_s^{-1/2}\tilde{U}^\dagger JV^{-1}Je_1 e_1^\dagger JV^{-1}J\tilde{U}R_s^{-1/2}}{(e_1^\dagger JV^{-1}Je_1)^2(q^\dagger R_s^{-1}q)}, \tag{2.F.8}$$

where the second equation is obtained by the fact that $JV^{-1}J$ and V^{-1} are statistically equivalent. Inserting (2.A.16) into (2.F.8) yields

$$\breve{w}\breve{w}^\dagger = \frac{R_s^{-1/2}\tilde{U}^\dagger Q^{-T}Q^{-1}e_1 e_1^\dagger Q^{-T}Q^{-1}\tilde{U}R_s^{-1/2}}{(e_1^\dagger Q^{-T}Q^{-1}e_1)^2(q^\dagger R_s^{-1}q)}. \tag{2.F.9}$$

Note that the real upper triangle matrix Q can be partitioned as

$$Q = \begin{bmatrix} q_{1,1} & q_{1,2}^T \\ 0 & Q_{2,2} \end{bmatrix}, \tag{2.F.10}$$

where $q_{1,1}$, $q_{1,2}$, and $Q_{2,2}$ are a scalar, an $(N-1)$-dimensional column vector, and an $(N-1) \times (N-1)$ matrix in the real domain, respectively. It is easy to show that [26, p. 92]

$$Q_{2,2}Q_{2,2}^T \sim \mathcal{W}_{N-1}(2K, I). \tag{2.F.11}$$

According to the partitioned matrix inversion theorem, we have

$$\mathbf{Q}^{-1} = \begin{bmatrix} q_{1,1}^{-1} & -q_{1,1}^{-1}\mathbf{q}_{1,2}^T\mathbf{Q}_{2,2}^{-1} \\ 0 & \mathbf{Q}_{2,2}^{-1} \end{bmatrix}. \tag{2.F.12}$$

Therefore,

$$(\mathbf{e}_1^\dagger\mathbf{Q}^{-T}\mathbf{Q}^{-1}\mathbf{e}_1)^2 = q_{1,1}^{-4}, \tag{2.F.13}$$

and

$$\begin{aligned} &\mathbf{Q}^{-T}\mathbf{Q}^{-1}\mathbf{e}_1\mathbf{e}_1^\dagger\mathbf{Q}^{-T}\mathbf{Q}^{-1} \\ &= q_{1,1}^{-2}\mathbf{Q}^{-T}\mathbf{e}_1\mathbf{e}_1^\dagger\mathbf{Q}^{-1} \\ &= q_{1,1}^{-2}\begin{bmatrix} q_{1,1}^{-1} \\ -(q_{1,1}^{-1}\mathbf{q}_{1,2}^T\mathbf{Q}_{2,2}^{-1})^T \end{bmatrix}\begin{bmatrix} q_{1,1}^{-1}, & -(q_{1,1}^{-1}\mathbf{q}_{1,2}^T\mathbf{Q}_{2,2}^{-1}) \end{bmatrix} \\ &= q_{1,1}^{-4}\begin{bmatrix} 1 & -\mathbf{q}_{1,2}^T\mathbf{Q}_{2,2}^{-1} \\ -(\mathbf{q}_{1,2}^T\mathbf{Q}_{2,2}^{-1})^T & \mathbf{Q}_{2,2}^{-T}\mathbf{q}_{1,2}\mathbf{q}_{1,2}^T\mathbf{Q}_{2,2}^{-1} \end{bmatrix}. \end{aligned} \tag{2.F.14}$$

Let

$$\mathbf{R}_s^{-1/2}\tilde{\mathbf{U}}^\dagger = [\mathbf{b}, \mathbf{B}], \tag{2.F.15}$$

where $\mathbf{b}$ is the first column vector, and $\mathbf{B}$ is an $N \times (N-1)$ matrix consisting of the other columns. Substituting (2.F.13), (2.F.14), and (2.F.15) into (2.F.9) leads to

$$\breve{\mathbf{w}}\breve{\mathbf{w}}^\dagger = \frac{[\mathbf{b}, \mathbf{B}]\begin{bmatrix} 1 & -\mathbf{q}_{1,2}^T\mathbf{Q}_{2,2}^{-1} \\ -(\mathbf{q}_{1,2}^T\mathbf{Q}_{2,2}^{-1})^T & \mathbf{Q}_{2,2}^{-T}\mathbf{q}_{1,2}\mathbf{q}_{1,2}^T\mathbf{Q}_{2,2}^{-1} \end{bmatrix}\begin{bmatrix} \mathbf{b}^\dagger \\ \mathbf{B}^\dagger \end{bmatrix}}{(\mathbf{q}^\dagger\mathbf{R}^{-1}\mathbf{q})}. \tag{2.F.16}$$

Since the entries of $\mathbf{Q}$ are independent, we get

$$E(\mathbf{q}_{1,2}^T\mathbf{Q}_{2,2}^{-1}) = \mathbf{0}. \tag{2.F.17}$$

Further, we have [27, Theorem 3.3.16]

$$E(\mathbf{Q}_{2,2}^{-T}\mathbf{q}_{1,2}\mathbf{q}_{1,2}^T\mathbf{Q}_{2,2}^{-1}) = \frac{1}{2K-N}\mathbf{I}_{N-1}. \tag{2.F.18}$$

Using (2.F.17) and (2.F.18), we have

$$\begin{aligned} E(\breve{\mathbf{w}}\breve{\mathbf{w}}^\dagger) &= \frac{[\mathbf{b}, \mathbf{B}]\begin{bmatrix} 1 & 0 \\ 0 & \frac{1}{2K-N}\mathbf{I}_{N-1} \end{bmatrix}\begin{bmatrix} \mathbf{b}^\dagger \\ \mathbf{B}^\dagger \end{bmatrix}}{(\mathbf{q}^\dagger\mathbf{R}_s^{-1}\mathbf{q})} \\ &= (\mathbf{q}^\dagger\mathbf{R}_s^{-1}\mathbf{q})^{-1}\left(\mathbf{b}\mathbf{b}^T + \frac{\mathbf{B}\mathbf{B}^\dagger}{2K-N}\right). \end{aligned} \tag{2.F.19}$$

It follows from (2.F.15) that

$$\mathbf{R}_s^{-1} = \mathbf{b}\mathbf{b}^\dagger + \mathbf{B}\mathbf{B}^\dagger, \tag{2.F.20}$$

and

$$R_s^{-1/2}\tilde{U}^\dagger e_1 = b. \tag{2.F.21}$$

In addition,

$$\begin{aligned}R_s^{-1}q &= R_s^{-1/2}\tilde{U}^\dagger\tilde{U}R_s^{-1/2}q \\ &= R_s^{-1/2}\tilde{U}^\dagger e_1(q^\dagger R_s^{-1}q)^{1/2} \\ &= b(q^\dagger R_s^{-1}q)^{1/2}, \end{aligned} \tag{2.F.22}$$

where the second and third equalities are obtained with (2.F.4) and (2.F.21), respectively. It means that

$$b = (q^\dagger R_s^{-1}q)^{-1/2}R_s^{-1}q. \tag{2.F.23}$$

Therefore, (2.F.19) becomes

$$\begin{aligned}E(\check{w}\check{w}^\dagger) &= (q^\dagger R_s^{-1}q)^{-1}\left(bb^\dagger + \frac{R_s^{-1} - bb^\dagger}{2K - N}\right) \\ &= \frac{1}{(2K - N)(q^\dagger R_s^{-1}q)}\left[R_s^{-1} + (2K - N - 1)\frac{R_s^{-1}qq^\dagger R_s^{-1}}{q^\dagger R_s^{-1}q}\right], \end{aligned} \tag{2.F.24}$$

where the first and second equalities are obtained with (2.F.20) and (2.F.23), respectively.

Bibliography

[1] I. S. Reed, J. D. Mallett, and L. E. Brennan, "Rapid convergence rate in adaptive arrays," *IEEE Transactions on Aerospace and Electronic Systems*, vol. 10, no. 6, pp. 853–863, 1974.

[2] C. D. Richmond, "Derived PDF of maximum-likelihood signal estimator which employs an estimated noise covariance," *IEEE Transactions on Signal Processing*, vol. 44, no. 2, pp. 305–315, February 1996.

[3] ——, "PDF's, confidence regions, and relevant statistics for a class of sample covariance-based array processors," *IEEE Transactions on Signal Processing*, vol. 44, no. 7, pp. 1779–1793, July 1996.

[4] ——, "Response of sample covariance based MVDR beamformer to imperfect look and inhomogeneities," *IEEE Signal Processing Letters*, vol. 5, no. 12, pp. 325–327, December 1998.

[5] ——, "Statistics of adaptive nulling and use of the generalized eigenrelation (GER) for modeling inhomogeneities in adaptive processing," *IEEE Transactions on Signal Processing*, vol. 48, no. 5, pp. 1263–1273, May 2000.

[6] F. Bandiera, D. Orlando, and G. Ricci, *Advanced Radar Detection Schemes under Mismatched Signal Models in Synthesis Lectures on Signal Processing*. San Rafael, CA, USA. Morgan & Claypool, 2009.

[7] J. Liu, D. Orlando, P. Addabbo, and W. Liu, "SINR distribution for the persymmetric SMI beamformer with steering vector mismatches," *IEEE Transactions on Signal Processing*, vol. 67, no. 5, pp. 1382–1392, March 1 2019.

[8] D. M. Boroson, "Sample size considerations for adaptive arrays," *IEEE Transactions on Aerospace and Electronic Systems*, vol. AES-16, no. 4, pp. 446–451, July 1980.

[9] J. Liu, W. Liu, H. Liu, C. Bo, X.-G. Xia, and F. Dai, "Average SINR calculation of a persymmetric sample matrix inversion beamformer," *IEEE Transactions on Signal Processing*, vol. 64, no. 8, pp. 2135–2145, April 15 2016.

[10] B. D. Carlson, "Covariance matrix estimation errors and diagonal loading in adaptive arrays," *IEEE Transactions on Aerospace and Electronic Systems*, vol. 24, no. 4, pp. 397–401, July 1988.

[11] J. Capon, "High resolution frequency-wavenumber spectrum analysis," *Proceedings of the IEEE*, vol. 57, no. 8, pp. 1408–1418, August 1969.

[12] E. J. Kelly, "Performance of an adaptive detection algorithm; rejection of unwanted signals," *IEEE Transactions on Aerospace and Electronic Systems*, no. 2, pp. 122–133, March 1989.

[13] F. C. Robey, D. R. Fuhrmann, E. J. Kelly, and R. Nitzberg, "A CFAR adaptive matched filter detector," *IEEE Transactions on Aerospace and Electronic Systems*, vol. 28, no. 1, pp. 208–216, January 1992.

[14] E. J. Kelly and K. Forsythe, "Adaptive detection and parameter estimation for multidimensional signal models," Lincoln Laboratory, MIT, Technical Report 848, 1989.

[15] H. L. Van Trees, *Optimum Array Processing, Part IV of Detection, Estimation, and Modulation Theory*. Wiley-Interscience, New York, 2002.

[16] R. Nitzberg, "Application of maximum likelihood estimation of persymmetric covariance matrices to adaptive processing," *IEEE Transactions on Aerospace and Electronic Systems*, vol. AES-16, no. 1, pp. 124–127, January 1980.

[17] L. Cai and H. Wang, "A persymmetric multiband GLR algorithm," *IEEE Transactions on Aerospace and Electronic Systems*, vol. 28, no. 3, pp. 806–816, July 1992.

[18] W. Liu, J. Liu, Q. Du, and Y. Wang, "Distributed target detection in partially homogeneous environment when signal mismatch occurs," *IEEE Transactions on Signal Processing*, vol. 66, no. 14, pp. 3918–3928, July 15 2018.

[19] J. Liu, W. Liu, H. Liu, B. Chen, X.-G. Xia, and D. Zhou, "Average SINR calculation of a persymmetric sample matrix inversion beamformer," *IEEE Transactions on Signal Processing*, vol. 64, no. 8, pp. 2135–2145, April 2016.

[20] I. S. Gradshteyn and I. M. Ryzhik, *Table of Integrals, Series, and Products*, 7th ed. San Diego: Academic Press, 2007.

[21] L. Yu, W. Liu, and R. Landley, "SINR analysis of the subtraction-based smi beamformer," *IEEE Transactions on Signal Processing*, vol. 58, no. 11, pp. 5926–5932, November 2010.

[22] F. Bandiera, D. Orlando, and G. Ricci, "A subspace-based adaptive sidelobe blanker," *IEEE Transactions on Signal Processing*, vol. 56, no. 9, pp. 4141–4151, September 2008.

[23] T. Anderson, *An Introduction to Multivariate Statistical Analysis*, 3rd ed. New York, USA: Wiley, 2003.

[24] M. V. Jambunathan, "Some properties of Beta and Gamma distributions," *The Annals of Mathematical Statistics*, vol. 25, no. 2, pp. 401–405, June 1954.

[25] J. Liu, Z.-J. Zhang, Y. Yang, and H. Liu, "A CFAR adaptive subspace detector for first-order or second-order Gaussian signals based on a single observation," *IEEE Transactions on Signal Processing*, vol. 59, no. 11, pp. 5126–5140, November 2011.

[26] R. J. Muirhead, *Aspects of Multivariate Statistical Theory*. New York: Wiley, 1982.

[27] A. K. Gupta and D. K. Nagar, *Matrix Variate Distributions*. Chapman and Hall/CRC, Boca Raton, Florida, 2000.

3

Invariance Issues under Persymmetry

This chapter frames the detection problem under persymmetry in the context of statistical invariance. Specifically, we show how the invariance principle allows us to design decision rules exhibiting some natural symmetries that become of practical interest for radar systems. As a matter of fact, the CFAR property can be viewed as a consequence of the invariance with respect to specific transformations. In this chapter, the so-called MISs are derived for both the homogeneous and partially homogeneous scenarios along with their statistical characterizations. Remarkably, the maximal invariant is the fundamental tile to build up invariant and, hence, CFAR detection architectures [1,2].

3.1 Preliminary Theory

Let us consider a binary hypothesis test

$$\begin{cases} H_0 : \boldsymbol{\theta} \in \Theta_0, \\ H_1 : \boldsymbol{\theta} \in \Theta_1, \end{cases} \tag{3.1}$$

where $\boldsymbol{\theta} \in \mathbb{C}^{N \times 1}$ is the distribution parameter vector, Θ_0 and Θ_1 form a disjoint covering of the parameter space Θ. Assuming that both the hypotheses are composite, there exist components of $\boldsymbol{\theta}$ that do not enter into the decision between H_0 and H_1, namely, that are nuisance parameters. In this case, we could be interested in decision rules that are somehow invariant to such parameters.

As first step toward this goal, we look for data transformations that introduce these parameters. Thus, let us define a group of such transformations $\mathcal{G}$ that leaves unaltered the formal structure of the hypothesis test (3.1), the parameter space, and the family of distribution of data. Otherwise stated, if $g(\cdot) \in \mathcal{G}$ and $\mathbf{z} \in \mathbb{C}^{N \times 1}$ is the observed data vector, we have that

- given $\mathbf{y} = g(\mathbf{z})$, then $F_{\boldsymbol{\theta}}(\mathbf{y}) = F_{\boldsymbol{\theta}}(g(\mathbf{z})) = F_{\bar{g}(\boldsymbol{\theta})}(\mathbf{y})$, where $F_{\boldsymbol{\theta}}(\cdot)$ is the CDF of $\mathbf{x}$ and $\bar{g}(\cdot)$ is the induced transformation in the parameter space;

- $\forall \boldsymbol{\theta} \in \Theta_i : \bar{g}(\boldsymbol{\theta}) \in \Theta_i, \ i = 0, 1.$

DOI: 10.1201/9781003340232-3

Under the above assumptions, the hypothesis testing problem is said invariant to $\mathcal{G}$ [3].

Now, let us focus on a decision rule, $\phi(\mathbf{z})$, for problem (3.1), we say that $\phi(\mathbf{z})$ is invariant to $\mathcal{G}$ if

$$\phi(\mathbf{z}) = \phi(g(\mathbf{z})), \quad \forall g \in \mathcal{G}. \tag{3.2}$$

A statistic (function of data) $\mathbf{m}(\mathbf{z})$ is said to be a MIS if

1. it is invariant: $\mathbf{m}(\mathbf{z}) = \mathbf{m}(g(\mathbf{z}))$, $\forall g \in \mathcal{G}$;

2. it is maximal: $\mathbf{m}(\mathbf{z}_1) = \mathbf{m}(\mathbf{z}_2) \Rightarrow \exists g \in \mathcal{G} : \mathbf{z}_2 = g(\mathbf{z}_1)$.

A MIS partitions the observation space into equivalence classes where the maximal invariant is constant and data are related through a specific transformation. From *Theorem 6.2.1* of [4], every invariant test can be written as a function of the MIS.

In the next sections, we provide the expressions of maximal invariant statistics for the conventional detection problem under persymmetry. More precisely, we start from the HE to conclude with the PHE.

3.2   Homogeneous Environment

The formal statement of the problem at hand is

$$
\begin{cases}
H_0 : \begin{cases} \mathbf{r} = \mathbf{m}, \\ \mathbf{r}_k = \mathbf{m}_k, & k = 1, \dots, K, \end{cases} \\[2ex]
H_1 : \begin{cases} \mathbf{r} = \alpha \mathbf{v} + \mathbf{m}, \\ \mathbf{r}_k = \mathbf{m}_k, & k = 1, \dots, K, \end{cases}
\end{cases}
\tag{3.3}
$$

where

- $\mathbf{v} \in \mathbb{C}^{N \times 1}$ with $\|\mathbf{v}\| = 1$ is the nominal steering vector [5] that exhibits a persymmetric structure, namely, it complies with the equation $\mathbf{v} = \mathbf{J}\mathbf{v}^*$ with $\mathbf{J}$ the $(N \times N)$-dimensional permutation matrix in (1.2);

- $\alpha \in \mathbb{C}$ is an unknown deterministic factor accounting for both target reflectivity and channel propagation effects;

- $\mathbf{m}$ and $\mathbf{m}_k \in \mathbb{C}^{N \times 1}$, $k = 1, \dots, K$, are IID circularly symmetric complex Gaussian random vectors with zero mean and positive definite covariance matrix $\mathbf{M}_0 \in \{\mathbf{R} \in \mathbb{C}^{N \times N} : \mathbf{R} = \mathbf{J}\mathbf{R}^*\mathbf{J}\}$.

Persymmetry of $\mathbf{M}_0$ and $\mathbf{v}$ allows us to transform data in order to transfer the problem to the real domain. In fact, let us define $\bar{\mathbf{z}} = \mathbf{Tr}$, $\bar{\mathbf{z}}_k = \mathbf{Tr}_k$, $k = 1, \ldots, K$, and observe that $\forall k = 1, \ldots, K$,

$$
\begin{aligned}
\mathrm{E}[(\bar{\mathbf{z}} - \mathrm{E}[\bar{\mathbf{z}}])(\bar{\mathbf{z}} - \mathrm{E}[\bar{\mathbf{z}}])^{\dagger}] &= \mathbf{T}\mathrm{E}[(\mathbf{r} - \mathrm{E}[\mathbf{r}])(\mathbf{r} - \mathrm{E}[\mathbf{r}])^{\dagger}]\mathbf{T}^{\dagger} \\
&= \mathbf{T}\mathbf{M}_0\mathbf{T}^{\dagger} = \mathbf{M}_r \in \mathbb{R}^{N \times N}, \quad (3.4)
\end{aligned}
$$

$$
\begin{aligned}
\mathrm{E}[(\bar{\mathbf{z}}_k - \mathrm{E}[\bar{\mathbf{z}}_k])(\bar{\mathbf{z}}_k - \mathrm{E}[\bar{\mathbf{z}}_k])^{\dagger}] &= \mathbf{T}\mathrm{E}[(\mathbf{r}_k - \mathrm{E}[\mathbf{r}_k])(\mathbf{r}_k - \mathrm{E}[\mathbf{r}_k])^{\dagger}]\mathbf{T}^{\dagger} \\
&= \mathbf{T}\mathbf{M}_0\mathbf{T}^{\dagger} = \mathbf{M}_r \in \mathbb{R}^{N \times N}, \quad (3.5)
\end{aligned}
$$

where $\mathbf{T} \in \mathbb{C}^{N \times N}$ is the unitary matrix given in (13) of [6]. Since $\mathbf{z}$ and the $\mathbf{z}_k$'s obey the complex Gaussian distribution with real covariance matrix, it follows that

$$
\begin{aligned}
&\mathrm{E}[(\mathfrak{Re}(\bar{\mathbf{z}}) - \mathrm{E}[\mathfrak{Re}(\bar{\mathbf{z}})])(\mathfrak{Im}(\bar{\mathbf{z}}) - \mathrm{E}[\mathfrak{Im}(\bar{\mathbf{z}})])^{\dagger}] \\
&= -\mathrm{E}[(\mathfrak{Im}(\bar{\mathbf{z}}) - \mathrm{E}[\mathfrak{Im}(\bar{\mathbf{z}})])(\mathfrak{Re}(\bar{\mathbf{z}}) - \mathrm{E}[\mathfrak{Re}(\bar{\mathbf{z}})])^{\dagger}] = \mathbf{0}, \quad (3.6)
\end{aligned}
$$

$$
\begin{aligned}
&\mathrm{E}[(\mathfrak{Re}(\bar{\mathbf{z}}_k) - \mathrm{E}[\mathfrak{Re}(\bar{\mathbf{z}}_k)])(\mathfrak{Im}(\bar{\mathbf{z}}_k) - \mathrm{E}[\mathfrak{Im}(\bar{\mathbf{z}}_k)])^{\dagger}] \\
&= -\mathrm{E}[(\mathfrak{Im}(\bar{\mathbf{z}}_k) - \mathrm{E}[\mathfrak{Im}(\bar{\mathbf{z}}_k)])(\mathfrak{Re}(\bar{\mathbf{z}}_k) - \mathrm{E}[\mathfrak{Re}(\bar{\mathbf{z}}_k)])^{\dagger}] = \mathbf{0}. \quad (3.7)
\end{aligned}
$$

Thus, $\mathfrak{Re}(\bar{\mathbf{z}})$, $\mathfrak{Im}(\bar{\mathbf{z}})$, $\mathfrak{Re}(\bar{\mathbf{z}}_k)$, and $\mathfrak{Im}(\bar{\mathbf{z}}_k)$ are IID Gaussian vectors with covariance matrix $\mathbf{M}_r/2$. Moreover, under H_1,

$$
\mathrm{E}[\mathfrak{Re}(\bar{\mathbf{z}})] = \alpha_1 \mathbf{v}_r \quad \text{and} \quad \mathrm{E}[\mathfrak{Im}(\bar{\mathbf{z}})] = \alpha_2 \mathbf{v}_r, \quad (3.8)
$$

where[1] $\mathbf{v}_r = \mathbf{Tv} \in \mathbb{R}^{N \times 1}$, $\alpha_1 = \mathfrak{Re}(\alpha)$, and $\alpha_2 = \mathfrak{Im}(\alpha)$ [6].

Finally, we apply another transformation to data by rotating the nominal steering vector into the first elementary vector $\mathbf{e}_1 = [1\ 0\ \ldots 0]^T \in \mathbb{R}^{N \times 1}$. This last transformation returns the so-called canonical form for the hypothesis testing problem [7]. To this end, let $\mathbf{V} \in \mathbb{R}^{N \times N}$ be an orthogonal matrix such that $\mathbf{V}\mathbf{v}_r = \mathbf{e}_1$ and rotate data to obtain $\mathbf{z}_1 = \mathbf{V}\mathfrak{Re}(\bar{\mathbf{z}})$, $\mathbf{z}_2 = \mathbf{V}\mathfrak{Im}(\bar{\mathbf{z}})$, $\mathbf{z}_{1k} = \mathbf{V}\mathfrak{Re}(\bar{\mathbf{z}}_k)$, $\mathbf{z}_{2k} = \mathbf{V}\mathfrak{Im}(\bar{\mathbf{z}}_k)$, $k = 1, \ldots, K$. Therefore, (3.3) is equivalent to the following problem (which is expressed in canonical form)

$$
\begin{cases}
H_0 : \begin{cases} \mathbf{z}_1 = \mathbf{n}_1, \ \mathbf{z}_2 = \mathbf{n}_2, \\ \mathbf{z}_{1k} = \mathbf{n}_{1k}, \ \mathbf{z}_{2k} = \mathbf{n}_{2k}, \ k = 1, \ldots, K, \end{cases} \\[1em]
H_1 : \begin{cases} \mathbf{z}_1 = \alpha_1 \mathbf{e}_1 + \mathbf{n}_1, \ \mathbf{z}_2 = \alpha_2 \mathbf{e}_1 + \mathbf{n}_2, \\ \mathbf{z}_{1k} = \mathbf{n}_{1k}, \ \mathbf{z}_{2k} = \mathbf{n}_{2k}, \ k = 1, \ldots, K, \end{cases}
\end{cases} \quad (3.9)
$$

where $\mathbf{n}_1 = \mathbf{V}\mathfrak{Re}(\mathbf{Tm})$, $\mathbf{n}_2 = \mathbf{V}\mathfrak{Im}(\mathbf{Tm})$, $\mathbf{n}_{1k} = \mathbf{V}\mathfrak{Re}(\mathbf{Tm}_k)$, $\mathbf{n}_{2k} = \mathbf{V}\mathfrak{Im}(\mathbf{Tm}_k)$, $k = 1, \ldots, K$, are IID real Gaussian vectors with zero mean and covariance matrix $\mathbf{M} = \frac{1}{2}\mathbf{V}\mathbf{M}_r\mathbf{V}^T$, namely, $\mathbf{n}_i, \mathbf{n}_{ik} \sim \mathcal{N}_N(\mathbf{0}, \mathbf{M})$, $i = 1, 2$, $k = 1, \ldots, K$. The sequence of transformations that lead to problem (3.9) is summarized in Figure 3.1.

---

[1] Note that $\|\mathbf{v}_r\| = 1$.

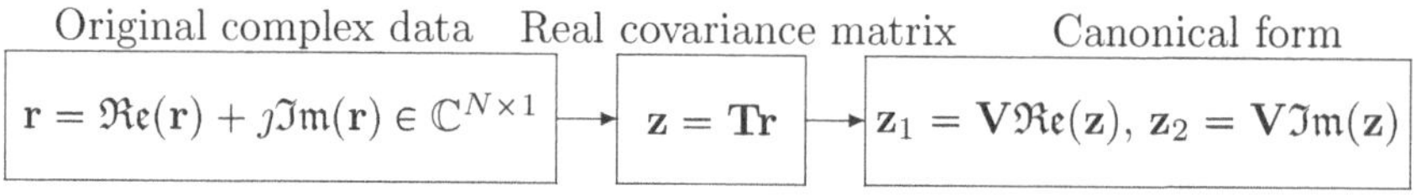

Original complex data	Real covariance matrix	Canonical form
$\mathbf{r} = \mathfrak{Re}(\mathbf{r}) + \jmath\mathfrak{Im}(\mathbf{r}) \in \mathbb{C}^{N \times 1}$	$\mathbf{z} = \mathbf{Tr}$	$\mathbf{z}_1 = \mathbf{V}\mathfrak{Re}(\mathbf{z}), \ \mathbf{z}_2 = \mathbf{V}\mathfrak{Im}(\mathbf{z})$

FIGURE 3.1
Sequence of transformations leading to the canonical form (3.9).

It is important to highlight that in (3.9), the relevant parameter vector is $\boldsymbol{\alpha} = [\alpha_1 \ \alpha_2]^T$, whereas the entries of $\mathbf{M}$ represent nuisance parameters. As a matter of fact, regardless the value of $\mathbf{M}$, if H_1 is true, then $\|\boldsymbol{\alpha}\| \neq 0$, whereas under the null hypothesis (H_0), $\|\boldsymbol{\alpha}\| = 0$. As a consequence, we can look for decision rules that are invariant to $\mathbf{M}$.

To this end, we can exploit the principle of invariance [4] and, as stated in Section 3.1, we find the transformation group that properly clusters data without altering

- the formal structure of the hypothesis testing problem given by $H_0 : \|\boldsymbol{\alpha}\| = 0$, $H_1 : \|\boldsymbol{\alpha}\| \neq 0$;

- the Gaussian model under the two hypotheses;

- the real symmetric structure of the covariance matrix.

Before defining such kind of transformations, let us partition data vectors as follows

$$\mathbf{Z}_p = [\mathbf{z}_1 \ \mathbf{z}_2] = \begin{bmatrix} \mathbf{z}_{1p} \\ \mathbf{Z}_{2p} \end{bmatrix}, \ \mathbf{S} = \mathbf{Z}_s\mathbf{Z}_s^T = \begin{bmatrix} s_{11} & \mathbf{s}_{12} \\ \mathbf{s}_{21} & \mathbf{S}_{22} \end{bmatrix}, \qquad (3.10)$$

where $\mathbf{Z}_s = [\mathbf{z}_{1k} \ \cdots \ \mathbf{z}_{1K} \ \mathbf{z}_{2k} \ \cdots \ \mathbf{z}_{2K}]$, $\mathbf{z}_{1p} \in \mathbb{R}^{1 \times 2}$, $\mathbf{Z}_{2p} \in \mathbb{R}^{(N-1) \times 2}$, $s_{11} \in \mathbb{R}$, $\mathbf{s}_{12} \in \mathbb{R}^{1 \times (N-1)}$, $\mathbf{s}_{21} \in \mathbb{R}^{(N-1) \times 1}$, and $\mathbf{S}_{22} \in \mathbb{R}^{(N-1) \times (N-1)}$. Moreover, we replace data $(\mathbf{Z}_p, \mathbf{Z}_s)$ by a sufficient statistic $(\mathbf{Z}_p, \mathbf{S})$ (reduction by sufficiency) [3, 8]. Problem (3.9) can be expressed in terms of the sufficient statistic as

$$\begin{cases} H_0 : \ \mathbf{Z}_p = [\mathbf{n}_1 \ \mathbf{n}_2], & \mathbf{S} = \mathbf{Z}_s\mathbf{Z}_s^T, \\ H_1 : \ \mathbf{Z}_p = \mathbf{e}_1\boldsymbol{\alpha}^T + [\mathbf{n}_1 \ \mathbf{n}_2], & \mathbf{S} = \mathbf{Z}_s\mathbf{Z}_s^T. \end{cases} \qquad (3.11)$$

Now, consider the group $\mathcal{G}_N$ of $N \times N$ non-singular matrices with the following structure[2]

$$\mathbf{G} = \begin{bmatrix} g_{11} & \mathbf{g}_{12} \\ \mathbf{0} & \mathbf{G}_{22} \end{bmatrix} \in \mathbb{R}^{N \times N}, \qquad (3.12)$$

---

[2]Hereafter, the dimensions of a submatrix indexed by the couple $(l, m) \in \{(1, 1), (1, 2), (2, 1), (2, 2)\}$ can be obtained replacing 1 and 2 with 1 and $N - 1$, respectively.

where $g_{11} \neq 0$ and $\det(\mathbf{G}_{22}) \neq 0$, the group $\mathcal{O}_2$ of 2×2 orthogonal matrices, and notice that the group

$$\mathcal{L} = \mathcal{G}_N \times \mathcal{O}_2 \tag{3.13}$$

with the operation $\circ$ defined by

$$(\mathbf{G}_1, \mathbf{U}_1) \circ (\mathbf{G}_2, \mathbf{U}_2) = (\mathbf{G}_2\mathbf{G}_1, \mathbf{U}_1\mathbf{U}_2), \quad \mathbf{G}_1, \mathbf{G}_2 \in \mathcal{G}_N, \; \mathbf{U}_1, \mathbf{U}_2 \in \mathcal{O}_2, \tag{3.14}$$

leaves the hypothesis testing problem invariant under the action l defined by

$$l(\mathbf{Z}_p, \mathbf{S}) = (\mathbf{G}\mathbf{Z}_p\mathbf{U}, \mathbf{G}\mathbf{S}\mathbf{G}^T), \quad \forall (\mathbf{G}, \mathbf{U}) \in \mathcal{L}. \tag{3.15}$$

In fact, it can be easily verified that

$$\begin{cases} H_0 : & \begin{cases} \mathbf{G}\mathbf{Z}_p\mathbf{U} = [\mathbf{G}\mathbf{n}_1 \; \mathbf{G}\mathbf{n}_2]\mathbf{U} = [\bar{\mathbf{n}}_1 \; \bar{\mathbf{n}}_2], \\ \mathbf{G}\mathbf{S}\mathbf{G}^T = \mathbf{G}\mathbf{Z}_s\mathbf{Z}_s^T\mathbf{G}^T = \bar{\mathbf{Z}}_s\bar{\mathbf{Z}}_s^T, \end{cases} \\ H_1 : & \begin{cases} \mathbf{G}\mathbf{Z}_p\mathbf{U} = \mathbf{e}_1 g_{11}\boldsymbol{\alpha}^T\mathbf{U} + [\bar{\mathbf{n}}_1 \; \bar{\mathbf{n}}_2] = \mathbf{e}_1\bar{\boldsymbol{\alpha}}^T + [\bar{\mathbf{n}}_1 \; \bar{\mathbf{n}}_2], \\ \mathbf{G}\mathbf{S}\mathbf{G}^T = \bar{\mathbf{Z}}_s\bar{\mathbf{Z}}_s^T, \end{cases} \end{cases} \tag{3.16}$$

where the columns of $\bar{\mathbf{Z}}_s$ are still Gaussian vectors with zero mean and unknown positive definite covariance matrix $\mathbf{G}\mathbf{M}\mathbf{G}^T$, $\bar{\boldsymbol{\alpha}} = g_{11}\mathbf{U}^T\boldsymbol{\alpha} \in \mathbb{R}^{2 \times 1}$ with $\|\bar{\boldsymbol{\alpha}}\| \neq \mathbf{0}$, and observe that the columns of $[\bar{\mathbf{n}}_1 \; \bar{\mathbf{n}}_2]^T = \mathbf{U}^T[\mathbf{n}_1 \; \mathbf{n}_2]^T$ are Gaussian vectors with identity covariance matrix, namely, $\bar{\mathbf{n}}_1$ and $\bar{\mathbf{n}}_2$ are IID Gaussian vectors with unknown positive definite covariance matrix $\mathbf{G}\mathbf{M}\mathbf{G}^T$. Moreover, notice that $\mathcal{L}$ preserves the family of distributions and, at the same time, includes those transformations which are relevant from the practical point of view such as the premultiplication by $\mathbf{G} \in \mathcal{G}_N$ that allow to claim the CFAR property as a consequence of the invariance.

Summarizing, we have identified a group of transformations which leaves unaltered the decision problem under consideration. As a consequence, it is reasonable to find decision rules that are invariant under the same group of transformations. The following theorem provides the expression of a maximal invariant for problem (3.9).

Theorem 3.2.1. *A MIS with respect to $\mathcal{L}$ for problem (3.9) is given by*

$$\mathbf{t}_1(\mathbf{Z}_p, \mathbf{S}) = [t_{GLRT} \; t_{P\text{-}AMF} \; \lambda_3 \; \lambda_4]^T, \tag{3.17}$$

where $\lambda_3 > \lambda_4 > 0$ are the eigenvalues of $\mathbf{Z}_{2p}^T\mathbf{S}_{22}^{-1}\mathbf{Z}_{2p}$,

$$t_{GLRT} = \frac{\det(\mathbf{I}_2 + \mathbf{Z}_p^T\mathbf{S}^{-1}\mathbf{Z}_p)}{\det\left(\mathbf{I}_2 + \mathbf{Z}_p^T\mathbf{S}^{-1}\mathbf{Z}_p - \dfrac{\mathbf{Z}_p^T\mathbf{S}^{-1}\mathbf{e}_1\mathbf{e}_1^T\mathbf{S}^{-1}\mathbf{Z}_p}{\mathbf{e}_1^T\mathbf{S}^{-1}\mathbf{e}_1}\right)} \tag{3.18}$$

is the statistic of the GLRT [9] for (3.9), and

$$t_{P\text{-}AMF} = \frac{\mathbf{e}_1^T\mathbf{S}^{-1}\mathbf{Z}_p\mathbf{Z}_p^T\mathbf{S}^{-1}\mathbf{e}_1}{\mathbf{e}_1^T\mathbf{S}^{-1}\mathbf{e}_1} \tag{3.19}$$

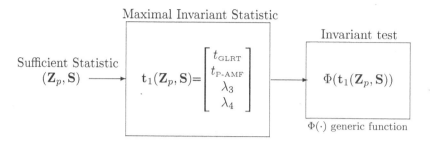

FIGURE 3.2
Construction of a generic invariant test starting from the maximal invariant statistic.

is the statistic of the persymmetric adaptive matched filter (P-AMF) that can be obtained by applying the two-step GLRT-based design procedure [10] to problem (3.9).

Proof. See Appendix 3.A.   □

Thus, the MIS is a four-dimensional vector whose the first two components are well-known decision schemes and the last two components are ancillary statistics independent of the true hypothesis. These last components arise from the persymmetric structure of the disturbance covariance matrix unlike the general Hermitian case that leads to a two-dimensional vector [11] formed by Kelly's GLRT [12] and the AMF [10]. As stated in Section 3.1, since every function of the MIS is an invariant test, we can build up invariant tests for problem 3.9 as shown in Figure 3.2.

In the next subsection, we provide a stochastic representation of $\mathbf{t}_1(\mathbf{Z}_p, \mathbf{S})$ defined in (3.17). Such a representation will be useful to find the PDF of the maximal invariant as well as to show that detectors obtained through suboptimum well-established design criteria can be expressed as a function of the MIS and characterized accordingly.

3.2.1   Stochastic Representation

Let us start by rewriting the first two components of $\mathbf{t}_1$, namely, the GLRT and the P-AMF, as follows [7]:

$$t_{\text{GLRT}} = 1 + \frac{\mathbf{z}(\mathbf{I}_2 + \mathbf{Z}_{2p}^T \mathbf{S}_{22}^{-1} \mathbf{Z}_{2p})^{-1}\mathbf{z}^T}{\zeta}, \qquad (3.20)$$

$$t_{\text{P-AMF}} = \frac{\mathbf{z}\mathbf{z}^T}{\zeta}, \qquad (3.21)$$

where $\mathbf{z} = \mathbf{z}_{1p} - \mathbf{s}_{12}\mathbf{S}_{22}^{-1}\mathbf{Z}_{2p}$ and $\zeta = s_{11} - \mathbf{s}_{12}\mathbf{S}_{22}^{-1}\mathbf{s}_{21}$. As a consequence, equation (3.17) becomes

$$\mathbf{t}_1(\mathbf{Z}_p, \mathbf{S}) = \begin{bmatrix} 1 + \mathbf{z}(\mathbf{I}_2 + \mathbf{Z}_{2p}^T\mathbf{S}_{22}^{-1}\mathbf{Z}_{2p})^{-1}\mathbf{z}^T/\zeta \\ \mathbf{z}\mathbf{z}^T/\zeta \\ \lambda_3 \\ \lambda_4 \end{bmatrix}. \tag{3.22}$$

Now, we apply a one-to-one transform to the above statistic to obtain

$$\mathbf{t}_2(\mathbf{Z}_p, \mathbf{S}) = \begin{bmatrix} (\mathbf{I}_2 + \mathbf{\Lambda})^{-1/2}\mathbf{U}\mathbf{z}^T/\sqrt{\zeta}] \\ \lambda_3 \\ \lambda_4 \end{bmatrix}, \tag{3.23}$$

where $\mathbf{U} \in \mathcal{O}_2$ is such that $\mathbf{Z}_{2p}^T\mathbf{S}_{22}^{-1}\mathbf{Z}_{2p} = \mathbf{U}^T\mathbf{\Lambda}\mathbf{U}$ with $\mathbf{\Lambda} \in \mathbb{R}^{2\times 2}$ a diagonal matrix whose non-zero entries are λ_3 and λ_4. In fact, given $\mathbf{t}_2(\mathbf{Z}_p, \mathbf{S})$, it is easy to evaluate $\mathbf{t}_1(\mathbf{Z}_p, \mathbf{S})$. On the other hand, given $\mathbf{t}_1(\mathbf{Z}_p, \mathbf{S})$, we denote by t_{1i} its ith entry and define

$$\mathbf{c} = \begin{bmatrix} c_1 \\ c_2 \end{bmatrix} = \frac{\mathbf{U}\mathbf{z}^T}{\sqrt{\zeta}}. \tag{3.24}$$

Thus, based upon the structure of t_{11} and t_{12}, we can build up the following system

$$\begin{cases} c_1^2/(1 + \lambda_3) + c_2^2/(1 + \lambda_4) = t_{11} - 1, \\ c_1^2 + c_2^2 = t_{12}, \end{cases} \tag{3.25}$$

that admits real solutions since $\lambda_3 \neq \lambda_4$ with probability one and the Rayleigh quotient conforms with[3]

$$\frac{1}{1 + \lambda_3} \leqslant \frac{t_{11} - 1}{t_{12}} = \frac{\mathbf{c}^T(\mathbf{I}_2 + \mathbf{\Lambda})^{-1}\mathbf{c}}{\mathbf{c}^T\mathbf{c}} \leqslant \frac{1}{1 + \lambda_4}. \tag{3.26}$$

Finally, from $\mathbf{c}$ it is possible to compute $\mathbf{t}_2(\mathbf{Z}_p, \mathbf{S})$ using λ_3 and λ_4.

Invoking invariance, the statistical characterization of the maximal invariant $\mathbf{t}_2(\mathbf{Z}_p, \mathbf{S})$ can be accomplished by whitening the original data through a matrix transformation $\mathbf{L}$ which belongs to the invariance group $\mathcal{G}_N$ and does not modify the useful signal direction $\mathbf{e}_1$. Specifically, partitioning $\mathbf{M}$ as

$$\mathbf{M} = \begin{bmatrix} m_{11} & \mathbf{m}_{12} \\ \mathbf{m}_{12}^T & \mathbf{M}_{22} \end{bmatrix}, \tag{3.27}$$

where m_{11} is a scalar. Then, the whitening matrix is

$$\mathbf{L} = \begin{bmatrix} l_{11} & -l_{11}\mathbf{m}_{12}\mathbf{M}_{22}^{-1} \\ \mathbf{0} & \mathbf{M}_{22}^{-1/2} \end{bmatrix} \tag{3.28}$$

---

[3]Condition (3.26) ensures that the solutions are real.

with $l_{11} = (m_{11} - \mathbf{m}_{12}\mathbf{M}_{22}^{-1}\mathbf{m}_{12}^T)^{-1/2}$. Thus, let $\mathbf{y} = \mathbf{x}_1 - \mathbf{s}_{0,12}\mathbf{S}_{0,22}^{-1}\mathbf{X}_2$ and $\xi = s_{0,11} - \mathbf{s}_{0,12}\mathbf{S}_{0,22}^{-1}\mathbf{s}_{0,21}$, where

$$\mathbf{L}\mathbf{Z}_p = \begin{bmatrix} \mathbf{x}_1 \\ \mathbf{X}_2 \end{bmatrix} = \mathbf{X}, \tag{3.29}$$

$$\mathbf{L}\mathbf{S}\mathbf{L}^T = \mathbf{L}\mathbf{Z}_s\mathbf{Z}_s^T\mathbf{L}^T = \mathbf{X}_s\mathbf{X}_s^T$$

$$= \begin{bmatrix} \mathbf{x}_{s,1} \\ \mathbf{X}_{s,2} \end{bmatrix}\begin{bmatrix} \mathbf{x}_{s,1} \\ \mathbf{X}_{s,2} \end{bmatrix}^T$$

$$= \begin{bmatrix} s_{0,11} & \mathbf{s}_{0,12} \\ \mathbf{s}_{0,21} & \mathbf{S}_{0,22} \end{bmatrix} = \mathbf{S}_0 \tag{3.30}$$

with $\mathbf{x}_1 \in \mathbb{R}^{1\times 2}$, $\mathbf{X}_2 \in \mathbb{R}^{(N-1)\times 2}$, $\mathbf{x}_{s,1} \in \mathbb{R}^{1\times 2K}$, $\mathbf{X}_{s,2} \in \mathbb{R}^{(N-1)\times 2K}$, $s_{0,11} \in \mathbb{R}$, $\mathbf{s}_{0,12} \in \mathbb{R}^{1\times(N-1)}$, $\mathbf{s}_{0,21} \in \mathbb{R}^{(N-1)\times 1}$, $\mathbf{S}_{0,22} \in \mathbb{R}^{(N-1)\times(N-1)}$. Since $\mathbf{t}_2$ is invariant to the above transformation (i.e., invariance principle), the following equality is true

$$\mathbf{y}_1 = \frac{(\mathbf{I}_2 + \mathbf{\Lambda}_0)^{-1/2}\mathbf{U}_0\mathbf{y}^T}{\sqrt{\xi}} = \frac{(\mathbf{I}_2 + \mathbf{\Lambda})^{-1/2}\mathbf{U}\mathbf{z}^T}{\sqrt{\zeta}} = \mathbf{q}, \tag{3.31}$$

where $\mathbf{U}_0 = \mathbf{U}$ is such that $\mathbf{X}_2^T\mathbf{S}_{0,22}^{-1}\mathbf{X}_2 = \mathbf{U}_0^T\mathbf{\Lambda}_0\mathbf{U}_0$ with $\mathbf{\Lambda}_0 = \mathbf{\Lambda}$ a diagonal matrix whose non-zero entries are the eigenvalues of $\mathbf{X}_2^T\mathbf{S}_{0,22}^{-1}\mathbf{X}_2$. Summarizing, in what follows we consider the following stochastic representation

$$\mathbf{t}_2(\mathbf{Z}_p, \mathbf{S}) = \mathbf{t}_2(\mathbf{L}\mathbf{Z}_p, \mathbf{L}\mathbf{S}\mathbf{L}^T) = \begin{bmatrix} (\mathbf{I}_2 + \mathbf{\Lambda})^{-1/2}\mathbf{U}\mathbf{y}^T/\sqrt{\xi} \\ \lambda_3 \\ \lambda_4 \end{bmatrix}. \tag{3.32}$$

Before providing the expression of the PDF of the maximal invariant as well as of detectors based upon this PDF, in the next subsection, we show that detection architectures devised by applying well-established design criteria are invariant.

3.2.2 Invariant Detectors

In Section 3.1, we noticed that any test that is function of the maximal invariant statistic is invariant to the same group of transformation. This property allows us to obtain invariant architectures selecting suitable function of the maximal invariant. A tangible example is provided by the GLRT and the P-AMF that can be written as

$$t_{\mathrm{GLRT}} = 1 + [\mathbf{t}_2(\mathbf{Z}_p, \mathbf{S})]^T \begin{bmatrix} \mathbf{I}_2 \\ \mathbf{0} \end{bmatrix} [\mathbf{I}_2 \; \mathbf{0}]\mathbf{t}_2(\mathbf{Z}_p, \mathbf{S})$$

$$= 1 + \mathbf{q}^T\mathbf{q} \underset{H_0}{\overset{H_1}{\gtrless}} \eta, \tag{3.33}$$

$$t_{\text{P-AMF}} = [\mathbf{t}_2(\mathbf{Z}_p, \mathbf{S})]^T \begin{bmatrix} \mathbf{I}_2 \\ \mathbf{0} \end{bmatrix} \begin{bmatrix} 1 + [0\ 0\ 1\ 0]\mathbf{t}_2(\mathbf{Z}_p, \mathbf{S}) & 0 \\ 0 & 1 + [0\ 0\ 0\ 1]\mathbf{t}_2(\mathbf{Z}_p, \mathbf{S}) \end{bmatrix}$$

$$\times [\mathbf{I}_2\ \mathbf{0}]\mathbf{t}_2(\mathbf{Z}_p, \mathbf{S}) = \mathbf{q}^T(\mathbf{I}_2 + \mathbf{\Lambda})\mathbf{q} \underset{H_0}{\overset{H_1}{\gtrless}} \eta, \tag{3.34}$$

where η is the detection threshold set according to the desired PFA; even though each decision rule has its own detection threshold, for simplicity, we denote by η the generic threshold. The above expressions highlight that both receivers are invariant with respect to $\mathcal{L}$.

An alternative procedure to solve (3.9) and based upon the estimate-and-plug paradigm consists in applying the GLRT assuming that the structure of $\mathbf{M}$ is known and, then, in replacing it with the SCM computed over the secondary data [13]. This strategy leads to the following invariant decision scheme [14]

$$\frac{\mathbf{e}_1^T \mathbf{S}^{-1} \mathbf{Z}_p (\mathbf{e}_1 \mathbf{S}^{-1} \mathbf{e}_1)^{-1} \mathbf{Z}_p^T \mathbf{S}^{-1} \mathbf{e}_1}{\text{tr}[\mathbf{Z}_p^T \mathbf{S}^{-1} \mathbf{Z}_p]} \underset{H_0}{\overset{H_1}{\gtrless}} \eta, \tag{3.35}$$

which is statistically equivalent to

$$t_{\text{P-ACE}} = \frac{t_{\text{P-AMF}}}{\lambda_3 + \lambda_4} \underset{H_0}{\overset{H_1}{\gtrless}} \eta. \tag{3.36}$$

In the sequel, we refer to the above detector as persymmetric adaptive coherence estimator (P-ACE). Observe that the P-ACE ensures the CFAR property in PHEs, where the interference covariance matrix of the primary data and that of secondary data coincide only up to a scale factor.

A design approach alternative to the GLRT (and its modifications) is the Rao test. In this respect, the authors of [15] provided the expression of the Rao test for the problem at hand, whose decision statistic can be written as

$$t_{\text{RAO}} = \frac{[\mathbf{e}_1^T (\mathbf{S} + \mathbf{Z}_p \mathbf{Z}_p^T)^{-1} \mathbf{z}_1]^2 + [\mathbf{e}_1^T (\mathbf{S} + \mathbf{Z}_p \mathbf{Z}_p^T)^{-1} \mathbf{z}_2]^2}{\mathbf{e}_1^T (\mathbf{S} + \mathbf{Z}_p \mathbf{Z}_p^T)^{-1} \mathbf{e}_1}. \tag{3.37}$$

In Appendix 3.B, we show that the above statistic can be recast as

$$t_{\text{RAO}} = \frac{\|(\mathbf{I}_2 + \mathbf{\Lambda})^{-1/2}(\mathbf{I}_2 + \mathbf{q}\mathbf{q}^T)^{-1}\mathbf{q}\|^2}{1 - \mathbf{q}^T(\mathbf{I}_2 + \mathbf{q}\mathbf{q}^T)^{-1}\mathbf{q}} \underset{H_0}{\overset{H_1}{\gtrless}} \eta. \tag{3.38}$$

Finally, another decision scheme is the energy detector, given by

$$t_{\text{ED}} = \text{tr}[\mathbf{Z}_p^T \mathbf{S}^{-1} \mathbf{Z}_p] = t_{\text{P-AMF}} + [0\ 0\ 1\ 1]\mathbf{t}_2(\mathbf{Z}_p, \mathbf{S})$$

$$= \mathbf{q}^T(\mathbf{I}_2 + \mathbf{\Lambda})\mathbf{q} + \lambda_3 + \lambda_4 \underset{H_0}{\overset{H_1}{\gtrless}} \eta. \tag{3.39}$$

In the next subsection, we provide the PDF of the maximal invariant statistic under each hypothesis exploiting the whitened stochastic representation (3.31).

3.2.3 Statistical Characterization

The aim of this last subsection is twofold. Specifically, besides providing the distribution of $\mathbf{t}_2$, it uses the corresponding PDFs to build up decision schemes based upon the LRT.

The PDFs of $\mathbf{t}_2$ are derived by exploiting the fact that λ_3 and λ_4 are ancillary statistics whose distribution is independent of the hypothesis. Thus, for each hypothesis H_i, $i = 1, 2$, we can write that

$$f_i(\mathbf{t}_2) = f_i(\mathbf{q}, \lambda_3, \lambda_4) = p_i(\mathbf{q}|\lambda_3, \lambda_4)p(\lambda_3, \lambda_4), \qquad (3.40)$$

where $p_i(\mathbf{q}|\lambda_3, \lambda_4)$ is the conditional PDF of $\mathbf{q}$ given (λ_3, λ_4) under H_i, while $p(\lambda_3, \lambda_4)$ is the joint PDF of λ_3 and λ_4 under both hypotheses. The next theorems provide the expressions of $p_i(\mathbf{q}|\lambda_3, \lambda_4)$, $i = 1, 2$, and $p(\lambda_3, \lambda_4)$.

Theorem 3.2.2. *Under the H_0 hypothesis, the conditional PDF of $\mathbf{q}$ given λ_3 and λ_4 is*

$$p_0(\mathbf{q}|\lambda_3, \lambda_4) = p_0(\mathbf{q}) = \frac{c}{\pi}(1 + \|\mathbf{q}\|^2)^{-(c+1)}, \qquad (3.41)$$

where $c = (2K - N + 1)/2$.

Proof. See Appendix 3.C. □

It is important to highlight that under H_0 the conditional PDF of $\mathbf{q}$ given λ_3 and λ_4 is independent of λ_3 and λ_4 and, hence, it is the unconditional one. It follows that $\mathbf{q}$ is statistically independent of λ_3 and λ_4 under H_0.

Theorem 3.2.3. *Under H_1, the conditional PDF of $\mathbf{q}$ given λ_3 and λ_4 has the following expression*

$$
\begin{aligned}
p_1(\mathbf{q}|\lambda_3, \lambda_4) = {}& \frac{c}{2\pi^2(1 + \|\mathbf{q}\|^2)^{c+1}} \exp\left\{-\frac{\rho}{2(1 + \lambda_4)}\right\} \\
& \times \int_0^{2\pi} \exp\left\{-\frac{1}{2}\frac{\rho(\cos\beta)^2(\lambda_4 - \lambda_3)}{(1 + \lambda_3)(1 + \lambda_4)}\right\} \\
& \times {}_1F_1\left(c + 1, \frac{1}{2}, \frac{\rho(\mathbf{q}^T(\mathbf{I} + \boldsymbol{\Lambda}_0)^{-1/2}\mathbf{u}(\beta))^2}{2(1 + \|\mathbf{q}\|^2)}\right) d\beta, \quad (3.42)
\end{aligned}
$$

where $\mathbf{u}(\beta) = [\cos\beta \ \sin\beta]^T$, $\rho = \|\boldsymbol{\alpha}\|^2 \mathbf{e}_1^T \mathbf{M}^{-1} \mathbf{e}_1$, and ${}_1F_1(\cdot, \cdot, \cdot)$ is the confluent hypergeometric function.[4]

Proof. See Appendix 3.D. □

---

[4] Notice that if N is an even number, then the hypergeometric function which appears in the integral (3.42) can be written in terms of elementary functions. Precisely, applying the Kummer's transformation ${}_1F_1\left(\frac{2K-N+3}{2}, \frac{1}{2}, z\right) = e^z {}_1F_1\left(-\frac{2K-N+2}{2}, \frac{1}{2}, -z\right)$. Hence,

Finally, the following theorem provides the last tile to complete the statistical characterization of $\mathbf{t}_2$.

Theorem 3.2.4. *If $N \geq 2$, the joint PDF of λ_3 and λ_4 $(\lambda_3 \geq \lambda_4)$ is given by*

$$
p(\lambda_3, \lambda_4) = \begin{cases} \omega_1(N, K) \dfrac{(\lambda_3 - \lambda_4) s(\lambda_3) s(\lambda_4)}{(1 + \lambda_3)^{K+1} (1 + \lambda_4)^{K+1}}, & N = 2, \\[3mm] \omega_2(N, K) \dfrac{\lambda_3^{\frac{N-4}{2}} \lambda_4^{\frac{N-4}{2}} (\lambda_3 - \lambda_4) s(\lambda_3) s(\lambda_4)}{(1 + \lambda_3)^{K+1} (1 + \lambda_4)^{K+1}}, & N \geq 3, \end{cases} \tag{3.43}
$$

where $\omega_1(N, K)$ and $\omega_2(N, K)$ are normalization constants while $s(\cdot)$ denotes the unit-step function.

Proof. To obtain (3.43), we notice that the ordered eigenvalues of $\mathbf{Z}_{2p}^T \mathbf{S}_{22}^{-1} \mathbf{Z}_{2p}$ are the same as those of $\mathbf{Z}_{2p} \mathbf{Z}_{2p}^T \mathbf{S}_{22}^{-1}$ whose PDF can be obtained through *Corollary 10.4.6* in [8] if $N \geq 3$, and via *Corollary 10.4.3* in [8] if $N = 2$. □

The last subsection shows how to obtain invariant detectors based upon the LRT.

3.2.3.1 LRT-Based Decision Schemes

The expressions of the PDFs provided by **Theorems 3.2.2** and **3.2.3** suggest that the LRT can be constructed neglecting the distribution of the ancillary part of the maximal invariant. Moreover, the induced maximal invariant can be easily derived from the fact that the PDF of $\mathbf{t}_1(\mathbf{Z}_p, \mathbf{S})$ depends on the parameters through ρ which plays the role of the SINR.

Thus, as a starting point of our analysis, we consider the MPID that, according to the Neyman-Pearson criterion, is given by the LRT after data compression. Specifically, the MPID has the following expression

$$
\begin{aligned}
t_{\mathrm{MPID}} ={}& \frac{p_1(\mathbf{q}|\lambda_3, \lambda_4)}{p_0(\mathbf{q}|\lambda_3, \lambda_4)} = \frac{1}{2\pi} \exp\left\{ -\frac{\rho}{2(1 + \lambda_4)} \right\} \\
&\times \int_0^{2\pi} \exp\left\{ -\frac{1}{2} \frac{\rho (\cos\beta)^2 (\lambda_4 - \lambda_3)}{(1 + \lambda_3)(1 + \lambda_4)} \right\} \\
&\times {}_1F_1\left(c + 1, \frac{1}{2}, \frac{\rho(\mathbf{q}^T (\mathbf{I}_2 + \mathbf{\Lambda}_0)^{-1/2} \mathbf{u}(\beta))^2}{2(1 + \|\mathbf{q}\|^2)} \right) d\beta \underset{H_0}{\overset{H_1}{\gtrless}} \eta. \tag{3.44}
\end{aligned}
$$

---

using [16, 13.6.9, 6.1.22], we obtain

$$
{}_1F_1\left(-\frac{2K - N + 2}{2}, \frac{1}{2}, -z \right) = \frac{\left(\frac{2K - N + 2}{2} \right)! \Gamma(1/2)}{\Gamma\left(\frac{2K - N + 3}{2} \right)} L_{\frac{2K - N + 2}{2}}^{-\frac{1}{2}}(-z),
$$

where $L_n^\alpha(z)$, $\alpha > -1$, is the generalized (associated) Laguerre polynomial given by

$$
L_n^\alpha(z) = \sum_{k=0}^n \frac{\Gamma(n + \alpha + 1)}{\Gamma(n - k + 1)\Gamma(\alpha + k + 1)k!} (-z)^k.
$$

It is important to observe that receiver (3.44) is clairvoyant and, as a consequence, it does not have a practical value in radar applications since it requires the knowledge of the induced maximal invariant. Nevertheless, it can be used as a detection performance benchmark.

The second detector based upon the LRT considered here is the LMPID that can be obtained from the LRT in the limit of zero SINR. From a practical point of view, the LMPID can be useful in applications where the performance becomes critical for low SINR values.

Following the lead of [17, 18], the LMPID is given by

$$t_{\text{LMPID}} = \frac{\left.\dfrac{\delta p_1(\mathbf{q}|\lambda_3, \lambda_4)}{\delta \rho}\right|_{\rho=0}}{p_0(\mathbf{q})} \underset{H_0}{\overset{H_1}{\gtrless}} \eta. \tag{3.45}$$

The expression of t_{LMPID} is provided by the following theorem.

Theorem 3.2.5. *The LMPID is*

$$t_{LMPID} = \frac{(c+1)}{2} \frac{\mathbf{q}^T (\mathbf{I}_2 + \mathbf{\Lambda}_0)^{-1} \mathbf{q}}{(1 + \|\mathbf{q}\|^2)} - \frac{2 + \lambda_3 + \lambda_4}{4(1 + \lambda_3)(1 + \lambda_4)} \underset{H_0}{\overset{H_1}{\gtrless}} \eta. \tag{3.46}$$

Proof. See Appendix 3.E.   □

3.3   Partially Homogeneous Environment

Let us consider again problem (3.3) with the difference that, now, primary data $\mathbf{r}$ and secondary data $\mathbf{r}_k$s share the same structure of the interference covariance structure but different power levels. Specifically, the distributions of the interference components are

$$\mathbf{n} \sim \mathcal{CN}_N(\mathbf{0}, \mathbf{M}_0), \tag{3.47}$$

$$\mathbf{n}_k \sim \mathcal{CN}_N(\mathbf{0}, \gamma \mathbf{M}_0), \quad k = 1, \ldots, K, \tag{3.48}$$

where $\gamma > 0$ is an unknown scaling factor.

Exploiting persymmetry, we can apply the same transformations as for the homogeneous case to come up with a detection problem involving real vectors whose canonical form is that of (3.9) but with the difference that, in this case, secondary data are characterized as follows:

$$\mathbf{z}_{ik} \sim \mathcal{N}_N(\mathbf{0}, \gamma \mathbf{M}), \quad i = 1, 2, \quad k = 1, \ldots, K. \tag{3.49}$$

Let $\mathbb{R}^+$ be the set of positive real numbers, then, after reduction by sufficiency, which replaces original data with the sufficient statistic $(\mathbf{Z}_p, \mathbf{S})$, it is not difficult to show that the group of transformations

$$\mathcal{L}_1 = \mathcal{G}_N \times \mathcal{O}_2 \times \mathbb{R}^+ \tag{3.50}$$

with the operation ∘ defined by

$$(\mathbf{G}_1, \mathbf{U}_1, \gamma_1) \circ (\mathbf{G}_2, \mathbf{U}_2, \gamma_2) = (\mathbf{G}_2\mathbf{G}_1, \mathbf{U}_1\mathbf{U}_2, \gamma_1\gamma_2), \tag{3.51}$$
$$\mathbf{G}_1, \mathbf{G}_2 \in \mathcal{G}_N, \quad \mathbf{U}_1, \mathbf{U}_2 \in \mathcal{O}_2, \quad \gamma_1, \gamma_2 \in \mathbb{R}^+,$$

leaves the hypothesis testing problem invariant under the action ℓ defined by

$$\ell(\mathbf{Z}_p, \mathbf{S}) = (\mathbf{G}\mathbf{Z}_p\mathbf{U}, \gamma\mathbf{G}\mathbf{S}\mathbf{G}^T), \quad \forall (\mathbf{G}, \mathbf{U}, \gamma) \in \mathcal{L}_1. \tag{3.52}$$

Exploiting data partition (3.10), a MIS for the PHE is provided by the following theorem.

Theorem 3.3.1. *A MIS with respect to $\mathcal{L}_1$ for problem (3.9) in PHE is given by*

$$\mathbf{t}_3(\mathbf{Z}_p, \mathbf{S}) \triangleq \begin{bmatrix} t_1 \\ t_2 \\ t_3 \end{bmatrix} = \begin{bmatrix} \lambda_1/\lambda_4 \\ \lambda_2/\lambda_4 \\ \lambda_3/\lambda_4 \end{bmatrix}, \tag{3.53}$$

where λ_i, $i = 1, 2$, and λ_j, $j = 3, 4$, are the eigenvalues of $\boldsymbol{\Psi}_0 \triangleq \mathbf{Z}_p^T\mathbf{S}^{-1}\mathbf{Z}_p$ and $\boldsymbol{\Psi}_1 \triangleq \mathbf{Z}_{2p}^T\mathbf{S}_{22}^{-1}\mathbf{Z}_{2p}$, respectively (see also Appendix 3.E).

Proof. See Appendix 3.F.   □

Notice that the MIS is given by a three-dimensional vector, where the third component (t_3) represents the ancillary part. Moreover, observe that the PDF of $\mathbf{t}_3(\mathbf{Z}_p, \mathbf{S})$ does not depend on γ, given the normalization by λ_4. Thus, the induced maximal invariant is the same as in the case of the homogeneous environment and corresponds to the SINR. Another way to claim the CFAR property consists in noticing that, when the hypothesis H_0 is in force, the SINR equals zero and thus the PDF of $\mathbf{t}(\mathbf{Z}_p, \mathbf{S})$ does not depend on any unknown parameter. This implies that every function of the MIS satisfies the CFAR property.

3.3.1 Invariant Detectors for Partially Homogeneous Scenarios

In this subsection, we show that detectors based on theoretically founded criteria can be expressed in terms of the MIS. Specifically, we focus on the GLRT (including its two-step version), Rao test, and Wald test [19].

Before proceeding, we first report the explicit expressions of the MLEs of the scale parameters under both hypotheses ($\hat{\gamma}_i$, $i = 0, 1$):

$$\hat{\gamma}_i \triangleq \frac{\beta_i - (K + 1 - N)\operatorname{tr}[\boldsymbol{\Psi}_i]}{2(2K + 2 - N)\det(\boldsymbol{\Psi}_i)}, \tag{3.54}$$
$$\beta_i \triangleq \sqrt{\operatorname{tr}[\boldsymbol{\Psi}_i]^2(K + 1 - N)^2 + 4N(2K + 2 - N)\det(\boldsymbol{\Psi}_i)}. \tag{3.55}$$

Such estimates can be further expressed as

$$\hat{\gamma}_0 = \frac{1}{\lambda_1}\, g_\gamma\left(\frac{\lambda_1}{\lambda_2}\right), \tag{3.56}$$

$$\hat{\gamma}_1 = \frac{1}{\lambda_3}\, g_\gamma\left(\frac{\lambda_3}{\lambda_4}\right), \tag{3.57}$$

where

$$g_\gamma(x) \triangleq (1+x)$$
$$\times \frac{\left\{\sqrt{(K+1-N)^2 + 4N(2K+2-N)\frac{1}{x+(1/x)+2}} - (K+1-N)\right\}}{2\,(2K+2-N)}.$$

The definition in (3.54) will be exploited to provide compact expressions of the considered detectors. In fact, the GLRT is expressed as [20]:

$$t_{\mathrm{glr}} \triangleq \frac{\hat{\gamma}_0^{-\frac{N}{K+1}} \det\left(\mathbf{I}_2 + \hat{\gamma}_0 \boldsymbol{\Psi}_0\right)}{\hat{\gamma}_1^{-\frac{N}{K+1}} \det\left(\mathbf{I}_2 + \hat{\gamma}_1 \boldsymbol{\Psi}_1\right)}, \tag{3.58}$$

while the two-step GLRT (that is the P-ACE) is [14]:

$$t_{\mathrm{2s-glr}} \triangleq \mathrm{tr}[\boldsymbol{\Psi}_0]\,/\,\mathrm{tr}[\boldsymbol{\Psi}_1]. \tag{3.59}$$

Differently, the Rao statistic is given by [21]:

$$t_{\mathrm{rao}} \triangleq \frac{\hat{\gamma}_0\, \mathrm{tr}\left[(\boldsymbol{\Psi}_0 - \boldsymbol{\Psi}_1)\,(\mathbf{I}_2 + \hat{\gamma}_0 \boldsymbol{\Psi}_0)^{-2}\right]}{1 - \hat{\gamma}_0\, \mathrm{tr}\left[(\boldsymbol{\Psi}_0 - \boldsymbol{\Psi}_1)\,(\mathbf{I}_2 + \hat{\gamma}_0 \boldsymbol{\Psi}_0)^{-1}\right]}. \tag{3.60}$$

Finally, the Wald statistic is [21]:

$$t_{\mathrm{wald}} \triangleq \hat{\gamma}_1 \left\{\mathrm{tr}[\boldsymbol{\Psi}_0] - \mathrm{tr}[\boldsymbol{\Psi}_1]\right\}. \tag{3.61}$$

It is possible to show that all the aforementioned statistics are functions of $\mathbf{t}_3(\mathbf{Z}_p, \mathbf{S})$. Indeed, t_{glr} in (3.58) can be rewritten as:

$$t_{\mathrm{glr}} = \left\{\frac{t_3\, g_\gamma(t_1/t_2)}{t_1\, g_\gamma(t_3)}\right\}^{-\frac{N}{K+1}} \frac{[1 + g_\gamma(t_1/t_2)]\,[1 + (t_2/t_1)\, g_\gamma(t_1/t_2)]}{[1 + g_\gamma(t_3)]\,[1 + (1/t_3)\, g_\gamma(t_3)]}. \tag{3.62}$$

The P-ACE can be expressed as:

$$t_{\mathrm{2s-glr}} = \frac{(t_1 + t_2)}{(1 + t_3)}. \tag{3.63}$$

Proving this property for the Rao statistic is much more difficult and it relies on showing that the terms

$$\hat{\gamma}_0\, \mathrm{tr}\left[(\boldsymbol{\Psi}_0 - \boldsymbol{\Psi}_1)\,(\mathbf{I}_2 + \hat{\gamma}_0 \boldsymbol{\Psi}_0)^{-2}\right] \tag{3.64}$$

and

$$\hat{\gamma}_0 \operatorname{tr}\left[\left(\boldsymbol{\Psi}_0 - \boldsymbol{\Psi}_1\right)\left(\mathbf{I}_2 + \hat{\gamma}_0 \boldsymbol{\Psi}_0\right)^{-1}\right] \qquad (3.65)$$

are both invariant. Finally, the Wald statistic can be rewritten as:

$$t_{\text{wald}} \;=\; g_\gamma(t_3)\left\{(t_1/t_3) + (t_2/t_3) - (1 + 1/t_3)\right\}, \qquad (3.66)$$

which proves its invariance.

Acknowledgments

This chapter gathers and re-organizes in a systematic way the results of [1, 2] that arise from the collaboration with Dr. Domenico Ciuonzo and Dr. Antonio De Maio.

3.A   Proof of Theorem 3.2.1

In this Appendix, we proceed according to the following rationale:

1. we find the maximal invariant with respect to the group $\mathcal{G}_N$;

2. we exploit the result at the previous step to obtain the maximal invariant with respect to $\mathcal{L} = \mathcal{G}_N \times \mathcal{O}_2$;

3. we apply a one-to-one transformation to obtain (3.17).

Let us recall that a statistic $\mathbf{T}(\mathbf{Z}_p, \mathbf{S})$ is said to be MIS with respect to a group of transformation $\bar{\mathcal{L}}$ if and only if [3, 4]

$$\mathbf{T}(\mathbf{Z}_p, \mathbf{S}) = \mathbf{T}[l(\mathbf{Z}_p, \mathbf{S})], \quad \forall l \in \bar{\mathcal{L}}, \qquad (3.A.1)$$

and

$$\mathbf{T}(\mathbf{Z}_{p,1}, \mathbf{S}_1) = \mathbf{T}(\mathbf{Z}_{p,2}, \mathbf{S}_2) \Rightarrow \exists\, l \in \bar{\mathcal{L}} \;:\; (\mathbf{Z}_{p,1}, \mathbf{S}_1) = l(\mathbf{Z}_{p,2}, \mathbf{S}_2). \qquad (3.A.2)$$

Exploiting *Proposition 2.1* of [22], we can claim that the maximal invariant statistics with respect to $\mathcal{G}_N$ are given by[5]

$$\mathbf{T}_1(\mathbf{Z}_p, \mathbf{S}) = \mathbf{Z}_p^T \mathbf{S}^{-1} \mathbf{Z}_p \in \mathbb{R}^{2\times 2}, \qquad (3.A.3)$$

$$\mathbf{T}_2(\mathbf{Z}_p, \mathbf{S}) = \mathbf{Z}_{2p}^T \mathbf{S}_{22}^{-1} \mathbf{Z}_{2p} \in \mathbb{R}^{2\times 2}. \qquad (3.A.4)$$

---

[5]Actually, Proposition 2.1 of [22] deals with a more general case and group that include those under consideration. Thus, (3.A.3) and (3.A.4) can be obtained by suitably setting some parameters of Proposition 2.1 to specific values.

Now, we notice that the eigenvalues of a square matrix are invariant to the orthogonal transformations and, hence, a possible candidate as maximal invariant with respect to $\mathcal{L}$ (defined by (3.13)) can be written as

$$t(\mathbf{Z}_p, \mathbf{S}) = \begin{bmatrix} \lambda(\mathbf{Z}_p^T \mathbf{S}^{-1} \mathbf{Z}_p) \\ \lambda(\mathbf{Z}_{2p}^T \mathbf{S}_{22}^{-1} \mathbf{Z}_{2p}) \end{bmatrix}, \tag{3.A.5}$$

where $\lambda(\cdot) \in \mathbb{R}^{2\times 1}$ is a vector whose entries are the eigenvalues of the matrix argument arranged in decreasing order.

The first property of the MIS (3.A.1) can be verified by letting $l = (\mathbf{G}, \mathbf{U}) \in \mathcal{L}$ and evaluating $t(\cdot, \cdot)$ at $l(\mathbf{Z}_p, \mathbf{S})$

$$
\begin{aligned}
t[l(\mathbf{Z}_p, \mathbf{S})] &= t(\mathbf{G}\mathbf{Z}_p\mathbf{U}, \mathbf{G}\mathbf{S}\mathbf{G}^T) \\
&= \begin{bmatrix} \lambda(\mathbf{U}^T\mathbf{Z}_p^T\mathbf{G}^T(\mathbf{G}\mathbf{S}\mathbf{G}^T)^{-1}\mathbf{G}\mathbf{Z}_p\mathbf{U}) \\ \lambda(\mathbf{U}^T\mathbf{Z}_{2p}^T\mathbf{G}_{22}^T(\mathbf{G}_{22}\mathbf{S}_{22}\mathbf{G}_{22}^T)^{-1}\mathbf{G}_{22}\mathbf{Z}_{2p}\mathbf{U}) \end{bmatrix} \\
&= \begin{bmatrix} \lambda(\mathbf{U}^T\mathbf{Z}_p^T\mathbf{S}^{-1}\mathbf{Z}_p\mathbf{U}) \\ \lambda(\mathbf{U}^T\mathbf{Z}_{2p}^T\mathbf{S}_{22}^{-1}\mathbf{Z}_{2p}\mathbf{U}) \end{bmatrix} = \begin{bmatrix} \lambda(\mathbf{Z}_p^T\mathbf{S}^{-1}\mathbf{Z}_p) \\ \lambda(\mathbf{Z}_{2p}^T\mathbf{S}_{22}^{-1}\mathbf{Z}_{2p}) \end{bmatrix} \\
&= t(\mathbf{Z}_p, \mathbf{S}). \tag{3.A.6}
\end{aligned}
$$

As to the second property, we assume that

$$t(\mathbf{Z}_{p,1}, \mathbf{S}_1) = t(\mathbf{Z}_{p,2}, \mathbf{S}_2) \tag{3.A.7}$$

and we find the transformation $\bar{l}$ such that $(\mathbf{Z}_{p,1}, \mathbf{S}_1) = \bar{l}(\mathbf{Z}_{p,2}, \mathbf{S}_2)$. To this end, observe that (3.A.7) can be written as

$$
\begin{cases} \lambda(\mathbf{Z}_{p,1}^T\mathbf{S}_1^{-1}\mathbf{Z}_{p,1}) = \lambda(\mathbf{Z}_{p,2}^T\mathbf{S}_2^{-1}\mathbf{Z}_{p,2}), \\ \lambda(\mathbf{Z}_{2p,1}^T\mathbf{S}_{22,1}^{-1}\mathbf{Z}_{2p,1}) = \lambda(\mathbf{Z}_{2p,2}^T\mathbf{S}_{22,2}^{-1}\mathbf{Z}_{2p,2}). \end{cases} \tag{3.A.8}
$$

Now, resorting to the formulas for the inverse of a partitioned matrix [23], it is tedious but not difficult to show that

$$\mathbf{Z}_{p,i}^T\mathbf{S}_i^{-1}\mathbf{Z}_{p,i} = \mathbf{B}_i + \mathbf{v}_i\mathbf{v}_i^T, \quad i = 1, 2, \tag{3.A.9}$$

where $\mathbf{B}_i = \mathbf{Z}_{2p,i}^T\mathbf{S}_{22,i}^{-1}\mathbf{Z}_{2p,i}$, $\mathbf{v}_i = F_i^{-1/2}(\mathbf{z}_{1p,i} - \mathbf{s}_{12,i}\mathbf{S}_{22,i}^{-1}\mathbf{Z}_{2p,i})^T$, $i = 1, 2$, with $F_i = s_{11,i} - \mathbf{s}_{12,i}\mathbf{S}_{22,i}^{-1}\mathbf{s}_{21,i}$, $i = 1, 2$. As a consequence, (3.A.8) can be recast as

$$
\begin{cases} \lambda(\mathbf{B}_1 + \mathbf{v}_1\mathbf{v}_1^T) = \lambda(\mathbf{B}_2 + \mathbf{v}_2\mathbf{v}_2^T), \\ \lambda(\mathbf{B}_1) = \lambda(\mathbf{B}_2). \end{cases} \tag{3.A.10}
$$

Using the second equality of (3.A.10) and the eigendecomposition of matrices $\mathbf{B}_i$, $i = 1, 2$, given by

$$\mathbf{B}_1 = \mathbf{U}_1\mathbf{\Lambda}_1\mathbf{U}_1^T, \quad \mathbf{B}_2 = \mathbf{U}_2\mathbf{\Lambda}_2\mathbf{U}_2^T, \tag{3.A.11}$$

where $\mathbf{U}_i \in \mathcal{O}_2$, $i = 1, 2$, while $\mathbf{\Lambda}_1$ and $\mathbf{\Lambda}_2$ are diagonal matrices whose non-zero entries are the eigenvalues of $\mathbf{B}_1$ and $\mathbf{B}_2$, respectively, we obtain that

$$\mathbf{\Lambda}_1 = \mathbf{\Lambda}_2 \Leftrightarrow \mathbf{U}_1^T \mathbf{B}_1 \mathbf{U}_1 = \mathbf{U}_2^T \mathbf{B}_2 \mathbf{U}_2 = \mathbf{Q} \mathbf{U}_2^T \mathbf{B}_2 \mathbf{U}_2 \mathbf{Q}^T$$
$$\Leftrightarrow \mathbf{B}_1 = \mathbf{U}_1 \mathbf{Q} \mathbf{U}_2^T \mathbf{B}_2 \mathbf{U}_2 \mathbf{Q}^T \mathbf{U}_1^T, \tag{3.A.12}$$

where $\mathbf{Q} \in \mathbb{R}^{2 \times 2}$ is any diagonal matrix belonging[6] to $\mathcal{O}_2$ (namely, such that $\mathbf{Q}^T \mathbf{Q} = \mathbf{Q} \mathbf{Q}^T = \mathbf{I}_2$). On the other hand, from the first equality of (3.A.10), it follows that

$$\lambda[\mathbf{U}_1^T(\mathbf{B}_1 + \mathbf{v}_1 \mathbf{v}_1^T)\mathbf{U}_1] = \lambda[\mathbf{U}_2^T(\mathbf{B}_2 + \mathbf{v}_2 \mathbf{v}_2^T)\mathbf{U}_2] \tag{3.A.13}$$
$$\Leftrightarrow \lambda(\mathbf{\Lambda}_1 + \mathbf{p}_1 \mathbf{p}_1^T) = \lambda(\mathbf{\Lambda}_2 + \mathbf{p}_2 \mathbf{p}_2^T), \tag{3.A.14}$$

where

$$\mathbf{p}_i = \mathbf{U}_i^T \mathbf{v}_i = \begin{bmatrix} p_{1,i} \\ p_{2,i} \end{bmatrix}, \quad i = 1, 2. \tag{3.A.15}$$

By *Lemma 1* of [24], the eigenvalues of $\mathbf{\Lambda}_i + \mathbf{p}_i \mathbf{p}_i^T$, $i = 1, 2$, are the roots of the following equations

$$1 + \frac{p_{1,i}^2}{\lambda_3 - x} + \frac{p_{2,i}^2}{\lambda_4 - x} = 0, \quad i = 1, 2, \tag{3.A.16}$$

that can be recast as

$$x^2 - x(\lambda_3 + \lambda_4 + p_{1,i}^2 + p_{2,i}^2) + \lambda_3 \lambda_4 + p_{1,i}^2 \lambda_4 + p_{2,i}^2 \lambda_3 = 0, \tag{3.A.17}$$

$i = 1, 2$, where λ_i, $i = 3, 4$, are the eigenvalues of $\mathbf{B}_1$ (or $\mathbf{B}_2$ which are equal due to (3.A.10)). From (3.A.14), it follows that the above equations have the same roots. Moreover, the coefficient of the term with the highest power is unity. Thus, imposing the equality between the coefficients of the terms with power less than 2 yields the following system of equations

$$\begin{cases} p_{1,1}^2 + p_{2,1}^2 = p_{1,2}^2 + p_{2,2}^2, \\ p_{1,1}^2 \lambda_4 + p_{2,1}^2 \lambda_3 = p_{1,2}^2 \lambda_4 + p_{2,2}^2 \lambda_3. \end{cases} \tag{3.A.18}$$

After simple algebraic manipulations, system (3.A.18) can be rewritten as

$$\begin{cases} h_1 + h_2 = 0, \\ h_1 \lambda_4 + h_2 \lambda_3 = 0, \end{cases} \tag{3.A.19}$$

where $h_1 = p_{1,1}^2 - p_{1,2}^2$ and $h_2 = p_{2,1}^2 - p_{2,2}^2$. Since $\lambda_3 \neq \lambda_4$ with probability one, the considered system admits the unique solution $h_1 = h_2 = 0$ which leads to

$$\begin{cases} p_{1,1}^2 = p_{1,2}^2, \\ p_{2,1}^2 = p_{2,2}^2, \end{cases} \Rightarrow \begin{cases} p_{1,1} = \pm p_{1,2}, \\ p_{2,1} = \pm p_{2,2}. \end{cases} \tag{3.A.20}$$

---

[6]Observe that the non-zero entries of $\mathbf{Q}$ can be either $+1$ or -1.

Thus, there exists an orthogonal and diagonal matrix $\bar{\mathbf{Q}} \in \mathbb{R}^{2 \times 2}$ such that

$$\mathbf{p}_1 = \bar{\mathbf{Q}}\mathbf{p}_2. \tag{3.A.21}$$

Replacing (3.A.15) in (3.A.21) yields

$$\mathbf{v}_1 = \mathbf{U}_1 \bar{\mathbf{Q}} \mathbf{U}_2^T \mathbf{v}_2. \tag{3.A.22}$$

Equations (3.A.9), (3.A.12) with $\mathbf{Q} = \bar{\mathbf{Q}}$, and (3.A.22) can be used to obtain the following equality

$$
\begin{aligned}
\mathbf{Z}_{p,1}^T \mathbf{S}_1^{-1} \mathbf{Z}_{p,1} &= \mathbf{B}_1 + \mathbf{v}_1 \mathbf{v}_1^T \\
&= \bar{\mathbf{U}}^T \mathbf{B}_2 \bar{\mathbf{U}} + \bar{\mathbf{U}}^T \mathbf{v}_2 \mathbf{v}_2^T \bar{\mathbf{U}} \\
&= \bar{\mathbf{U}}^T (\mathbf{B}_2 + \mathbf{v}_2 \mathbf{v}_2^T) \bar{\mathbf{U}} = \bar{\mathbf{U}}^T \mathbf{Z}_{p,2}^T \mathbf{S}_2^{-1} \mathbf{Z}_{p,2} \bar{\mathbf{U}};
\end{aligned} \tag{3.A.23}
$$

where $\bar{\mathbf{U}} = \mathbf{U}_2 \bar{\mathbf{Q}} \mathbf{U}_1^T$.

Summarizing, (3.A.12) and (3.A.23) lead to the following equalities

$$\mathbf{Z}_{p,1}^T \mathbf{S}_1^{-1} \mathbf{Z}_{p,1} = \bar{\mathbf{Z}}_{p,2}^T \mathbf{S}_2^{-1} \bar{\mathbf{Z}}_{p,2}, \tag{3.A.24}$$

$$\mathbf{Z}_{2p,1}^T \mathbf{S}_{22,1}^{-1} \mathbf{Z}_{2p,1} = \bar{\mathbf{Z}}_{2p,2}^T \mathbf{S}_{22,2}^{-1} \bar{\mathbf{Z}}_{2p,2}, \tag{3.A.25}$$

where $\bar{\mathbf{Z}}_{p,2} = \mathbf{Z}_{p,2}\bar{\mathbf{U}}$ and $\bar{\mathbf{Z}}_{2p,2} = \mathbf{Z}_{2p,2}\bar{\mathbf{U}}$. Since (3.A.3) and (3.A.4) are the maximal invariants with respect to $\mathcal{G}_N$, exploiting property (3.A.2), there exists a transformation $\bar{\mathbf{G}} \in \mathcal{G}_N$ such that

$$\mathbf{Z}_{p,1} = \bar{\mathbf{G}}\bar{\mathbf{Z}}_{p,2} = \bar{\mathbf{G}}\mathbf{Z}_{p,2}\bar{\mathbf{U}} \quad \text{and} \quad \mathbf{S}_1 = \bar{\mathbf{G}}\mathbf{S}_2\bar{\mathbf{G}}^T. \tag{3.A.26}$$

Gathering the above results, we have found a transformation $\bar{l} = (\bar{\mathbf{G}}, \bar{\mathbf{U}}) \in \mathcal{L}$ such that $(\mathbf{Z}_{p,1}, \mathbf{S}_1) = \bar{l}(\mathbf{Z}_{p,2}, \mathbf{S}_2)$ and, hence, the MIS with respect to $\mathcal{L}$ is given by (3.A.5). In order to write (3.A.5) in terms of the statistic of the GLRT and that of the P-AMF, observe that

$$
\begin{aligned}
t_{\text{GLRT}} &= \frac{\det(\mathbf{I}_2 + \mathbf{Z}_p^T \mathbf{S}^{-1} \mathbf{Z}_p)}{\det\left(\mathbf{I}_2 + \mathbf{Z}_p^T \mathbf{S}^{-1}\mathbf{Z}_p - \frac{\mathbf{Z}_p^T \mathbf{S}^{-1}\mathbf{e}_1 \mathbf{e}_1^T \mathbf{S}^{-1}\mathbf{Z}_p}{\mathbf{e}_1^T \mathbf{S}^{-1}\mathbf{e}_1}\right)} \\
&= \frac{\det(\mathbf{I}_2 + \mathbf{Z}_p^T \mathbf{S}^{-1} \mathbf{Z}_p)}{\det(\mathbf{I}_2 + \mathbf{Z}_{2p}^T \mathbf{S}_{22}^{-1} \mathbf{Z}_{2p})} \\
&= \frac{\prod\limits_{i=1}^{2}(1 + \lambda_i)}{\prod\limits_{i=3}^{4}(1 + \lambda_i)},
\end{aligned} \tag{3.A.27}
$$

and

$$t_{\text{P-AMF}} = \frac{\mathbf{e}_1^T \mathbf{S}^{-1} \mathbf{Z}_p \mathbf{Z}_p^T \mathbf{S}^{-1} \mathbf{e}_1}{\mathbf{e}_1^T \mathbf{S}^{-1} \mathbf{e}_1}$$

$$= \text{tr}(\mathbf{Z}_p^T \mathbf{S}^{-1} \mathbf{Z}_p - \mathbf{Z}_{2p}^T \mathbf{S}_{22}^{-1} \mathbf{Z}_{2p})$$

$$= \sum_{i=1}^{2} (\lambda_i - \lambda_{i+2}). \tag{3.A.28}$$

Finally, it is not difficult to show that there exists a one-to-one transformation which allows to obtain $\mathbf{t}_1(\mathbf{Z}_p, \mathbf{S})$ starting from $\mathbf{t}(\mathbf{Z}_p, \mathbf{S})$ and vice versa (in the latter case, the fact is exploited that $\lambda_1 > \lambda_2$).

3.B   Derivation of (3.38)

The test statistic of the Rao detector for problem (3.9) can be easily obtained leveraging the results in [15] to come up with

$$t_8 = \frac{[\mathbf{e}_1^T (\mathbf{S} + \mathbf{Z}_p \mathbf{Z}_p^T)^{-1} \mathbf{z}_1]^2 + [\mathbf{e}_1^T (\mathbf{S} + \mathbf{Z}_p \mathbf{Z}_p^T)^{-1} \mathbf{z}_2]^2}{\mathbf{e}_1^T (\mathbf{S} + \mathbf{Z}_p \mathbf{Z}_p^T)^{-1} \mathbf{e}_1}$$

$$= \frac{\|\mathbf{Z}_p^T (\mathbf{S} + \mathbf{Z}_p \mathbf{Z}_p^T)^{-1} \mathbf{e}_1\|^2}{\mathbf{e}_1^T (\mathbf{S} + \mathbf{Z}_p \mathbf{Z}_p^T)^{-1} \mathbf{e}_1}. \tag{3.B.1}$$

Using the Woodbury identity [23], the above equation can be recast as

$$t_8 = \frac{\|[\mathbf{I}_2 - \mathbf{Z}_p^T \mathbf{S}^{-1} \mathbf{Z}_p (\mathbf{I}_2 + \mathbf{Z}_p^T \mathbf{S}^{-1} \mathbf{Z}_p)^{-1}] \mathbf{Z}_p^T \mathbf{S}^{-1} \mathbf{e}_1\|^2}{\mathbf{e}_1^T \mathbf{S}^{-1} \mathbf{e}_1 - \mathbf{e}_1^T \mathbf{S}^{-1} \mathbf{Z}_p (\mathbf{I}_2 + \mathbf{Z}_p^T \mathbf{S}^{-1} \mathbf{Z}_p)^{-1} \mathbf{Z}_p^T \mathbf{S}^{-1} \mathbf{e}_1}.$$

Now, observe that

$$(\mathbf{I}_2 + \mathbf{Z}_p^T \mathbf{S}^{-1} \mathbf{Z}_p)(\mathbf{I}_2 + \mathbf{Z}_p^T \mathbf{S}^{-1} \mathbf{Z}_p)^{-1} = \mathbf{I}$$

$$\Rightarrow (\mathbf{I}_2 + \mathbf{Z}_p^T \mathbf{S}^{-1} \mathbf{Z}_p)^{-1} = \mathbf{I}_2 - \mathbf{Z}_p^T \mathbf{S}^{-1} \mathbf{Z}_p (\mathbf{I}_2 + \mathbf{Z}_p^T \mathbf{S}^{-1} \mathbf{Z}_p)^{-1}, \tag{3.B.2}$$

it follows that

$$t_8 = \frac{\|(\mathbf{I}_2 + \mathbf{Z}_p^T \mathbf{S}^{-1} \mathbf{Z}_p)^{-1} \mathbf{Z}_p^T \mathbf{S}^{-1} \mathbf{e}_1\|^2}{\mathbf{e}_1^T \mathbf{S}^{-1} \mathbf{e}_1 - \mathbf{e}_1^T \mathbf{S}^{-1} \mathbf{Z}_p (\mathbf{I}_2 + \mathbf{Z}_p^T \mathbf{S}^{-1} \mathbf{Z}_p)^{-1} \mathbf{Z}_p^T \mathbf{S}^{-1} \mathbf{e}_1}. \tag{3.B.3}$$

Using (3.31) in conjunction with

$$\mathbf{Z}_p^T \mathbf{S}^{-1} \mathbf{Z}_p - \mathbf{Z}_p^T \mathbf{S}^{-1} \mathbf{e}_1 (\mathbf{e}_1^T \mathbf{S}^{-1} \mathbf{e}_1)^{-1} \mathbf{e}_1^T \mathbf{S}^{-1} \mathbf{Z}_p = \mathbf{Z}_{2p}^T \mathbf{S}_{22}^{-1} \mathbf{Z}_{2p}, \tag{3.B.4}$$

$$\mathbf{z}^T = \frac{\mathbf{Z}_p^T \mathbf{S}^{-1} \mathbf{e}_1}{\mathbf{e}_1^T \mathbf{S}^{-1} \mathbf{e}_1}, \tag{3.B.5}$$

$$\zeta = (\mathbf{e}_1^T \mathbf{S}^{-1} \mathbf{e}_1)^{-1}, \tag{3.B.6}$$

it is possible to show that (3.B.3) can be rewritten as

$$
\begin{aligned}
t_8 &= \frac{\|[\mathbf{U}^T(\mathbf{I}_2 + \boldsymbol{\Lambda})^{1/2}(\mathbf{I}_2 + \mathbf{q}\mathbf{q}^T)(\mathbf{I}_2 + \boldsymbol{\Lambda})^{1/2}\mathbf{U}]^{-1}\mathbf{z}^T/\sqrt{\xi}\|^2}{(1 - \mathbf{q}^T(\mathbf{I}_2 + \mathbf{q}\mathbf{q}^T)^{-1}\mathbf{q})} \\
&= \frac{\|\mathbf{U}^T(\mathbf{I}_2 + \boldsymbol{\Lambda})^{-1/2}(\mathbf{I}_2 + \mathbf{q}\mathbf{q}^T)^{-1}\mathbf{q}\|^2}{1 - \mathbf{q}^T(\mathbf{I}_2 + \mathbf{q}\mathbf{q}^T)^{-1}\mathbf{q}} \\
&= \frac{\|(\mathbf{I}_2 + \boldsymbol{\Lambda})^{-1/2}(\mathbf{I}_2 + \mathbf{q}\mathbf{q}^T)^{-1}\mathbf{q}\|^2}{1 - \mathbf{q}^T(\mathbf{I}_2 + \mathbf{q}\mathbf{q}^T)^{-1}\mathbf{q}} .
\end{aligned}
\tag{3.B.7}
$$

3.C   Proof of Theorem 3.2.2

Let us focus on $\mathbf{y}_1$ defined in (3.31) and assumes that, with reference to equations (3.29) and (3.30), the "2-components" are given, namely, $\mathbf{X}_2$ and $\mathbf{X}_{s,2}$ are assigned. As a consequence, $\mathbf{S}_{0,22}$, some components of $\mathbf{s}_{0,12}$ and $\mathbf{s}_{0,21}$, $\boldsymbol{\Lambda}_0$, and $\mathbf{U}_0$ are deterministic.

Now, observe that, under the H_0 hypothesis and given the two-components, $\mathbf{y}^T$ is a Gaussian random vector with zero mean and covariance matrix $\mathbf{I}_2 + \mathbf{X}_2^T\mathbf{S}_{0,22}^{-1}\mathbf{X}_2$ and, hence, the following transformation

$$
\mathbf{y}_2 = \mathbf{y}_1\sqrt{\xi} = (\mathbf{I}_2 + \boldsymbol{\Lambda}_0)^{-1/2}\mathbf{U}_0\mathbf{y}^T
\tag{3.C.1}
$$

obeys the Gaussian distribution with zero mean and covariance matrix

$$
\begin{aligned}
\mathrm{E}[\mathbf{y}_2\mathbf{y}_2^T] &= (\mathbf{I}_2 + \boldsymbol{\Lambda}_0)^{-1/2}\mathbf{U}_0\mathrm{E}[\mathbf{y}^T\mathbf{y}]\mathbf{U}_0^T(\mathbf{I}_2 + \boldsymbol{\Lambda}_0)^{-1/2} \\
&= (\mathbf{I}_2 + \boldsymbol{\Lambda}_0)^{-1/2}\mathbf{U}_0(\mathbf{I}_2 + \mathbf{X}_2^T\mathbf{S}_{0,22}^{-1}\mathbf{X}_2)\mathbf{U}_0^T(\mathbf{I}_2 + \boldsymbol{\Lambda}_0)^{-1/2} \\
&= \mathbf{I}_2.
\end{aligned}
\tag{3.C.2}
$$

It clearly turns out that the conditional distribution of $\mathbf{y}_2$ given the 2-components is independent of the two-components and, hence, it is also the unconditional one. As to the statistical characterization of ξ, by *Theorem 3.2.10* of [8], it is a central chi-square random variable with $2K - N + 1$ degrees of freedom, i.e., $\xi \sim \chi^2_{2K-N+1}$, and is independent of $\{\mathbf{s}_{0,12}, \mathbf{S}_{0,22}\}$. It follows that ξ is also independent of $\mathbf{X}_2^T\mathbf{S}_{0,22}^{-1}\mathbf{X}_2$ and $\mathbf{y}_2$. Given ξ, we have that

$$
\mathbf{y}_1 = \frac{\mathbf{y}_2}{\sqrt{\xi}} \sim \mathcal{N}_2\left(0, \frac{1}{\xi}\mathbf{I}_2\right)
\tag{3.C.3}
$$

and is independent of λ_3 and λ_4, i.e.,

$$
p_0(\mathbf{y}_1|\xi, \lambda_3, \lambda_4) = p_0(\mathbf{y}_1|\xi).
\tag{3.C.4}
$$

Finally, the PDF of $\mathbf{y}_1 = \mathbf{q}$ can be obtained by integrating the conditional PDF with respect to ξ, namely,

$$p_0(\mathbf{q}) = p_0(\mathbf{y}_1) = \int_0^{+\infty} p_0(\mathbf{y}_1|\xi = u) f_{\chi^2}(u)\, du \qquad (3.\text{C}.5)$$

$$= \frac{1}{2\pi} \left(\frac{1}{2}\right)^c \frac{1}{\Gamma(c)} \int_0^{+\infty} u^c \exp\left[-u\frac{1+\|\mathbf{y}_1\|^2}{2}\right] du \qquad (3.\text{C}.6)$$

$$= \frac{1}{\pi} \frac{\Gamma(c+1)}{\Gamma(c)} (1 + \|\mathbf{y}_1\|^2)^{-(c+1)} \qquad (3.\text{C}.7)$$

$$= \frac{c}{\pi} (1 + \|\mathbf{y}_1\|^2)^{-(c+1)} \qquad (3.\text{C}.8)$$

$$= \frac{c}{\pi} (1 + \|\mathbf{q}\|^2)^{-(c+1)}, \qquad (3.\text{C}.9)$$

where $f_{\chi^2}(\cdot)$ is the PDF of ξ, $c = (2K - N + 1)/2$, and $\Gamma(\cdot)$ is the Gamma function [16]. The last equality is due to the following property of the Gamma function

$$\Gamma(c+1) = c\Gamma(c). \qquad (3.\text{C}.10)$$

3.D   Proof of Theorem 3.2.3

Following the same reasoning as in the proof of **Theorem 3.2.2**, we focus on $\mathbf{y}_1$ and define

$$\mathbf{y}_2 = \mathbf{y}_1 \sqrt{\xi} = (\mathbf{I}_2 + \mathbf{\Lambda}_0)^{-1/2} \mathbf{U}_0 \mathbf{y}^T. \qquad (3.\text{D}.1)$$

Under H_1 and given the two-components, $\mathbf{y}^T$ is ruled by the Gaussian distribution with mean

$$\sqrt{\mathbf{e}_1^T \mathbf{M}^{-1} \mathbf{e}_1}\, \alpha \qquad (3.\text{D}.2)$$

and covariance matrix

$$\mathbf{I}_2 + \mathbf{X}_2^T \mathbf{S}_{0,22}^{-1} \mathbf{X}_2. \qquad (3.\text{D}.3)$$

Then, ξ and $\mathbf{y}_2$ are statistically independent and characterized as follows:

$$\xi \sim \chi^2_{2K-N+1}, \qquad (3.\text{D}.4)$$

$$\mathbf{y}_2 \sim \mathcal{N}_2 \left((\mathbf{I}_2 + \mathbf{\Lambda}_0)^{-1/2} \mathbf{U}_0 \sqrt{\mathbf{e}_1^T \mathbf{M}^{-1} \mathbf{e}_1}\, \alpha, \mathbf{I}_2 \right). \qquad (3.\text{D}.5)$$

It follows that the conditional distribution of $\mathbf{y}_1$ given the two-components and ξ is Gaussian with mean and covariance matrix given by

$$\mathbf{m}_{\mathbf{y}_1}(\xi, \mathbf{\Lambda}_0, \mathbf{U}_0) = \frac{1}{\sqrt{\xi}} (\mathbf{I}_2 + \mathbf{\Lambda}_0)^{-1/2} \mathbf{U}_0 \sqrt{\mathbf{e}_1^T \mathbf{M}^{-1} \mathbf{e}_1}\, \alpha \quad \text{and} \quad \frac{1}{\xi} \mathbf{I}_2, \qquad (3.\text{D}.6)$$

respectively. Now, observe that random matrix $\mathbf{X}_2^T \mathbf{S}_{0,22}^{-1} \mathbf{X}_2$ is unitarily invariant [25, *Definition 2.6*]. As a consequence, $\mathbf{U}_0$ obeys the Haar invariant distribution and is independent of $\mathbf{\Lambda}_0$ [25, *Lemma 2.6*]. As a consequence, we can integrate the conditional PDF of $\mathbf{y}_1$ over $\mathbf{U}_0$ and ξ to obtain the PDF of $\mathbf{y}_1$ given λ_3 and λ_4 under H_1, namely,

$$p_1(\mathbf{y}_1|\lambda_3, \lambda_4) = \int_{\mathcal{O}_2} \int_0^{+\infty} \frac{u}{2\pi} \exp\left\{-\frac{u}{2}\|\mathbf{y}_1 - \mathbf{m}_{y_1}(u, \mathbf{\Lambda}_0, \mathbf{U})\|^2\right\} f_{\chi^2}(u) du (d\mathbf{U}),$$
$$(3.D.7)$$

where [8]

$$(d\mathbf{U}) = \frac{1}{4\pi}(\mathbf{U}^T d\mathbf{U}), \quad \mathbf{U} \in \mathcal{O}_2. \qquad (3.D.8)$$

In (3.D.8), $(\mathbf{U}^T d\mathbf{U})$ is the exterior product[7] of the subdiagonal elements of the skew-symmetric matrix $\mathbf{U}^T d\mathbf{U}$ with $d\mathbf{U}$ being the matrix of differentials,[8] i.e.,

$$d\mathbf{U} = \begin{bmatrix} du_{11} & du_{12} \\ du_{21} & du_{22} \end{bmatrix}. \qquad (3.D.9)$$

In order to simplify (3.D.7), we express $\mathbf{U}$ in a proper manner to obtain an integral that can be solved at least by means of numerical routines. To this end, observe that

$$\mathcal{O}_2 = \mathcal{O}_2^+ \cup \mathcal{O}_2^-, \qquad (3.D.10)$$

where $\cup$ denotes the union of sets,

- $\mathcal{O}_2^+$ is the subgroup of $\mathcal{O}_2$ consisting of 2×2 orthogonal matrices, $\mathbf{U}^+$ say, with $\det(\mathbf{U}^+) = 1$;

- $\mathcal{O}_2^-$ is the subgroup of $\mathcal{O}_2$ consisting of 2×2 orthogonal matrices, $\mathbf{U}^-$ say, with $\det(\mathbf{U}^-) = -1$.

It is not difficult to show that the elements of $\mathcal{O}_2^+$ can be written as

$$\mathbf{U}_\theta^+ = \begin{bmatrix} \cos\theta & -\sin\theta \\ \sin\theta & \cos\theta \end{bmatrix}, \quad 0 < \theta \leqslant 2\pi, \qquad (3.D.11)$$

while the elements of $\mathcal{O}_2^-$ have the following form

$$\mathbf{U}_\theta^- = \begin{bmatrix} \cos\theta & -\sin\theta \\ -\sin\theta & -\cos\theta \end{bmatrix}, \quad 0 < \theta \leqslant 2\pi. \qquad (3.D.12)$$

Substituting $\mathbf{U}$ with $\mathbf{U}_\theta^+$ and $\mathbf{U}_\theta^-$ in (3.D.8) yields

$$\frac{1}{4\pi}([\mathbf{U}_\theta^+]^T d\mathbf{U}_\theta^+) = \frac{1}{4\pi} d\theta, \qquad (3.D.13)$$

and

$$\frac{1}{4\pi}([\mathbf{U}_\theta^-]^T d\mathbf{U}_\theta^-) = \frac{1}{4\pi} d\theta, \qquad (3.D.14)$$

---

[7]The interested reader is referred to Chapter 2 of [8] for further details.
[8]We denote by u_{ij}, $i = 1, 2$, $j = 1, 2$, the elements of $\mathbf{U}$.

respectively. Thus, equation (3.D.7) becomes

$$p_1(\mathbf{y}_1|\lambda_3, \lambda_4) = \frac{1}{4\pi}\left\{\int_{\mathbb{O}_2^+}\int_0^{+\infty}\frac{u}{2\pi}\exp\left\{-\frac{u}{2}\|\mathbf{y}_1 - \mathbf{m}_{y_1}(u, \boldsymbol{\Lambda}_0, \mathbf{U}_\theta^+)\|^2\right\}\right.$$

$$\tag{3.D.15}$$

$$\times f_{\chi^2}(u)du([\mathbf{U}_\theta^+]^T d\mathbf{U}_\theta^+)$$

$$+ \int_{\mathbb{O}_2^-}\int_0^{+\infty}\frac{u}{2\pi}\exp\left\{-\frac{u}{2}\|\mathbf{y}_1 - \mathbf{m}_{y_1}(u, \boldsymbol{\Lambda}_0, \mathbf{U}_\theta^-)\|^2\right\} \quad (3.D.16)$$

$$\left.\times f_{\chi^2}(u)du([\mathbf{U}_\theta^-]^T d\mathbf{U}_\theta^-)\right\}$$

$$= \frac{1}{4\pi}\left\{\int_0^{2\pi}\int_0^{+\infty}\frac{u}{2\pi}\exp\left\{-\frac{u}{2}\|\mathbf{y}_1 - \mathbf{m}_{y_1}(u, \boldsymbol{\Lambda}_0, \mathbf{U}_\theta^+)\|^2\right\}\right.$$

$$\tag{3.D.17}$$

$$\times f_{\chi^2(u)}\,dud\theta$$

$$+ \int_0^{2\pi}\int_0^{+\infty}\frac{u}{2\pi}\exp\left\{-\frac{u}{2}\|\mathbf{y}_1 - \mathbf{m}_{y_1}(u, \boldsymbol{\Lambda}_0, \mathbf{U}_\theta^-)\|^2\right\} \quad (3.D.18)$$

$$\left.\times f_{\chi^2}(u)\,dud\theta\right\}. \tag{3.D.19}$$

The above equation can be further simplified by recasting $\mathbf{m}_{y_1}(u, \boldsymbol{\Lambda}_0, \mathbf{U}_\theta^+)$ and $\mathbf{m}_{y_1}(u, \boldsymbol{\Lambda}_0, \mathbf{U}_\theta^-)$ as follows:

$$\mathbf{m}_{y_1}^+(u, \boldsymbol{\Lambda}_0, \theta) = \frac{(\mathbf{I}_2 + \boldsymbol{\Lambda}_0)^{-1/2}}{\sqrt{u}}\mathbf{U}_\theta^+\sqrt{\rho}\mathbf{u}(\varphi)$$

$$= \frac{1}{\sqrt{u}}(\mathbf{I}_2 + \boldsymbol{\Lambda}_0)^{-1/2}\sqrt{\rho}\begin{bmatrix}\cos(\theta + \varphi)\\\sin(\theta + \varphi)\end{bmatrix}$$

$$= \frac{1}{\sqrt{u}}(\mathbf{I}_2 + \boldsymbol{\Lambda}_0)^{-1/2}\sqrt{\rho}\mathbf{u}(\theta + \varphi), \tag{3.D.20}$$

$$\mathbf{m}_{y_1}^-(u, \boldsymbol{\Lambda}_0, \theta) = \frac{(\mathbf{I}_2 + \boldsymbol{\Lambda}_0)^{-1/2}}{\sqrt{u}}\mathbf{U}_\theta^-\sqrt{\rho}\mathbf{u}(\varphi)$$

$$= -\frac{1}{\sqrt{u}}(\mathbf{I}_2 + \boldsymbol{\Lambda}_0)^{-1/2}\sqrt{\rho}\begin{bmatrix}\cos(\theta + \varphi)\\\sin(\theta + \varphi)\end{bmatrix}$$

$$= -\frac{1}{\sqrt{u}}(\mathbf{I}_2 + \boldsymbol{\Lambda}_0)^{-1/2}\sqrt{\rho}\mathbf{u}(\theta + \varphi), \tag{3.D.21}$$

where $\mathbf{u}(\psi) = [\cos\psi \ \sin\psi]^T$, φ is such that $\boldsymbol{\alpha} = \|\boldsymbol{\alpha}\|\mathbf{u}(\varphi)$, and $\rho = \mathbf{e}_1^T \mathbf{M}^{-1}\mathbf{e}_1 \|\boldsymbol{\alpha}\|^2$. It follows that (3.D.19) can be rewritten as

$$p_1(\mathbf{y}_1|\lambda_3, \lambda_4) = \frac{1}{4\pi} \int_0^{+\infty} \int_0^{2\pi} \frac{u}{2\pi} \tag{3.D.22}$$

$$\times \left[\exp\left\{ -\frac{u}{2}\left\| \mathbf{y}_1 - \frac{1}{\sqrt{u}}(\mathbf{I}_2 + \boldsymbol{\Lambda}_0)^{-1/2}\sqrt{\rho}\,\mathbf{u}(\theta + \varphi) \right\|^2 \right\} \right.$$

$$\left. + \exp\left\{ -\frac{u}{2}\left\| \mathbf{y}_1 + \frac{1}{\sqrt{u}}(\mathbf{I}_2 + \boldsymbol{\Lambda}_0)^{-1/2}\sqrt{\rho}\,\mathbf{u}(\theta + \varphi) \right\|^2 \right\} \right] \tag{3.D.23}$$

$$\times d\theta f_{\chi^2}(u)du. \tag{3.D.24}$$

Let $\beta = \varphi + \theta$, then

$$p_1(\mathbf{y}_1|\lambda_3, \lambda_4) = \frac{1}{4\pi} \int_0^{+\infty} \frac{u}{2\pi} \exp\left\{ -\frac{u}{2}\left(\|\mathbf{y}_1\|^2 + \frac{\rho}{u(1 + \lambda_4)} \right) \right\}$$

$$\times \int_0^{2\pi} \exp\left\{ -\frac{1}{2}\frac{\rho(\cos\beta)^2(\lambda_4 - \lambda_3)}{(1 + \lambda_3)(1 + \lambda_4)} \right\}$$

$$\times \left[\exp\left\{ \sqrt{u\rho}\,\mathbf{y}_1^T(\mathbf{I} + \boldsymbol{\Lambda}_0)^{-1/2}\mathbf{u}(\beta) \right\} \right.$$

$$\left. + \exp\left\{ -\sqrt{u\rho}\,\mathbf{y}_1^T(\mathbf{I} + \boldsymbol{\Lambda}_0)^{-1/2}\mathbf{u}(\beta) \right\} \right] d\beta f_{\chi^2}(u)du$$

$$= \frac{1}{4\pi} \int_0^{+\infty} \frac{u}{\pi} \exp\left\{ -\frac{u}{2}\left(\|\mathbf{y}_1\|^2 + \frac{\rho}{u(1 + \lambda_4)} \right) \right\}$$

$$\times \int_0^{2\pi} \exp\left\{ -\frac{1}{2}\frac{\rho(\cos\beta)^2(\lambda_4 - \lambda_3)}{(1 + \lambda_3)(1 + \lambda_4)} \right\}$$

$$\times \cosh(\sqrt{u\rho}\,\mathbf{y}_1^T(\mathbf{I} + \boldsymbol{\Lambda}_0)^{-1/2}\mathbf{u}(\beta))d\beta f_{\chi^2}(u)du. \tag{3.D.25}$$

The last integrals can be further recast by defining

$$A = \|\mathbf{y}_1\|^2, \quad B = -\frac{\rho}{2(1 + \lambda_4)}, \tag{3.D.26}$$

$$C = -\frac{1}{2}\frac{\rho(\lambda_4 - \lambda_3)}{(1 + \lambda_3)(1 + \lambda_4)}, \tag{3.D.27}$$

$$D = \sqrt{\rho}\,\mathbf{y}_1^T(\mathbf{I} + \boldsymbol{\Lambda}_0)^{-1/2}\mathbf{u}(\beta). \tag{3.D.28}$$

Thus, (3.D.25) can be further simplified as below

$$(3.D.25) = \frac{1}{4\pi^2} \int_0^{2\pi} e^{B+C\cos^2\beta} \qquad (3.D.29)$$

$$\times \left[\int_0^{+\infty} u e^{-uA/2} \cosh(\sqrt{u}D) \frac{1}{\Gamma(c)} \frac{1}{2^c} u^{c-1} e^{-u/2} du \right] d\beta$$

$$= \frac{2}{4\pi^2\Gamma(c)} \int_0^{2\pi} e^{B+C\cos^2\beta} \left[\int_0^{+\infty} x^c e^{-xA_1} \cosh(\sqrt{x}D_1) dx \right] d\beta,$$
$$(3.D.30)$$

where $A_1 = A + 1$, $D_1 = \sqrt{2}D$ and $c = (2K - N + 1)/2$.

Now, we recast the hyperbolic cosine in terms of Bessel functions

$$\cosh(\sqrt{x}D_1) = \sqrt{\frac{\pi\sqrt{x}D_1}{2}} I_{-1/2}(\sqrt{x}D_1)$$

$$= \sqrt{\frac{j\pi\sqrt{x}D_1}{2}} J_{-1/2}(j\sqrt{x}D_1), \qquad (3.D.31)$$

where $I_{-1/2}(\cdot)$ and $J_{-1/2}(\cdot)$ are the modified Bessel function and the Bessel function of the first kind, respectively. In (3.D.31), we have used equations (10.1.14) and (9.6.3) of [16]. As a consequence, (3.D.30) can be written as

$$(3.D.30) = \frac{e^B c}{2\pi^2 A_1^{c+1}} \int_0^{2\pi} e^{C\cos^2\beta} {}_1F_1\left(c+1, 1/2, \frac{D_1^2}{4A_1}\right) d\beta,$$

where the last equality comes from equation (11.4.28) of [16] and ${}_1F_1(\cdot, \cdot, \cdot)$ is the confluent hypergeometric function. Gathering the above results, under H_1 the conditional PDF of $\mathbf{y}_1$ given λ_3 and λ_4 exhibits the following expression

$$p_1(\mathbf{y}_1|\lambda_3, \lambda_4) = \frac{\exp\left\{-\frac{\rho}{2(1+\lambda_4)}\right\} c}{2\pi^2(1 + \|\mathbf{y}_1\|^2)^{(2K-N+3)/2}}$$

$$\times \int_0^{2\pi} \exp\left\{-\frac{1}{2} \frac{\rho(\cos\beta)^2(\lambda_4 - \lambda_3)}{(1+\lambda_3)(1+\lambda_4)}\right\}$$

$$\times {}_1F_1\left(c+1, \frac{1}{2}, \frac{\rho(\mathbf{y}_1^T(\mathbf{I} + \mathbf{\Lambda}_0)^{-1/2}\mathbf{u}(\beta))^2}{2(1 + \|\mathbf{y}_1\|^2)}\right) d\beta. \qquad (3.D.32)$$

Finally, recall that the conditional PDF of $\mathbf{q}$ given λ_3 and λ_4 can be obtained replacing $\mathbf{y}_1$ with $\mathbf{q}$ in the last equation.

3.E   Proof of Theorem 3.2.5

In order to find the expression of t_{LMPID}, we refer to equation (3.D.25), wherein[9] we replace $\mathbf{y}_1$ with $\mathbf{q}$ and exploit the Leibniz's rule for differentiation under the integral sign, which allows us to write

$$
\frac{\delta p_1(\mathbf{q}|\lambda_3, \lambda_4)}{\delta \rho} = \int_0^{+\infty} \frac{u}{4\pi^2} \int_0^{2\pi} \left\{ \cosh(\sqrt{u\rho}\mathbf{q}^T(\mathbf{I} + \boldsymbol{\Lambda}_0)^{-1/2}\mathbf{u}(\beta)) \right.
$$

$$
\times \exp\left\{ -\frac{u}{2}\left(\|\mathbf{q}\|^2 + \frac{\rho}{u(1+\lambda_4)} \right) - \frac{1}{2}\frac{\rho(\cos\beta)^2(\lambda_4 - \lambda_3)}{(1+\lambda_3)(1+\lambda_4)} \right\}
$$

$$
\times \left[-\frac{1}{2(1+\lambda_4)} - \frac{(\cos\beta)^2(\lambda_4 - \lambda_3)}{2(1+\lambda_3)(1+\lambda_4)} \right]
$$

$$
+ \exp\left\{ -\frac{u}{2}\left(\|\mathbf{q}\|^2 + \frac{\rho}{u(1+\lambda_4)} \right) - \frac{1}{2}\frac{\rho(\cos\beta)^2(\lambda_4 - \lambda_3)}{(1+\lambda_3)(1+\lambda_4)} \right\}
$$

$$
\times \sinh(\sqrt{u\rho}\mathbf{q}^T(\mathbf{I} + \boldsymbol{\Lambda}_0)^{-1/2}\mathbf{u}(\beta))
$$

$$
\left. \times \frac{\sqrt{u}\mathbf{q}^T(\mathbf{I} + \boldsymbol{\Lambda}_0)^{-1/2}\mathbf{u}(\beta)}{2\sqrt{\rho}} \right\} d\beta f_{\chi^2}(u)du. \tag{3.E.1}
$$

It follows that

$$
\lim_{\rho \to 0} \frac{\delta p_1(\mathbf{q}|\lambda_3, \lambda_4)}{\delta \rho} = \int_0^{+\infty} \frac{u}{4\pi^2} \int_0^{2\pi} \left\{ \exp\left\{ -\frac{u}{2}\|\mathbf{q}\|^2 \right\} \right. \tag{3.E.2}
$$

$$
\times \left[-\frac{1}{2(1+\lambda_4)} - \frac{(\cos\beta)^2(\lambda_4 - \lambda_3)}{2(1+\lambda_3)(1+\lambda_4)} \right]
$$

$$
+ \frac{1}{2}\exp\left\{ -\frac{u}{2}\|\mathbf{q}\|^2 \right\}
$$

$$
\left. \times u\left[\mathbf{q}^T(\mathbf{I} + \boldsymbol{\Lambda}_0)^{-1/2}\mathbf{u}(\beta) \right]^2 \right\} d\beta f_{\chi^2}(u)du. \tag{3.E.3}
$$

It can be easily shown that

$$
\int_0^{+\infty} \left(\frac{u}{2} \right)^c \exp\left\{ -\frac{u}{2}(1 + \|\mathbf{q}\|^2) \right\} du = \frac{2\Gamma(c+1)}{(1+\|\mathbf{q}\|^2)^{(c+1)}} \tag{3.E.4}
$$

and, hence, the right-hand side of (3.E.3) can be recast as

$$
\frac{c}{2\pi^2}(1 + \|\mathbf{q}\|^2)^{-(c+1)}\left[-\frac{\pi}{1+\lambda_4} - \frac{(\lambda_4 - \lambda_3)\int_0^{2\pi}(\cos\beta)^2 d\beta}{2(1+\lambda_3)(1+\lambda_4)} \right]
$$

$$
+ \frac{c(c+1)}{2\pi^2}(1 + \|\mathbf{q}\|^2)^{-(c+2)}\int_0^{2\pi}\left[\mathbf{q}^T(\mathbf{I} + \boldsymbol{\Lambda}_0)^{-1/2}\mathbf{u}(\beta) \right]^2 d\beta
$$

$$
= \frac{c(c+1)}{2\pi}\frac{\mathbf{q}^T(\mathbf{I} + \boldsymbol{\Lambda}_0)^{-1}\mathbf{q}}{(1+\|\mathbf{q}\|^2)^{c+2}} - \frac{c}{2\pi(1+\|\mathbf{q}\|^2)^{c+1}}\frac{2+\lambda_3+\lambda_4}{2(1+\lambda_3)(1+\lambda_4)}. \tag{3.E.5}
$$

---

[9]See equation (3.31).

Finally, substituting (3.41) and (3.E.5) in (3.45) leads to

$$t_{\text{LMPID}} = \frac{(c+1)}{2} \frac{\mathbf{q}^T(\mathbf{I}+\boldsymbol{\Lambda}_0)^{-1}\mathbf{q}}{(1+\|\mathbf{q}\|^2)} - \frac{2+\lambda_3+\lambda_4}{4(1+\lambda_3)(1+\lambda_4)} \overset{H_1}{\underset{H_0}{\gtrless}} \eta. \tag{3.E.6}$$

3.F Proof of Theorem 3.3.1

The proof is based on the key observation that the action $\ell(\cdot,\cdot)$ can be re-interpreted as the sequential application of the following sub-actions:

$$l(\mathbf{Z}_p, \mathbf{S}) = (\mathbf{G}\,\mathbf{Z}_p\mathbf{U}, \mathbf{G}\,\mathbf{S}\,\mathbf{G}^T) \quad \forall(\mathbf{G}, \mathbf{U}) \in \mathcal{L}$$
$$\ell_2(\mathbf{Z}_p, \mathbf{S}) = (\mathbf{Z}_p, \varphi\mathbf{S}) \quad \forall \varphi \in \mathcal{L}_2, \tag{3.F.1}$$

where $\mathcal{L}$ is the group of transformation for the HE (see (3.13)) and $\mathcal{L}_2 \triangleq \{\mathbb{R}^+, \text{`` }\times\text{ ''}\}$ (i.e., the composition operator for $\mathcal{L}_1$ simply corresponds to the product). Then, a MIS for the sub-action $\ell_1(\cdot,\cdot)$ is that for the HE, namely, the four-dimensional statistic

$$\mathbf{t}(\mathbf{Z}_p, \mathbf{S}) \triangleq \begin{bmatrix} \lambda_1 \\ \lambda_2 \\ \lambda_3 \\ \lambda_4 \end{bmatrix}, \tag{3.F.2}$$

where $\lambda_1 \geqslant \lambda_2$ are the two eigenvalues of $\mathbf{Z}_p^T\mathbf{S}^{-1}\mathbf{Z}_p$ and $\lambda_3 \geqslant \lambda_4$ denote the two eigenvalues of $\mathbf{Z}_{2p}^T\mathbf{S}_{22}^{-1}\mathbf{Z}_{2p}$, respectively.

Now, define the action $\ell_2^*(\cdot)$ acting on the couple of positive-valued scalars a_i collected in the vector $\mathbf{a} \triangleq \begin{bmatrix} a_1 & a_2 & a_3 & a_4 \end{bmatrix}^T$ (with a_i corresponding to λ_i) as:

$$\ell_2^*(\mathbf{a}) = \varphi^{-1}\mathbf{a}, \quad \forall \varphi \in \mathcal{L}_2. \tag{3.F.3}$$

It is not difficult to show that a MIS for the elementary operation $\ell_2^*(\cdot)$ in (3.F.3) is given by

$$\mathbf{t}^*(\mathbf{a}) \triangleq \begin{bmatrix} a_1/a_4 \\ a_2/a_4 \\ a_3/a_4 \end{bmatrix}. \tag{3.F.4}$$

Indeed invariance follows from

$$\mathbf{t}^*(\varphi^{-1}\mathbf{a}) = \begin{bmatrix} \frac{\varphi^{-1}a_1}{\varphi^{-1}a_4} \\ \frac{\varphi^{-1}a_2}{\varphi^{-1}a_4} \\ \frac{\varphi^{-1}a_3}{\varphi^{-1}a_4} \end{bmatrix} = \mathbf{t}_3(\mathbf{a}), \tag{3.F.5}$$

while maximality can be proved as follows. Suppose that $\mathbf{t}^*(\mathbf{a}) = \mathbf{t}^*(\bar{\mathbf{a}})$ holds, which implies

$$\bar{a}_i = \frac{\bar{a}_4}{a_4} a_i, \quad i = 1, 2, 3. \tag{3.F.6}$$

Thus, there exists a $\varphi = a_4/\bar{a}_4 \in \mathcal{L}_2$ such that $(\varphi^{-1}\mathbf{a}) = \bar{\mathbf{a}}$. Additionally, we notice that

$$\mathbf{t}(\bar{\mathbf{Z}}_p, \bar{\mathbf{S}}) = \mathbf{t}(\mathbf{Z}_p, \mathbf{S}) \Rightarrow$$
$$\mathbf{t}(\bar{\mathbf{Z}}_p, \varphi\bar{\mathbf{S}}) = \mathbf{t}(\mathbf{Z}_p, \varphi\mathbf{S}), \ \forall \varphi \in \mathcal{L}_2, \tag{3.F.7}$$

since $\mathbf{t}(\mathbf{Z}_p, \varphi\mathbf{S}) = \frac{1}{\varphi}\mathbf{t}(\mathbf{Z}_p, \mathbf{S})$ holds for the problem at hand. Therefore, exploiting [4, p. 217, Theorem. 6.2.2], it follows that a MIS for the action $\ell(\cdot, \cdot)$ is given by the composite function

$$\mathbf{t}_3(\mathbf{Z}_p, \mathbf{S}) \triangleq \mathbf{t}^*(\mathbf{t}(\mathbf{Z}_p, \mathbf{S})) = \begin{bmatrix} \lambda_1/\lambda_4 \\ \lambda_2/\lambda_4 \\ \lambda_3/\lambda_4 \end{bmatrix}. \tag{3.F.8}$$

Bibliography

[1] A. De Maio and D. Orlando, "An invariant approach to adaptive radar detection under covariance persymmetry," *IEEE Transactions on Signal Processing*, vol. 63, no. 5, pp. 1297–1309, 2015.

[2] D. Ciuonzo, D. Orlando, and L. Pallotta, "On the maximal invariant statistic for adaptive radar detection in partially homogeneous disturbance with persymmetric covariance," *IEEE Signal Processing Letters*, vol. 23, no. 12, pp. 1830–1834, 2016.

[3] L. Scharf and C. Demeure, *Statistical Signal Processing: Detection, Estimation, and Time Series Analysis*, ser. Addison-Wesley series in electrical and computer engineering. Reading, MA, Addison-Wesley Publishing Company, 1991.

[4] E. L. Lehmann, J. P. Romano, and G. Casella, *Testing Statistical Hypotheses*. New York: Wiley, 1986, vol. 150.

[5] F. Bandiera, D. Orlando, and G. Ricci, *Advanced Radar Detection Schemes Under Mismatched Signal Models*. San Rafael, US: Synthesis Lectures on Signal Processing No. 8, Morgan & Claypool Publishers, 2009.

[6] G. Pailloux, P. Forster, J. P. Ovarlez, and F. Pascal, "Persymmetric adaptive radar detectors," *IEEE Transactions on Aerospace and Electronic Systems*, vol. 47, no. 4, pp. 2376–2390, 2011.

[7] E. J. Kelly and K. Forsythe, "Adaptive detection and parameter estimation for multidimensional signal models," Lincoln Lab, MIT, Lexington, US, Technical Report 848, 1989.

[8] R. J. Muirhead, *Aspects of multivariate statistical theory*. Hoboken, New Jersey, John Wiley & Sons, 2009, vol. 197.

[9] L. Cai and H. Wang, "A persymmetric multiband GLR algorithm," *IEEE Transactions on Aerospace and Electronic Systems*, vol. 28, no. 3, pp. 806–816, 1992.

[10] F. C. Robey, D. R. Fuhrmann, E. J. Kelly, and R. Nitzberg, "A CFAR Adaptive Matched Filter Detector," *IEEE Transactions on Aerospace and Electronic Systems*, vol. 28, no. 1, pp. 208–216, 1992.

[11] S. Bose and A. O. Steinhardt, "A Maximal Invariant Framework for Adaptive Detection with Structured and Unstructured Covariance Matrices," *IEEE Transactions on Signal Processing*, vol. 43, no. 9, pp. 2164–2175, September 1995.

[12] E. J. Kelly, "An Adaptive Detection Algorithm," *IEEE Transactions on Aerospace and Electronic Systems*, vol. 22, no. 2, pp. 115–127, 1986.

[13] S. Kraut and L. L. Scharf, "The CFAR Adaptive Subspace Detector is a Scale-invariant GLRT," *IEEE Transactions on Signal Processing*, vol. 47, no. 9, pp. 2538–2541, 1999.

[14] Y. Gao, G. Liao, S. Zhu, X. Zhang, and D. Yang, "Persymmetric Adaptive Detectors in Homogeneous and Partially Homogeneous Environments," *IEEE Transactions on Signal Processing*, vol. 62, no. 2, pp. 331–342, 2014.

[15] C. Hao, D. Orlando, X. Ma, S. Yan, and C. Hou, "Persymmetric Detectors with Enhanced Rejection Capabilities," *IET Radar, Sonar & Navigation*, vol. 8, pp. 557–563(6), June 2014.

[16] M. Abramowitz and I. Stegun, *Handbook of Mathematical Functions: With Formulas, Graphs, and Mathematical Tables*, ser. Applied mathematics series. Washington, DC, Dover Publications, 1965.

[17] H. Poor, *An Introduction to Signal Detection and Estimation*, ser. Springer Texts in Electrical Engineering. New York: Springer, 1998.

[18] J. Via and L. Vielva, "Locally Most Powerful Invariant Tests for the Properness of Quaternion Gaussian Vectors," *IEEE Transactions on Signal Processing*, vol. 60, no. 3, pp. 997–1009, 2012.

[19] S. M. Kay, *Fundamentals of Statistical Signal Processing: Detection Theory*, New Jersey, USA, P. Hall, Ed., 1998, vol. 2.

[20] M. Casillo, A. De Maio, S. Iommelli, and L. Landi, "A Persymmetric GLRT for Adaptive Detection in Partially-homogeneous Environment," *IEEE Signal Processing Letters*, vol. 14, no. 12, pp. 1016–1019, 2007.

[21] C. Hao, D. Orlando, X. Ma, and C. Hou, "Persymmetric Rao and Wald Tests for Partially Homogeneous Environment," *IEEE Signal Processing Letters*, vol. 19, no. 9, pp. 587–590, September 2012.

[22] T. Kariya, "The General MANOVA Problem," *The Annals of Statistics*, vol. 6, no. 1, pp. 200–214, 1978.

[23] R. A. Horn and C. R. Johnson, *Matrix Analysis*, New York C. U. Press, Ed., 1985.

[24] M. Gu and S. C. Eisenstat, "A Stable and Efficient Algorithm for the Rank-One Modification of the Symmetric Eigenproblem," *SIAM Journal on Matrix Analysis and Applications*, vol. 15, no. 4, pp. 1266–1276, 1994.

[25] A. Tulino, S. Verdú, and S. Verdu, *Random Matrix Theory and Wireless Communications*, ser. Foundations and Trends in Comm. Delft, The Netherlands, 2004.

4

Persymmetric Adaptive Subspace Detector

The previous two chapters assume the target signal is rank-one, namely, the target steering vector is assumed known up to a scalar. In practice, the target signal may be multi-rank, e.g., due to beam-pointing errors [1] and multipath [2]. In such cases, a subspace model is widely used in the open literature [2–11]. More specifically, in the subspace model, the target signal is expressed as the product of a known full-column-rank matrix and an unknown column vector. It means that the target signal lies in a known subspace spanned by the columns of a matrix, but its exact location is unknown since the coordinate vector is unknown. More detailed explanations about the subspace model can be found in [2, 4, 7].

In [2], several matched subspace detectors were developed in Gaussian clutter, where the clutter covariance matrix is assumed known. The authors in [12] designed a one-step GLRT detector for detecting subspace signals when the clutter covariance matrix is unknown. For ease of reference, this detector is called SGLRT detector hereafter. The SGLRT detector was also applied to polarimetric target detection [13]. A two-step GLRT detector for subspace signal, also referred to as subspace adaptive matched filter (SAMF), was proposed in [14, 15]. The analytical performance of the SGLRT and SAMF detectors was provided in [15]. In [9, 10], several adaptive detectors were designed for detecting double subspace signals.

This chapter addresses the subspace detection problem in the HE by exploiting persymmetry. An adaptive detector is designed according to the criterion of one-step GLRT [11]. Its PFA is derived in closed form, which indicates that it has CFAR property with respect to the clutter covariance matrix. Numerical examples show the superiority over its counterparts.

4.1 Problem Formulation

Assume that the data under test are collected from Q (temporal, spatial, or spatial-temporal) channels. We examine the problem of detecting the presence of a point-like target. The echoes from the range cell under test, called primary or test data, are denoted by

$$\mathbf{x} = \Sigma \mathbf{a} + \mathbf{c}, \tag{4.1}$$

DOI: 10.1201/9781003340232-4

where $\boldsymbol{\Sigma} \in \mathbb{C}^{Q \times q}$ is a known target subspace matrix[1], $\mathbf{a} \in \mathbb{C}^{q \times 1}$ is an unknown complex coordinate vector accounting for the target reflectivity and channel propagation effects, and $\mathbf{c}$ denotes clutter in the range cell under test. Suppose that the clutter $\mathbf{c}$ has a circularly symmetric, complex Gaussian distribution with zero mean, and covariance matrix $\mathbf{R}_p \in \mathbb{C}^{Q \times Q}$, i.e., $\mathbf{c} \sim \mathcal{CN}_Q(\mathbf{0}, \mathbf{R}_p)$.

As customary, a set of secondary (training) data $\{\mathbf{y}_k\}_{k=1}^K$ free of the target echoes is assumed available, i.e.,

$$\mathbf{y}_k = \mathbf{c}_k \sim \mathcal{CN}_Q(\mathbf{0}, \mathbf{R}_p), \tag{4.2}$$

for $k = 1, 2, \ldots, K$. Suppose further that the clutter $\mathbf{c}$ and $\mathbf{c}_k$s for $k = 1, 2, \ldots, K$ are IID. In radar applications, these training data are collected from the cells adjacent to the CUT.

The target detection problem at hand can be cast to the following binary hypothesis test

$$H_0 : \begin{cases} \mathbf{x} \sim \mathcal{CN}_Q(\mathbf{0}, \mathbf{R}_p), \\ \mathbf{y}_k \sim \mathcal{CN}_Q(\mathbf{0}, \mathbf{R}_p), \ k = 1, 2, \ldots, K, \end{cases} \tag{4.3a}$$

$$H_1 : \begin{cases} \mathbf{x} \sim \mathcal{CN}_Q(\boldsymbol{\Sigma}\mathbf{a}, \mathbf{R}_p), \\ \mathbf{y}_k \sim \mathcal{CN}_Q(\mathbf{0}, \mathbf{R}_p), \ k = 1, 2, \ldots, K, \end{cases} \tag{4.3b}$$

where $\mathbf{a}$ and $\mathbf{R}_p$ are both unknown.

4.2   Persymmetric One-Step GLRT

In practice, the covariance matrix $\mathbf{R}_p$ has a persymmetric structure, when the receiver uses a symmetrically spaced linear array and/or symmetrically spaced pulse trains [17, 18]. When $\mathbf{R}_p$ is persymmetric, it satisfies

$$\mathbf{R}_p = \mathbf{J}\mathbf{R}_p^*\mathbf{J}. \tag{4.4}$$

We assume that the target subspace matrix $\boldsymbol{\Sigma}$ is persymmetric, which means that $\boldsymbol{\Sigma} = \mathbf{J}\boldsymbol{\Sigma}^*$. The persymmetric subspace model can be used in the following cases:

- First, target detection in practice is performed on a grid of spatial and/or Doppler frequencies, while the true target frequency may be located between the grid. In this mismatched case, the persymmetric structure in the steering vector is not destroyed. Hence, the persymmetric subspace model holds for this mismatched case.

---

[1]The subspace target model is widely used in two cases. First, the target signal comes from multipaths [2]; second, the subspace model can improve the robustness to signal mismatch [4, 6, 12, 16].

- Second, target echoes may occupy several Doppler and/or angular resolution cells in practice. This kind of target is referred to as Doppler and/or spatially distributed target. In this case, the subspace model can be adopted to describe the dispersions of the target echoes in the Doppler and/or angular domains. Note that no mismatch is assumed in this case, and the persymmetry structure of the target steering vector exists.

- Finally, when mismatches occur in the steering vector due to array calibration errors and/or waveform distortions, the persymmetry of the target steering vector is destroyed. However, in such mismatched cases, we can still use the persymmetric subspace model to *approximate* the true steering vector, even though the true steering vector does not have persymmetry. Note that the true steering vector is not guaranteed to belong to the chosen subspace, since the exact steering vector in the mismatched cases cannot be known in practice. Strictly speaking, the true steering vector is just *approximated* in the subspace model by using the linear combination of the column vectors of a carefully designed matrix. The aim of the subspace is to represent the true steering vector as well as possible, and this is accomplished through the additional directions.

In the following, a one-step GLRT detector is designed by exploiting the persymmetry. Moreover, an exact finite-sum expression for the PFA is derived. For mathematical tractability, we assume that $K \geqslant \lceil \frac{Q}{2} \rceil$ in the detection problem (4.3). This is a more interesting and also practically motivated case. This constraint is much less restrictive than the one (i.e., $K \geqslant Q$) required in [6, 8, 12].

We incorporate persymmetry to design the GLRT detector according to the one-step method, namely,

$$\Lambda_1 = \frac{\max_{\{\mathbf{a}, \mathbf{R}_p\}} f_1(\mathbf{x}, \mathbf{y}_1, \ldots, \mathbf{y}_K)}{\max_{\{\mathbf{R}_p\}} f_0(\mathbf{x}, \mathbf{y}_1, \ldots, \mathbf{y}_K)} \overset{H_1}{\underset{H_0}{\gtrless}} \lambda_1, \tag{4.5}$$

where λ_1 is a detection threshold, $f_1(\mathbf{x}, \mathbf{y}_1, \ldots, \mathbf{y}_K)$ and $f_0(\mathbf{x}, \mathbf{y}_1, \ldots, \mathbf{y}_K)$ denote the joint PDFs of the primary and secondary data under H_1 and H_0, respectively. Due to the independence among the primary and secondary data, $f_1(\mathbf{x}, \mathbf{y}_1, \ldots, \mathbf{y}_K)$ can be written as

$$f_1(\mathbf{x}, \mathbf{y}_1, \ldots, \mathbf{y}_K) = \left\{ \frac{1}{\pi^Q \det(\mathbf{R}_p)} \exp\left[-\mathrm{tr}(\mathbf{R}_p^{-1} \mathbf{T}_1)\right] \right\}^{K+1}, \tag{4.6}$$

where

$$\mathbf{T}_1 = \frac{1}{K+1} \left[(\mathbf{x} - \mathbf{\Sigma}\mathbf{a})(\mathbf{x} - \mathbf{\Sigma}\mathbf{a})^{\dagger} + \tilde{\mathbf{R}}_p \right] \tag{4.7}$$

with $\tilde{\mathbf{R}}_p = \sum_{k=1}^{K} \mathbf{y}_k \mathbf{y}_k^{\dagger}$. In addition, $f_0(\mathbf{x}, \mathbf{y}_1, \ldots, \mathbf{y}_K)$ can be expressed by

$$f_0(\mathbf{x}, \mathbf{y}_1, \ldots, \mathbf{y}_K) = \left\{ \frac{1}{\pi^Q \det(\mathbf{R}_p)} \exp\left[-\mathrm{tr}(\mathbf{R}_p^{-1} \mathbf{T}_0)\right] \right\}^{K+1}, \tag{4.8}$$

where

$$\mathbf{T}_0 = \frac{1}{K+1} \left(\mathbf{x}\mathbf{x}^\dagger + \tilde{\mathbf{R}}_p \right). \tag{4.9}$$

Using the persymmetry structure in the covariance matrix $\mathbf{R}_p$, we have

$$\begin{aligned}
\mathrm{tr}(\mathbf{R}_p^{-1}\mathbf{T}_j) &= \mathrm{tr}[\mathbf{J}(\mathbf{R}_\mathbf{p}^*)^{-1}\mathbf{J}\mathbf{T}_j] \\
&= \mathrm{tr}[(\mathbf{R}_\mathbf{p}^*)^{-1}\mathbf{J}\mathbf{T}_j\mathbf{J}] \\
&= \mathrm{tr}[\mathbf{R}_p^{-1}\mathbf{J}\mathbf{T}_j^*\mathbf{J}], \qquad j = 0, 1.
\end{aligned} \tag{4.10}$$

As a result, (4.6) can be rewritten as

$$f_1(\mathbf{x}, \mathbf{y}_1, \ldots, \mathbf{y}_K) = \left\{ \frac{1}{\pi^Q \det(\mathbf{R}_p)} \exp\left[-\mathrm{tr}(\mathbf{R}_p^{-1}\mathbf{T}_{1p}) \right] \right\}^{K+1}, \tag{4.11}$$

where

$$\mathbf{T}_{1p} = \frac{1}{2} \left(\mathbf{T}_1 + \mathbf{J}\mathbf{T}_1^*\mathbf{J} \right). \tag{4.12}$$

As derived in Appendix 4.A, we have

$$\mathbf{T}_{1p} = \frac{1}{K+1} \left[(\mathbf{X}_p - \boldsymbol{\Sigma}\mathbf{A})(\mathbf{X}_p - \boldsymbol{\Sigma}\mathbf{A})^\dagger + \hat{\mathbf{R}}_p \right], \tag{4.13}$$

where

$$\hat{\mathbf{R}}_p = \frac{1}{2} \left(\tilde{\mathbf{R}}_p + \mathbf{J}\tilde{\mathbf{R}}_p^*\mathbf{J} \right) \in \mathbb{C}^{Q \times Q}, \tag{4.14}$$

$$\mathbf{X}_p = [\mathbf{x}_e, \mathbf{x}_o] \in \mathbb{C}^{Q \times 2}, \tag{4.15}$$

and

$$\mathbf{A} = [\mathbf{a}_e, \mathbf{a}_o] \in \mathbb{C}^{q \times 2}, \tag{4.16}$$

with

$$\begin{cases} \mathbf{x}_e = \frac{1}{2}(\mathbf{x} + \mathbf{J}\mathbf{x}^*) \in \mathbb{C}^{Q \times 1}, \\ \mathbf{x}_o = \frac{1}{2}(\mathbf{x} - \mathbf{J}\mathbf{x}^*) \in \mathbb{C}^{Q \times 1}, \end{cases} \tag{4.17}$$

and

$$\begin{cases} \mathbf{a}_e = \frac{1}{2}(\mathbf{a} + \mathbf{a}^*) = \mathfrak{Re}(\mathbf{a}), \\ \mathbf{a}_o = \frac{1}{2}(\mathbf{a} - \mathbf{a}^*) = \jmath\,\mathfrak{Im}(\mathbf{a}). \end{cases} \tag{4.18}$$

Similarly,

$$f_0(\mathbf{y}, \mathbf{y}_1, \ldots, \mathbf{y}_K) = \left\{ \frac{1}{\pi^Q \det(\mathbf{R}_p)} \exp\left[-\mathrm{tr}(\mathbf{R}_p^{-1}\mathbf{T}_{0p}) \right] \right\}^{K+1}, \tag{4.19}$$

where

$$\mathbf{T}_{0p} = \frac{1}{K+1} \left(\mathbf{X}_p\mathbf{X}_p^\dagger + \hat{\mathbf{R}}_p \right). \tag{4.20}$$

It is easy to check that

$$
\begin{cases}
\max_{\{\mathbf{R}_p\}} f_1(\mathbf{y}, \mathbf{y}_1, \ldots, \mathbf{y}_K) = \left[(e\pi)^Q \det(\mathbf{T}_{1p}) \right]^{-(K+1)}, \\
\max_{\{\mathbf{R}_p\}} f_0(\mathbf{y}, \mathbf{y}_1, \ldots, \mathbf{y}_K) = \left[(e\pi)^Q \det(\mathbf{T}_{0p}) \right]^{-(K+1)}.
\end{cases}
\tag{4.21}
$$

As a result, the GLRT in (4.5) can be equivalently written as

$$
\Lambda_2 = \frac{\det(\mathbf{T}_{0p})}{\min_{\{\mathbf{A}\}} \det(\mathbf{T}_{1p})} \underset{H_0}{\overset{H_1}{\gtrless}} \lambda_2,
\tag{4.22}
$$

where λ_2 is a detection threshold. The minimization of $\det(\mathbf{T}_{1p})$ with respect to $\mathbf{A}$ can be achieved at

$$
\hat{\mathbf{A}} = \left(\boldsymbol{\Sigma}^\dagger \hat{\mathbf{R}}_p^{-1} \boldsymbol{\Sigma} \right)^{-1} \boldsymbol{\Sigma}^\dagger \hat{\mathbf{R}}_p^{-1} \mathbf{X}_p.
\tag{4.23}
$$

Hence, we have

$$
\min_{\{\mathbf{A}\}} \det(\mathbf{T}_{1p}) = \frac{1}{(K+1)^Q} \det(\hat{\mathbf{R}}_p) \det \left(\mathbf{I}_2 + \mathbf{X}_p^\dagger \mathbf{P}_p \mathbf{X}_p \right),
\tag{4.24}
$$

where

$$
\mathbf{P}_p = \hat{\mathbf{R}}_p^{-1} - \hat{\mathbf{R}}_p^{-1} \boldsymbol{\Sigma} \left(\boldsymbol{\Sigma}^\dagger \hat{\mathbf{R}}_p^{-1} \boldsymbol{\Sigma} \right)^{-1} \boldsymbol{\Sigma}^\dagger \hat{\mathbf{R}}_p^{-1} \in \mathbb{C}^{Q \times Q}.
\tag{4.25}
$$

It is obvious from (4.9) that

$$
\det(\mathbf{T}_{0p}) = \frac{1}{(K+1)^Q} \det(\hat{\mathbf{R}}_p) \det \left(\mathbf{I}_2 + \mathbf{X}_p^\dagger \hat{\mathbf{R}}_p^{-1} \mathbf{X}_p \right).
\tag{4.26}
$$

Substituting (4.24) and (4.26) into (4.22), we derive the one-step persymmetric GLRT detector as

$$
\Lambda = \frac{\det \left(\mathbf{I}_2 + \mathbf{X}_p^\dagger \hat{\mathbf{R}}_p^{-1} \mathbf{X}_p \right)}{\det \left(\mathbf{I}_2 + \mathbf{X}_p^\dagger \mathbf{P}_p \mathbf{X}_p \right)} \underset{H_0}{\overset{H_1}{\gtrless}} \lambda,
\tag{4.27}
$$

where λ is a detection threshold. This detector is referred to as persymmetric adaptive subspace detector (PASD), for ease of reference.

4.3   Threshold Setting

To complete the construction of the detection scheme in (4.27), a way is provided to set the detection threshold λ for a given PFA. To this end, a finite-sum expression is derived for the PFA in the following. At the end of this section, we will provide some discussions on the detection probability.

4.3.1   Transformation from Complex Domain to Real Domain

For convenience of statistical analysis, we first transform all quantities in the original data from the complex-valued domain to the real-valued domain. Let us start by defining a unitary matrix as

$$\mathbf{D} = \frac{1}{2}\left[(\mathbf{I}_Q + \mathbf{J}) + \jmath(\mathbf{I}_Q - \mathbf{J})\right] \in \mathbb{C}^{Q \times Q}. \tag{4.28}$$

Using (4.15), (4.17), and (4.28), we have

$$\mathbf{D}\mathbf{X}_p = [\mathbf{x}_{er}, \jmath\mathbf{x}_{or}], \tag{4.29}$$

where

$$\mathbf{x}_{er} = \mathbf{D}\mathbf{x}_e = \frac{1}{2}\left[(\mathbf{I}_Q + \mathbf{J})\mathfrak{Re}(\mathbf{x}) - (\mathbf{I}_Q - \mathbf{J})\mathfrak{Im}(\mathbf{x})\right] \in \mathbb{R}^{Q \times 1}, \tag{4.30}$$

and

$$\mathbf{x}_{or} = -\jmath\mathbf{D}\mathbf{x}_o = \frac{1}{2}\left[(\mathbf{I}_Q - \mathbf{J})\mathfrak{Re}(\mathbf{x}) + (\mathbf{I}_Q + \mathbf{J})\mathfrak{Im}(\mathbf{x})\right] \in \mathbb{R}^{Q \times 1}. \tag{4.31}$$

It is worth noting that $\mathbf{x}_{er}$ and $\mathbf{x}_{or}$ are now real-valued column vectors of dimension Q.

We proceed by defining another unitary matrix as

$$\mathbf{V}_2 = \begin{bmatrix} 1 & 0 \\ 0 & -\jmath \end{bmatrix} \in \mathbb{C}^{2 \times 2}. \tag{4.32}$$

Now we use the two unitary matrices $\mathbf{D}$ and $\mathbf{V}_2$ to transform all complex-valued quantities in (4.27) to be real-valued ones. Specifically, we define

$$\mathbf{M} \triangleq \mathbf{D}\hat{\mathbf{R}}_p\mathbf{D}^\dagger = \mathfrak{Re}(\hat{\mathbf{R}}_p) + \mathbf{J}\mathfrak{Im}(\hat{\mathbf{R}}_p) \in \mathbb{R}^{Q \times Q}, \tag{4.33}$$

$$\mathbf{\Omega} \triangleq \mathbf{D}\mathbf{\Sigma} = \mathfrak{Re}(\mathbf{\Sigma}) - \mathfrak{Im}(\mathbf{\Sigma}) \in \mathbb{R}^{Q \times q}, \tag{4.34}$$

and

$$\mathbf{X} \triangleq \mathbf{D}\mathbf{X}_p\mathbf{V}_2 = [\mathbf{x}_{er}, \mathbf{x}_{or}] \in \mathbb{R}^{Q \times 2}. \tag{4.35}$$

Then, the PASD in (4.27) can be recast to

$$\begin{aligned} \Lambda &= \frac{\det\left(\mathbf{I}_2 + \mathbf{V}_2\mathbf{X}^\dagger\mathbf{M}^{-1}\mathbf{X}\mathbf{V}_2^\dagger\right)}{\det\left(\mathbf{I}_2 + \mathbf{V}_2\mathbf{X}^\dagger\mathbf{P}\mathbf{X}\mathbf{V}_2^\dagger\right)} \\ &= \frac{\det\left(\mathbf{I}_2 + \mathbf{X}^T\mathbf{M}^{-1}\mathbf{X}\right)}{\det\left(\mathbf{I}_2 + \mathbf{X}^T\mathbf{P}\mathbf{X}\right)} \overset{H_1}{\underset{H_0}{\gtrless}} \lambda, \end{aligned} \tag{4.36}$$

where

$$\mathbf{P} \triangleq \mathbf{D}\mathbf{P}_p\mathbf{D}^\dagger = \mathbf{M}^{-1} - \mathbf{M}^{-1}\mathbf{\Omega}\left(\mathbf{\Omega}^T\mathbf{M}^{-1}\mathbf{\Omega}\right)^{-1}\mathbf{\Omega}^T\mathbf{M}^{-1} \in \mathbb{R}^{Q \times Q}. \tag{4.37}$$

It is worth pointing out that all quantities in (4.36) are real-valued. Next, we will perform statistical analysis on the PASD in the real-valued domain.

4.3.2 Statistical Characterizations

First, the statistical properties of $\mathbf{M}$ and $\mathbf{X}$ are provided. As derived in Appendix 4.B, we have

$$\begin{cases} \mathbf{M} \sim \mathcal{W}_Q(2K, \mathbf{R}), \\ \mathbf{X} \sim \mathcal{N}_{Q \times 2}(\mathbf{0}, \mathbf{R} \otimes \mathbf{I}_2), \end{cases} \tag{4.38}$$

where $\mathbf{R}$ is defined in (4.B.7).

4.3.2.1 Equivalent Form of $\mathbf{X}^T \mathbf{P} \mathbf{X}$

Now we turn to obtain an equivalent form of $\mathbf{X}^T \mathbf{P} \mathbf{X}$. To this end, we define

$$\mathbf{E} \triangleq \mathbf{\Omega}(\mathbf{\Omega}^T \mathbf{\Omega})^{-1/2} \in \mathbb{R}^{Q \times q}, \tag{4.39}$$

and hence

$$\mathbf{E}^T \mathbf{E} = \mathbf{I}_q. \tag{4.40}$$

Therefore, we can construct a $Q \times (Q - q)$ matrix $\mathbf{F}$ such that

$$\begin{cases} \mathbf{F}^T \mathbf{E} = \mathbf{0}_{(Q-q) \times q}, \\ \mathbf{F}^T \mathbf{F} = \mathbf{I}_{Q-q}. \end{cases} \tag{4.41}$$

Define

$$\mathbf{U} \triangleq [\mathbf{E}, \mathbf{F}] \in \mathbb{R}^{Q \times Q}. \tag{4.42}$$

Obviously, $\mathbf{U}$ is an orthogonal matrix, i.e., $\mathbf{U}\mathbf{U}^T = \mathbf{U}^T\mathbf{U} = \mathbf{I}_Q$.
   Define

$$\begin{cases} \mathbf{X}_1 \triangleq \mathbf{E}^T \mathbf{X} \in \mathbb{R}^{q \times 2}, \\ \mathbf{X}_2 \triangleq \mathbf{F}^T \mathbf{X} \in \mathbb{R}^{(Q-q) \times 2}, \end{cases} \tag{4.43}$$

and then we have

$$\begin{bmatrix} \mathbf{X}_1 \\ \mathbf{X}_2 \end{bmatrix} = \mathbf{U}^T \mathbf{X}, \tag{4.44}$$

and

$$\mathbf{X} = \mathbf{E}\mathbf{X}_1 + \mathbf{F}\mathbf{X}_2. \tag{4.45}$$

Using (4.37) and the definition of $\mathbf{E}$ in (4.39), we obtain

$$\begin{cases} \mathbf{P}\mathbf{E} = \mathbf{0}_{Q \times q}, \\ \mathbf{E}^T \mathbf{P} = \mathbf{0}_{q \times Q}, \end{cases} \tag{4.46}$$

and

$$\mathbf{P} = \mathbf{M}^{-1} - \mathbf{M}^{-1}\mathbf{E}\left(\mathbf{E}^T \mathbf{M}^{-1}\mathbf{E}\right)^{-1}\mathbf{E}^T\mathbf{M}^{-1}. \tag{4.47}$$

Employing (4.45) and (4.46), we have

$$\mathbf{X}^T \mathbf{P} \mathbf{X} = \mathbf{X}_2^T \mathbf{F}^T \mathbf{P} \mathbf{F} \mathbf{X}_2. \tag{4.48}$$

In the following, we make a transformation for the term $\mathbf{F}^T\mathbf{P}\mathbf{F}$. Write

$$\mathbf{U}^T\mathbf{M}\mathbf{U} = \begin{bmatrix} \mathbf{M}_{11} & \mathbf{M}_{12} \\ \mathbf{M}_{21} & \mathbf{M}_{22} \end{bmatrix}, \tag{4.49}$$

where $\mathbf{M}_{11} \in \mathbb{R}^{q \times q}$, $\mathbf{M}_{12} \in \mathbb{R}^{q \times (Q-q)}$, $\mathbf{M}_{21} \in \mathbb{R}^{(Q-q) \times q}$, and $\mathbf{M}_{22} \in \mathbb{R}^{(Q-q) \times (Q-q)}$. We note that

$$\begin{aligned} \mathbf{U}^T\mathbf{M}^{-1}\mathbf{U} &= (\mathbf{U}^T\mathbf{M}\mathbf{U})^{-1} \\ &\triangleq \begin{bmatrix} \bar{\mathbf{M}}_{11} & \bar{\mathbf{M}}_{12} \\ \bar{\mathbf{M}}_{21} & \bar{\mathbf{M}}_{22} \end{bmatrix}, \end{aligned} \tag{4.50}$$

where $\bar{\mathbf{M}}_{11} \in \mathbb{R}^{q \times q}$, $\bar{\mathbf{M}}_{12} \in \mathbb{R}^{q \times (Q-q)}$, $\bar{\mathbf{M}}_{21} \in \mathbb{R}^{(Q-q) \times q}$, and $\bar{\mathbf{M}}_{22} \in \mathbb{R}^{(Q-q) \times (Q-q)}$. According to the matrix inversion lemma [19, p. 99, theorem 8.5.11], we obtain

$$\mathbf{M}_{22}^{-1} = \bar{\mathbf{M}}_{22} - \bar{\mathbf{M}}_{21}\bar{\mathbf{M}}_{11}^{-1}\bar{\mathbf{M}}_{12}. \tag{4.51}$$

Using (4.42), we have

$$\mathbf{U}^T\mathbf{M}^{-1}\mathbf{U} = \begin{bmatrix} \mathbf{E}^T\mathbf{M}^{-1}\mathbf{E} & \mathbf{E}^T\mathbf{M}^{-1}\mathbf{F} \\ \mathbf{F}^T\mathbf{M}^{-1}\mathbf{E} & \mathbf{F}^T\mathbf{M}^{-1}\mathbf{F} \end{bmatrix}. \tag{4.52}$$

Combining (4.50) and (4.52), we obtain

$$\begin{bmatrix} \bar{\mathbf{M}}_{11} & \bar{\mathbf{M}}_{12} \\ \bar{\mathbf{M}}_{21} & \bar{\mathbf{M}}_{22} \end{bmatrix} = \begin{bmatrix} \mathbf{E}^T\mathbf{M}^{-1}\mathbf{E} & \mathbf{E}^T\mathbf{M}^{-1}\mathbf{F} \\ \mathbf{F}^T\mathbf{M}^{-1}\mathbf{E} & \mathbf{F}^T\mathbf{M}^{-1}\mathbf{F} \end{bmatrix}. \tag{4.53}$$

Using (4.47), we have

$$\begin{aligned} \mathbf{F}^T\mathbf{P}\mathbf{F} &= \mathbf{F}^T\mathbf{M}^{-1}\mathbf{F} - \mathbf{F}^T\mathbf{M}^{-1}\mathbf{E}\left(\mathbf{E}^T\mathbf{M}^{-1}\mathbf{E}\right)^{-1}\mathbf{E}^T\mathbf{M}^{-1}\mathbf{F} \\ &= \bar{\mathbf{M}}_{22} - \bar{\mathbf{M}}_{21}\bar{\mathbf{M}}_{11}^{-1}\bar{\mathbf{M}}_{12} \\ &= \mathbf{M}_{22}^{-1}, \end{aligned} \tag{4.54}$$

where the second and third lines are obtained from (4.53) and (4.51), respectively. Substituting (4.54) into (4.48) results in

$$\mathbf{X}^T\mathbf{P}\mathbf{X} = \mathbf{X}_2^T\mathbf{M}_{22}^{-1}\mathbf{X}_2, \tag{4.55}$$

which is only relative to the two-components of the data.

4.3.2.2 Equivalent Form of $\mathbf{X}^T\mathbf{M}^{-1}\mathbf{X}$

Now we turn our attention to $\mathbf{X}^T\mathbf{M}^{-1}\mathbf{X}$, which can be written as

$$\begin{aligned} \mathbf{X}^T\mathbf{M}^{-1}\mathbf{X} &= (\mathbf{U}^T\mathbf{X})^T(\mathbf{U}^T\mathbf{M}^{-1}\mathbf{U})\mathbf{U}^T\mathbf{X} \\ &= [\mathbf{X}_1^T, \mathbf{X}_2^T] \begin{bmatrix} \bar{\mathbf{M}}_{11} & \bar{\mathbf{M}}_{12} \\ \bar{\mathbf{M}}_{21} & \bar{\mathbf{M}}_{22} \end{bmatrix} \begin{bmatrix} \mathbf{X}_1 \\ \mathbf{X}_2 \end{bmatrix} \\ &= \mathbf{Z}^T\bar{\mathbf{M}}_{11}\mathbf{Z} + \mathbf{X}_2^T\mathbf{M}_{22}^{-1}\mathbf{X}_2, \end{aligned} \tag{4.56}$$

where the second and third lines are obtained using (4.44) and [20, p. 135, eq. (A1-9)], respectively, and

$$\mathbf{Z} \triangleq \mathbf{X}_1 - \mathbf{M}_{12}\mathbf{M}_{22}^{-1}\mathbf{X}_2 \in \mathbb{R}^{q \times 2}. \tag{4.57}$$

4.3.2.3 Statistical Distribution of Λ

Taking (4.55) and (4.56) into (4.36) produces

$$\Lambda = \frac{\det(\boldsymbol{\Delta} + \mathbf{Z}^T \bar{\mathbf{M}}_{11} \mathbf{Z})}{\det(\boldsymbol{\Delta})}$$

$$= \det(\mathbf{I}_2 + \mathbf{G}^T \bar{\mathbf{M}}_{11} \mathbf{G}) \underset{H_0}{\overset{H_1}{\gtrless}} \lambda, \tag{4.58}$$

where

$$\boldsymbol{\Delta} \triangleq \mathbf{I}_2 + \mathbf{X}_2^T \mathbf{M}_{22}^{-1} \mathbf{X}_2 \in \mathbb{R}^{2 \times 2}, \tag{4.59}$$

and

$$\mathbf{G} \triangleq \mathbf{Z} \boldsymbol{\Delta}^{-1/2} \in \mathbb{R}^{q \times 2}. \tag{4.60}$$

Now we can rewrite (4.58) as

$$\Lambda = \frac{\det(\bar{\mathbf{M}}_{11} + \mathbf{G}\mathbf{G}^T)}{\det(\bar{\mathbf{M}}_{11})} \underset{H_0}{\overset{H_1}{\gtrless}} \lambda, \tag{4.61}$$

where $\bar{\mathbf{M}}_{11}$ and $\mathbf{G}$ are independent, and their distributions under H_0 are given by (see the detailed derivations in Appendix 4.C)

$$\begin{cases} \bar{\mathbf{M}}_{11} \sim \mathcal{W}_q(2K - Q + q, \bar{\mathbf{R}}_{11}^{-1}), \\ \mathbf{G} \sim \mathcal{N}_{q \times 2}(\mathbf{0}, \bar{\mathbf{R}}_{11}^{-1} \otimes \mathbf{I}_2), \end{cases} \tag{4.62}$$

with $\bar{\mathbf{R}}_{11}$ defined in (4.C.4). According to [21, p. 305, Lemma 8.4.2], Λ^{-1} under H_0 has the central Wilks' Lambda distribution, written as

$$\Lambda^{-1} \sim U_{q, 2, 2K+q-Q}, \tag{4.63}$$

where $U_{n,m,j}$ denotes the central Wilks' Lambda distribution with n, m and j being the number of dimensions, the hypothesis degrees of freedom, and the error degrees of freedom, respectively.

4.3.3 Probability of False Alarm

According to [21, p. 311, Theorem 8.4.6], we have

$$\frac{1 - \sqrt{\Lambda^{-1}}}{\sqrt{\Lambda^{-1}}} \sim \frac{q}{2K - Q + 1} F_{2q, 2(2K-Q+1)}. \tag{4.64}$$

As a result,

$$\sqrt{\Lambda} - 1 \sim \frac{\chi_{2q}^2}{\chi_{2(2K-Q+1)}^2}. \tag{4.65}$$

Now we can equivalently write (4.61) as

$$\frac{t}{\tau} \underset{H_0}{\overset{H_1}{\gtrless}} \sqrt{\lambda} - 1, \tag{4.66}$$

where t and τ are independent real-valued random variables. Under H_0, these two random variables are distributed as

$$\begin{cases} t \sim \frac{1}{2}\chi^2_{2q}, \\ \tau \sim \frac{1}{2}\chi^2_{2(2K-Q+1)}. \end{cases} \tag{4.67}$$

Similar to the derivation of the PFA in [8], we can obtain an analytical expression for the PFA as

$$P_{\mathrm{FA}} = \sum_{j=1}^{q} \mathrm{C}^{q-j}_{2K+q-Q-j} \left(\sqrt{\lambda}-1\right)^{q-j} \lambda^{-\frac{2K+q+1-Q-j}{2}}. \tag{4.68}$$

It can be seen that the PASD exhibits the CFAR property against the clutter covariance matrix, since (4.68) is independent of the clutter covariance matrix.

As to the detection probability of the PASD, an analytical expression cannot be obtained. This is because that the test statistic Λ under H_1 is subject to a noncentral Wilks' Lambda distribution whose complementary cumulative distribution function is quite complicated [22], if not intractable.

4.4   Numerical Examples

In this section, numerical examples are provided to verify the above theoretical results. Assume that a pulsed Doppler radar is used, which transmits symmetrically spaced pulse trains. The number of pulses in a coherent processing interval is $Q = 8$. The (i,j)th element of the clutter covariance matrix is chosen as $\mathbf{R}(i,j) = \sigma^2 0.9^{|i-j|}$, where σ^2 is the clutter power. The target coordinate vector is $\mathbf{a} = \sigma_a[1,1,\ldots,1]^T$, where σ_a^2 is the target power. The SCR in decibel is defined by $\mathrm{SCR} = 10\log_{10}\frac{\sigma_a^2}{\sigma^2}$. The target subspace matrix $\boldsymbol{\Sigma}$ is chosen to be

$$\boldsymbol{\Sigma} = [\mathbf{p}(f_{\mathrm{d},1}), \mathbf{p}(f_{\mathrm{d},2}), \ldots, \mathbf{p}(f_{\mathrm{d},q})], \tag{4.69}$$

where $f_{\mathrm{d},i}$ is the i-th normalized Doppler frequency, and for $i = 1,2,\ldots,q$,

$$\mathbf{p}(f_{\mathrm{d},i}) = \frac{1}{\sqrt{Q}}\left[e^{-j2\pi f_{\mathrm{d},i}\frac{(Q-1)}{2}}, \ldots, e^{-j\,2\pi f_{\mathrm{d},i}\frac{1}{2}}, \quad e^{j\,2\pi f_{\mathrm{d},i}\frac{1}{2}}, \ldots, e^{j2\pi f_{\mathrm{d},i}\frac{(Q-1)}{2}}\right]^T, \tag{4.70}$$

for even Q, and

$$\mathbf{p}(f_{\mathrm{d},i}) = \frac{1}{\sqrt{Q}}\left[e^{-j2\pi f_{\mathrm{d},i}\frac{(Q-1)}{2}}, \ldots, e^{-j\,2\pi f_{\mathrm{d},i}}, \quad 1, \quad e^{j\,2\pi f_{\mathrm{d},i}}, \ldots, e^{j2\pi f_{\mathrm{d},i}\frac{(Q-1)}{2}}\right]^T, \tag{4.71}$$

for odd Q. Here, we select $q = 2$, $f_{d,1} = 0.07$, and $f_{d,2} = 0.1$. For comparison purposes, the SGLRT and SAMF in [15] are considered, i.e.,

$$\Lambda_{\text{SGLRT}} = \frac{\mathbf{x}^H \hat{\mathbf{R}}^{-1} \mathbf{\Sigma} (\mathbf{\Sigma}^H \hat{\mathbf{R}}^{-1} \mathbf{\Sigma})^{-1} \mathbf{\Sigma}^H \hat{\mathbf{R}}^{-1} \mathbf{x}}{1 + \mathbf{x}^H \hat{\mathbf{R}}^{-1} \mathbf{x}} \overset{H_1}{\underset{H_0}{\gtrless}} \lambda_{\text{SGLRT}}, \tag{4.72}$$

and

$$\Lambda_{\text{SAMF}} = \mathbf{x}^H \hat{\mathbf{R}}^{-1} \mathbf{\Sigma} (\mathbf{\Sigma}^H \hat{\mathbf{R}}^{-1} \mathbf{\Sigma})^{-1} \mathbf{\Sigma}^H \hat{\mathbf{R}}^{-1} \mathbf{x} \overset{H_1}{\underset{H_0}{\gtrless}} \lambda_{\text{SAMF}}, \tag{4.73}$$

where λ_{SGLRT} and λ_{SAMF} are detection thresholds, and

$$\hat{\mathbf{R}} = \sum_{k=1}^{K} \mathbf{y}_k \mathbf{y}_k^H. \tag{4.74}$$

In Fig. 4.1, the ROC curves are plotted with SCR = 8 dB for $K = 16$ and 8. It can be observed that the PASD obviously outperforms its counterparts, especially for small K. This is because the estimation accuracy of the clutter covariance matrix is significantly improved with the exploitation of persymmetry.

In Fig. 4.2, the ROC curves are provided with $K = 6$ for different SCRs. Note that for the case of small K (i.e., $K < Q$), the SGLRT and SAMF cannot work. We can see from Fig. 4.2 that the PASD can work for small K, and its performance improves as the SCR increases.

The detection probability as a function of SCR is presented in Fig. 4.3, where $P_{\text{FA}} = 10^{-3}$. These results highlight that the PASD has better performance than the SGLRT and SAMF. It can be easily explained, because the former exploits the persymmetry of the clutter covariance matrix.

In Fig. 4.4, we depict the detection probability as a function of K, where SCR = 12 dB and the PFA is 10^{-3}. It can be seen that the PASD performs the best, especially when the training data size is small. Here, it should be emphasized again that the PASD can work for $\lceil \frac{Q}{2} \rceil \leqslant K < Q$, but the SGLRT and SAMF cannot.

4.A   Derivations of (4.13)

Define

$$\begin{cases} \tilde{\mathbf{T}}_1 = (\mathbf{x} - \mathbf{\Sigma} \mathbf{a})(\mathbf{x} - \mathbf{\Sigma} \mathbf{a})^\dagger + \tilde{\mathbf{R}}_p, \\ \tilde{\mathbf{T}}_{1p} = (\mathbf{X}_p - \mathbf{\Sigma} \mathbf{A})(\mathbf{X}_p - \mathbf{\Sigma} \mathbf{A})^\dagger + \hat{\mathbf{R}}_p. \end{cases} \tag{4.A.1}$$

Then, (4.13) holds true if the following equation is valid:

$$\tilde{\mathbf{T}}_{1p} = \frac{1}{2} \left(\tilde{\mathbf{T}}_1 + \mathbf{J} \tilde{\mathbf{T}}_1^* \mathbf{J} \right), \tag{4.A.2}$$

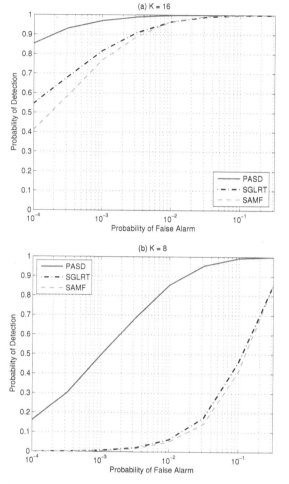

FIGURE 4.1
Receiver operating characteristic curves with SCR = 8 dB.

which is derived as follows. Using (4.A.1), we have

$$\tilde{\mathbf{T}}_1 + \mathbf{J}\tilde{\mathbf{T}}_1^*\mathbf{J} = \tilde{\mathbf{R}}_p + \mathbf{J}\tilde{\mathbf{R}}_p^*\mathbf{J} + \mathbf{H}, \qquad (4.A.3)$$

where

$$
\begin{aligned}
\mathbf{H} =& (\mathbf{x} - \mathbf{\Sigma}\mathbf{a})(\mathbf{x} - \mathbf{\Sigma}\mathbf{a})^\dagger + \mathbf{J}(\mathbf{x} - \mathbf{\Sigma}\mathbf{a})^*(\mathbf{x} - \mathbf{\Sigma}\mathbf{a})^T\mathbf{J} \\
=& \mathbf{x}\mathbf{x}^\dagger + \mathbf{J}\mathbf{x}^*\mathbf{x}^T\mathbf{J} - \left(\mathbf{x}\mathbf{a}^\dagger\mathbf{\Sigma}^\dagger + \mathbf{J}\mathbf{x}^*\mathbf{a}^T\mathbf{\Sigma}^T\mathbf{J}\right) \\
& - \left(\mathbf{\Sigma}\mathbf{a}\mathbf{x}^\dagger + \mathbf{J}\mathbf{\Sigma}^*\mathbf{a}^*\mathbf{x}^T\mathbf{J}\right) + \left(\mathbf{\Sigma}\mathbf{a}\mathbf{a}^\dagger\mathbf{\Sigma}^\dagger + \mathbf{J}\mathbf{\Sigma}^*\mathbf{a}^*\mathbf{a}^T\mathbf{\Sigma}^T\mathbf{J}\right).
\end{aligned}
\qquad (4.A.4)
$$

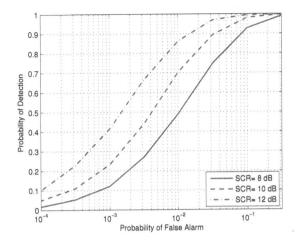

FIGURE 4.2
Receiver operating characteristic curves with $K = 6$.

It is easy to check that

$$\begin{cases} 2\mathbf{X}_p \mathbf{X}_p^\dagger = \mathbf{x}\mathbf{x}^\dagger + \mathbf{J}\mathbf{x}^*\mathbf{x}^T\mathbf{J}, \\ 2\mathbf{\Sigma}\mathbf{A}\mathbf{A}^\dagger\mathbf{\Sigma}^\dagger = \mathbf{\Sigma}\mathbf{a}\mathbf{a}^\dagger\mathbf{\Sigma}^\dagger + \mathbf{J}\mathbf{\Sigma}^*\mathbf{a}^*\mathbf{a}^T\mathbf{\Sigma}^T\mathbf{J}, \end{cases} \tag{4.A.5}$$

and

$$\begin{cases} 2\mathbf{X}_p\mathbf{A}^\dagger\mathbf{\Sigma}^\dagger = \mathbf{x}\mathbf{a}^\dagger\mathbf{\Sigma}^\dagger + \mathbf{J}\mathbf{x}^*\mathbf{a}^T\mathbf{\Sigma}^T\mathbf{J}, \\ 2\mathbf{\Sigma}\mathbf{A}\mathbf{X}_p^\dagger = \mathbf{\Sigma}\mathbf{a}\mathbf{x}^\dagger + \mathbf{J}\mathbf{\Sigma}^*\mathbf{a}^*\mathbf{x}^T\mathbf{J}, \end{cases} \tag{4.A.6}$$

where $\mathbf{\Sigma}^\dagger = \mathbf{\Sigma}^T\mathbf{J}$ is used. Taking (4.A.5) and (4.A.6) into (4.A.4) produces

$$\mathbf{H} = 2\left(\mathbf{X}_p - \mathbf{\Sigma}\mathbf{A}\right)\left(\mathbf{X}_p - \mathbf{\Sigma}\mathbf{A}\right)^\dagger. \tag{4.A.7}$$

Substituting (4.A.7) into (4.A.3), and using (4.A.1), we obtain (4.A.2). That is to say, (4.13) holds true.

4.B   Derivations of (4.38)

Define $\tilde{\mathbf{Y}} = [\mathbf{y}_{e1}, \ldots, \mathbf{y}_{eK}, \mathbf{y}_{o1}, \ldots, \mathbf{y}_{oK}] \in \mathbb{C}^{Q \times 2K}$, where

$$\begin{cases} \mathbf{y}_{ek} \triangleq \frac{1}{2}\left(\mathbf{y}_k + \mathbf{J}\mathbf{y}_k^*\right), \\ \mathbf{y}_{ok} \triangleq \frac{1}{2}\left(\mathbf{y}_k - \mathbf{J}\mathbf{y}_k^*\right). \end{cases} \tag{4.B.1}$$

Define $\mathbf{Y} \triangleq \mathbf{D}\tilde{\mathbf{Y}}\mathbf{V}_{2K}$, where

$$\mathbf{V}_{2K} = \begin{bmatrix} \mathbf{I}_K & \mathbf{0} \\ \mathbf{0} & -\jmath\mathbf{I}_K \end{bmatrix} \in \mathbb{C}^{2K \times 2K}. \tag{4.B.2}$$

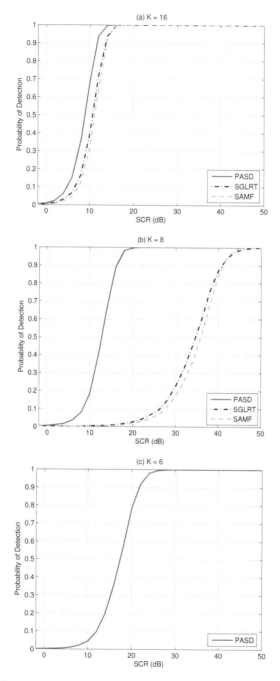

FIGURE 4.3
Detection probability versus SCR for $P_{\mathrm{FA}} = 10^{-3}$. (a) $K = 16$; (b) $K = 8$;
(c)$K = 6$.

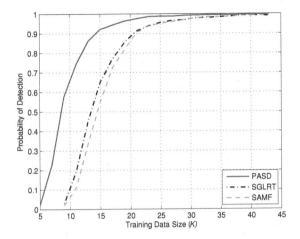

FIGURE 4.4
Detection probability versus K for SCR $= 12$ dB and $P_{\mathrm{FA}} = 10^{-3}$.

Then, we have

$$\mathbf{Y} = [\mathbf{y}_{er1}, \ldots, \mathbf{y}_{erK}, \mathbf{y}_{or1}, \ldots, \mathbf{y}_{orK}] \in \mathbb{R}^{Q \times 2K}, \qquad (4.\mathrm{B}.3)$$

where

$$\mathbf{y}_{erk} \triangleq \mathbf{D}\mathbf{y}_{ek} = \frac{1}{2}\left[(\mathbf{I}_Q + \mathbf{J})\Re\mathfrak{e}(\mathbf{y}_k) - (\mathbf{I}_Q - \mathbf{J})\Im\mathfrak{m}(\mathbf{y}_k)\right] \in \mathbb{R}^{Q \times 1}, \qquad (4.\mathrm{B}.4)$$

and

$$\mathbf{y}_{ork} \triangleq -\jmath\mathbf{D}\mathbf{y}_{ok} = \frac{1}{2}\left[(\mathbf{I}_Q - \mathbf{J})\Re\mathfrak{e}(\mathbf{y}_k) + (\mathbf{I}_Q + \mathbf{J})\Im\mathfrak{m}(\mathbf{y}_k)\right] \in \mathbb{R}^{Q \times 1}, \qquad (4.\mathrm{B}.5)$$

for $k = 1, 2, \ldots, K$. Moreover,

$$\begin{cases} \mathbf{y}_{erk} \sim \mathcal{N}_Q(\mathbf{0}, \mathbf{R}), \\ \mathbf{y}_{ork} \sim \mathcal{N}_Q(\mathbf{0}, \mathbf{R}), \end{cases} \qquad (4.\mathrm{B}.6)$$

where $k = 1, 2, \ldots, K$, they are independent of each other, and

$$\mathbf{R} = \frac{1}{2}\mathbf{D}\mathbf{R}_p\mathbf{D}^{\dagger} = \frac{1}{2}\left[\Re\mathfrak{e}(\mathbf{R}_p) + \mathbf{J}\Im\mathfrak{m}(\mathbf{R}_p)\right] \in \mathbb{R}^{Q \times Q}. \qquad (4.\mathrm{B}.7)$$

That is to say, $\mathbf{Y} \sim \mathcal{N}_{Q \times 2K}(\mathbf{0}, \mathbf{R} \otimes \mathbf{I}_{2K})$. According to [17, eq. (B11)], we have

$$\mathbf{M} = \mathbf{Y}\mathbf{Y}^T. \qquad (4.\mathrm{B}.8)$$

Thus,

$$\mathbf{M} \sim \mathcal{W}_Q(2K, \mathbf{R}). \qquad (4.\mathrm{B}.9)$$

Similar to (4.B.6), we have

$$\begin{cases} \mathbf{x}_{er} \sim \mathcal{N}_Q(\mathbf{0}, \mathbf{R}), \\ \mathbf{x}_{or} \sim \mathcal{N}_Q(\mathbf{0}, \mathbf{R}). \end{cases} \tag{4.B.10}$$

It follows from (4.35) and (4.B.10) that under H_0,

$$\mathbf{X} \sim \mathcal{N}_{Q \times 2}(\mathbf{0}, \mathbf{R} \otimes \mathbf{I}_2). \tag{4.B.11}$$

Until now we have derived the distributions of $\mathbf{M}$ and $\mathbf{X}$ as shown in (4.B.9) and (4.B.11), respectively. In summary, (4.38) holds true.

4.C   Derivations of (4.62)

Here we aim to derive the distributions of $\mathbf{G}$ and $\bar{\mathbf{M}}_{11}$. Let us begin by writing

$$\mathbf{U}^T \mathbf{R} \mathbf{U} = \begin{bmatrix} \mathbf{R}_{11} & \mathbf{R}_{12} \\ \mathbf{R}_{21} & \mathbf{R}_{22} \end{bmatrix} \tag{4.C.1}$$

and

$$\mathbf{U}^T \mathbf{R}^{-1} \mathbf{U} = \begin{bmatrix} \bar{\mathbf{R}}_{11} & \bar{\mathbf{R}}_{12} \\ \bar{\mathbf{R}}_{21} & \bar{\mathbf{R}}_{22} \end{bmatrix}, \tag{4.C.2}$$

where $\mathbf{R}_{11} \in \mathbb{R}^{q \times q}$, $\mathbf{R}_{12} \in \mathbb{R}^{q \times (Q-q)}$, $\mathbf{R}_{21} \in \mathbb{R}^{(Q-q) \times q}$, $\mathbf{R}_{22} \in \mathbb{R}^{(Q-q) \times (Q-q)}$, $\bar{\mathbf{R}}_{11} \in \mathbb{R}^{q \times q}$, $\bar{\mathbf{R}}_{12} \in \mathbb{R}^{q \times (Q-q)}$, $\bar{\mathbf{R}}_{21} \in \mathbb{R}^{(Q-q) \times q}$, and $\bar{\mathbf{R}}_{22} \in \mathbb{R}^{(Q-q) \times (Q-q)}$. It is straightforward to show that

$$\begin{bmatrix} \mathbf{R}_{11} & \mathbf{R}_{12} \\ \mathbf{R}_{21} & \mathbf{R}_{22} \end{bmatrix}^{-1} = \begin{bmatrix} \bar{\mathbf{R}}_{11} & \bar{\mathbf{R}}_{12} \\ \bar{\mathbf{R}}_{21} & \bar{\mathbf{R}}_{22} \end{bmatrix}, \tag{4.C.3}$$

and

$$\bar{\mathbf{R}}_{11}^{-1} = \mathbf{R}_{11} - \mathbf{R}_{12} \mathbf{R}_{22}^{-1} \mathbf{R}_{21}. \tag{4.C.4}$$

According to [21, p. 262, Theorem 7.3.6], we have

$$\bar{\mathbf{M}}_{11} \sim \mathcal{W}_q(2K - Q + q, \bar{\mathbf{R}}_{11}^{-1}). \tag{4.C.5}$$

Define

$$\begin{cases} \mathbf{Y}_1 \triangleq \mathbf{E}^T \mathbf{Y} \in \mathbb{R}^{q \times 2K}, \\ \mathbf{Y}_2 \triangleq \mathbf{F}^T \mathbf{Y} \in \mathbb{R}^{(Q-q) \times 2K}, \end{cases} \tag{4.C.6}$$

and we have

$$\begin{bmatrix} \mathbf{Y}_1 \\ \mathbf{Y}_2 \end{bmatrix} = \mathbf{U}^T \mathbf{Y}. \tag{4.C.7}$$

Then,

$$\mathbf{U}^T\mathbf{M}\mathbf{U} = \mathbf{U}^T\mathbf{Y}\mathbf{Y}^T\mathbf{U} = \begin{bmatrix} \mathbf{Y}_1\mathbf{Y}_1^T & \mathbf{Y}_1\mathbf{Y}_2^T \\ \mathbf{Y}_2\mathbf{Y}_1^T & \mathbf{Y}_2\mathbf{Y}_2^T \end{bmatrix}, \tag{4.C.8}$$

where we have used (4.B.8) and (4.C.7). Comparing (4.49) and (4.C.8), we can obtain that

$$\mathbf{M}_{12} = \mathbf{Y}_1\mathbf{Y}_2^T, \tag{4.C.9}$$

and

$$\mathbf{M}_{22} = \mathbf{Y}_2\mathbf{Y}_2^T. \tag{4.C.10}$$

Substituting (4.C.9) into (4.57) leads to

$$\mathbf{Z} = \mathbf{X}_1 - \mathbf{Y}_1\mathbf{Y}_2^T\mathbf{M}_{22}^{-1}\mathbf{X}_2. \tag{4.C.11}$$

Due to (4.B.11), the PDF of $\mathbf{X}$ can be written as

$$f(\mathbf{X}) = \frac{1}{(2\pi)^Q \det(\mathbf{R})} \exp\left[-\frac{1}{2}\operatorname{tr}\left(\mathbf{X}^T\mathbf{R}^{-1}\mathbf{X}\right)\right]. \tag{4.C.12}$$

According to [19, p. 188, Theorem 13.3.8], we have

$$\begin{aligned} \det(\mathbf{R}) &= \det(\mathbf{R}_{22}) \det(\mathbf{R}_{11} - \mathbf{R}_{12}\mathbf{R}_{22}^{-1}\mathbf{R}_{21}) \\ &= \det(\mathbf{R}_{22}) \det(\bar{\mathbf{R}}_{11}^{-1}). \end{aligned} \tag{4.C.13}$$

It follows from (4.44) that

$$\begin{aligned} \mathbf{X}^T\mathbf{R}^{-1}\mathbf{X} &= \begin{bmatrix} \mathbf{X}_1^T, \mathbf{X}_2^T \end{bmatrix} \mathbf{U}^T\mathbf{R}^{-1}\mathbf{U} \begin{bmatrix} \mathbf{X}_1 \\ \mathbf{X}_2 \end{bmatrix} \\ &= \begin{bmatrix} \mathbf{X}_1^T, \mathbf{X}_2^T \end{bmatrix} \begin{bmatrix} \bar{\mathbf{R}}_{11} & \bar{\mathbf{R}}_{12} \\ \bar{\mathbf{R}}_{21} & \bar{\mathbf{R}}_{22} \end{bmatrix} \begin{bmatrix} \mathbf{X}_1 \\ \mathbf{X}_2 \end{bmatrix} \\ &= (\mathbf{X}_1 - \bar{\mathbf{X}}_1)^T\bar{\mathbf{R}}_{11}(\mathbf{X}_1 - \bar{\mathbf{X}}_1) + \mathbf{X}_2^T\mathbf{R}_{22}^{-1}\mathbf{X}_2, \end{aligned} \tag{4.C.14}$$

where the second and third equations are obtained using (4.C.2) and [20, p. 135, eq. (A1-9)], respectively, and $\bar{\mathbf{X}}_1 = \mathbf{R}_{12}\mathbf{R}_{22}^{-1}\mathbf{X}_2$. Substituting (4.C.13) and (4.C.14) into (4.C.12) yields

$$f(\mathbf{X}) = f(\mathbf{X}_1|\mathbf{X}_2)f(\mathbf{X}_2), \tag{4.C.15}$$

where

$$f(\mathbf{X}_1|\mathbf{X}_2) = \frac{\exp\left\{-\frac{1}{2}\operatorname{tr}\left[\bar{\mathbf{R}}_{11}(\mathbf{X}_1 - \bar{\mathbf{X}}_1)(\mathbf{X}_1 - \bar{\mathbf{X}}_1)^T\right]\right\}}{(2\pi)^q \det(\bar{\mathbf{R}}_{11}^{-1})}, \tag{4.C.16}$$

and

$$f(\mathbf{X}_2) = \frac{1}{(2\pi)^{Q-q} \det(\mathbf{R}_{22})} \exp\left[-\frac{1}{2}\operatorname{tr}\left(\mathbf{R}_{22}^{-1}\mathbf{X}_2\mathbf{X}_2^T\right)\right]. \tag{4.C.17}$$

We fix temporarily the two-components of the data. It can be seen from (4.C.16) that the PDF of $\mathbf{X}_1$ conditioned on $\mathbf{X}_2$ is

$$\mathbf{X}_1 \sim \mathcal{N}_{q \times 2}(\bar{\mathbf{X}}_1, \bar{\mathbf{R}}_{11}^{-1} \otimes \mathbf{I}_2). \qquad (4.C.18)$$

That means that the covariance matrix of $\mathbf{X}_1$ conditioned on the two-components is

$$\text{Cov}_2(\mathbf{X}_1) = \bar{\mathbf{R}}_{11}^{-1} \otimes \mathbf{I}_2, \qquad (4.C.19)$$

where $\bar{\mathbf{R}}_{11}^{-1}$ is given in (4.C.4). Similarly, we can derive the covariance matrix of $\mathbf{Y}_1$ conditioned on the two-components as

$$\text{Cov}_2(\mathbf{Y}_1) = \bar{\mathbf{R}}_{11}^{-1} \otimes \mathbf{I}_{2K}. \qquad (4.C.20)$$

From (4.C.11), we can obtain the covariance matrix of $\mathbf{Z}$ conditioned on the two-components as

$$
\begin{aligned}
\text{Cov}_2(\mathbf{Z}) &= \text{Cov}_2(\mathbf{X}_1) + \text{Cov}_2(\mathbf{Y}_1)\mathbf{Y}_2^T\mathbf{M}_{22}^{-1}\mathbf{X}_2 \\
&= \bar{\mathbf{R}}_{11}^{-1} \otimes \mathbf{I}_2 + \bar{\mathbf{R}}_{11}^{-1} \otimes (\mathbf{X}_2^T\mathbf{M}_{22}^{-1}\mathbf{Y}_2\mathbf{Y}_2^T\mathbf{M}_{22}^{-1}\mathbf{X}_2) \\
&= \bar{\mathbf{R}}_{11}^{-1} \otimes \mathbf{I}_2 + \bar{\mathbf{R}}_{11}^{-1} \otimes (\mathbf{X}_2^T\mathbf{M}_{22}^{-1}\mathbf{X}_2) \\
&= \bar{\mathbf{R}}_{11}^{-1} \otimes \boldsymbol{\Delta},
\end{aligned}
\qquad (4.C.21)
$$

where the third and last lines are obtained using (4.C.10) and (4.59), respectively. Using (4.60) and (4.C.21), we can obtain that under H_0,

$$\mathbf{G} \sim \mathcal{N}_{q \times 2}(\mathbf{0}, \bar{\mathbf{R}}_{11}^{-1} \otimes \mathbf{I}_2). \qquad (4.C.22)$$

Combining (4.C.5) and (4.C.22) achieves (4.62).

Bibliography

[1] F. Bandiera, D. Orlando, and G. Ricci, *Advanced Radar Detection Schemes under Mismatched Signal Models in Synthesis Lectures on Signal Processing*. San Rafael, CA, USA. Morgan & Claypool, 2009.

[2] L. L. Scharf and B. Friedlander, "Matched subspace detectors," *IEEE Transactions on Signal Processing*, vol. 42, no. 8, pp. 2146–2157, August 1994.

[3] D. Ciuonzo, A. De Maio, and D. Orlando, "On the statistical invariance for adaptive radar detection in partially homogeneous disturbance plus structured interference," *IEEE Transaction on Signal Processing*, vol. 5, no. 65, pp. 1222–1234, March 1 2017.

[4] F. Gini and A. Farina, "Vector subspace detection in compound-Gaussian clutter. part I: survey and new results," *IEEE Transactions on Aerospace and Electronic Systems*, vol. 38, no. 4, pp. 1295–1311, October 2002.

[5] J. Liu, W. Liu, B. Chen, H. Liu, H. Li, and C. Hao, "Modified Rao test for multichannel adaptive signal detection," *IEEE Transactions on Signal Processing*, vol. 64, no. 3, pp. 714–725, February 1 2016.

[6] S. Kraut, L. L. Scharf, and L. T. McWhorter, "Adaptive subspace detectors," *IEEE Transactions on Signal Processing*, vol. 49, no. 1, pp. 1–16, January 2001.

[7] O. Besson, L. L. Scharf, and F. Vincent, "Matched direction detectors and estimators for array processing with subspace steering vector uncertainties," *IEEE Transactions on Signal Processing*, vol. 53, no. 12, pp. 4453–4463, December 2005.

[8] J. Liu, Z.-J. Zhang, Y. Yang, and H. Liu, "A CFAR adaptive subspace detector for first-order or second-order Gaussian signals based on a single observation," *IEEE Transactions on Signal Processing*, vol. 59, no. 11, pp. 5126–5140, November 2011.

[9] W. Liu, W. Xie, J. Liu, and Y. Wang, "Adaptive double subspace signal detection in Gaussian background–Part I: Homogeneous environments," *IEEE Transactions on Signal Processing*, vol. 62, no. 9, pp. 2345–2357, May 2014.

[10] ——, "Adaptive double subspace signal detection in Gaussian background–Part II: Partially homogeneous environments," *IEEE Transactions on Signal Processing*, vol. 62, no. 9, pp. 2358–2369, May 2014.

[11] J. Liu, S. Sun, and W. Liu, "One-step persymmetric GLRT for subspace signals," *IEEE Transaction on Signal Processing*, vol. 14, no. 67, pp. 3639–3648, July 15 2019.

[12] R. S. Raghavan, N. Pulsone, and D. J. McLaughlin, "Performance of the GLRT for adaptive vector subspace detection," *IEEE Transactions on Aerospace and Electronic Systems*, vol. 32, no. 4, pp. 1473–1487, October 1996.

[13] D. Pastina, P. Lombardo, and T. Bucciarelli, "Adaptive polarimetric target detection with coherent radar Part I: Detection against gaussian background," *IEEE Transactions on Aerospace and Electronic Systems*, vol. 37, no. 4, pp. 1194–1206, October 2001.

[14] A. De Maio and G. Ricci, "A polarimetric adaptive matched filter," *Signal Processing*, vol. 81, no. 12, pp. 2583–2589, December 2001.

[15] J. Liu, Z.-J. Zhang, and Y. Yun, "Optimal waveform design for generalized likelihood ratio and adaptive matched filter detectors using a diversely polarized antenna," *Signal Processing*, vol. 92, no. 4, pp. 1126–1131, April 2012.

[16] F. Gini and A. Farina, "Vector subspace detection in compound-Gaussian clutter. part II: performance analysis," *IEEE Transactions on Aerospace and Electronic Systems*, vol. 38, no. 4, pp. 1312–1323, October 2002.

[17] L. Cai and H. Wang, "A persymmetric multiband GLR algorithm," *IEEE Transactions on Aerospace and Electronic Systems*, vol. 28, no. 3, pp. 806–816, July 1992.

[18] A. De Maio, D. Orlando, C. Hao, and G. Foglia, "Adaptive detection of point-like targets in spectrally symmetric interference," *IEEE Transactions on Signal Processing*, vol. 64, no. 12, pp. 3207–3220, December 2016.

[19] D. A. Harville, *Matrix Algebra for a Statistician's Perspective*. New York: Springer-Verlag, 1997.

[20] E. J. Kelly and K. Forsythe, "Adaptive detection and parameter estimation for multidimensional signal models," Lincoln Laboratory, MIT, Technical Report 848, 1989.

[21] T. Anderson, *An Introduction to Multivariate Statistical Analysis*, 3rd ed. New York, USA: Wiley, 2003.

[22] A. M. Mathai and P. N. Rathie, "The exact distribution of Wilks' criterion," *The Annals of Mathematical Statistics*, vol. 42, no. 3, pp. 1010–1019, 1971.

5

Persymmetric Detectors with Enhanced Rejection Capabilities

In radar applications, the region to be searched will be scanned by successively pointing the beam in various directions. The goal of the search is not only to detect a target but to report its approximate direction, which is the beam pointing direction. The search must be considered unsuccessful if the target is "detected" while the beam is pointing elsewhere. This is because realistic radar beams have "sidelobes", which are directions with relatively high gain outside the "main" beam, and a strong target can sometimes trigger a "detection" when it is located in a sidelobe direction, therefore appearing just like a somewhat weaker target in the main beam. This is a false alarm, with system consequences not very different from reporting a target when there is only noise [1].

With the above remark in mind, in this chapter, we exploit, at the design stage, the persymmetry of the noise covariance matrix to design two selective adaptive receivers capable of rejecting with high probability signals whose signatures are unlikely to correspond to that of interest [2–5]. To this end, we apply the design criterion of Rao test (results not reported here highlight that the Wald test design criterion for the same problem leads to a not selective receiver. For this reason, it has not been considered), which can lead to selective receivers (see, for instance, [6,7]) and the GLRT assuming that under the null hypothesis the collected returns contain a signal component orthogonal to the nominal one [1,8,9].

5.1 Problem Formulation

The detection problem at hand can be formulated as a conventional binary hypothesis testing problem

$$H_0 : \begin{cases} \mathbf{y} = \mathbf{n}, \\ \mathbf{y}_k = \mathbf{n}_k, \ k = 2,\ldots,K+1, \end{cases} \tag{5.1a}$$

DOI: 10.1201/9781003340232-5

$$H_1 : \begin{cases} \mathbf{y} = a\,\mathbf{p} + \mathbf{n}, \\ \mathbf{y}_k = \mathbf{n}_k, \;\; k = 2,\ldots,K+1, \end{cases} \tag{5.1b}$$

where $\mathbf{y} = \mathbf{y}_1 \in \mathbb{C}^{N\times 1}$ denotes the primary data, while $\mathbf{Y}_1 = [\mathbf{y}_2, \mathbf{y}_3, \ldots, \mathbf{y}_{K+1}] \in \mathbb{C}^{N\times K}$ denotes the secondary data. The complex-valued scalar a is an unknown quantity accounting for the target reflectivity and channel propagation effects, $\mathbf{p} \in \mathbb{C}^{N\times 1}$ is the target signal steering vector. Assume that the clutter vectors $\{\mathbf{n}, \mathbf{n}_k\}$ are independent of each other, and each has a circularly symmetric, complex Gaussian distribution with zero-mean and unknown covariance matrix $\mathbf{R} \in \mathbb{C}^{N\times N}$, i.e., $\mathbf{n}, \mathbf{n}_k \sim \mathcal{CN}_N(\mathbf{0}, \mathbf{R})$ for $k = 2,\ldots,K+1$.

Now, we are interested in deciding whether or not $\mathbf{y}$ contains useful target components. In this respect, we modify the H_0 hypothesis so that the data under test contain a signal component orthogonal to the nominal target signature, that is

$$H_0 : \begin{cases} \mathbf{y} = \mathbf{p}_\perp + \mathbf{n}, \\ \mathbf{y}_k = \mathbf{n}_k, \;\; k = 2,\ldots,K+1, \end{cases} \tag{5.2a}$$

$$H_1 : \begin{cases} \mathbf{y} = a\,\mathbf{p} + \mathbf{n}, \\ \mathbf{y}_k = \mathbf{n}_k, \;\; k = 2,\ldots,K+1, \end{cases} \tag{5.2b}$$

where $\mathbf{p}_\perp \in \mathbb{C}^{N\times 1}$ is a fictitious signal vector orthogonal to $\mathbf{p}$ in the whitened observation space such that $\mathbf{p}_\perp^\dagger \mathbf{R}^{-1}\mathbf{p} = 0$.

In addition, we assume that $\mathbf{R}$ is a (positive-definite Hermitian) persymmetric matrix, namely, $\mathbf{R} = \mathbf{J}\mathbf{R}^*\mathbf{J}$ with $\mathbf{J}$ the permutation matrix, and the nominal steering vector is a persymmetric vector, namely, $\mathbf{p} = \mathbf{J}\mathbf{p}^*$. Generally speaking, for such a system configuration, we expect that received data exhibit a preassigned symmetry and, hence, we assume also that the fictitious signal under H_0 has a persymmetric structure, namely, $\mathbf{p}_\perp = \mathbf{J}\mathbf{p}_\perp^*$.

It follows that the PDF of $\mathbf{y}$ and $\mathbf{Y}_1$ can be written as

$$\begin{cases} f_0(\mathbf{y}, \mathbf{Y}_1 | \mathbf{w}, \mathbf{R}) = \frac{1}{[\pi^N \det(\mathbf{R})]^{K+1}} \exp\left[-\operatorname{tr}\left(\mathbf{R}^{-1}\mathbf{T}_0\right) \right], \\ f_1(\mathbf{y}, \mathbf{Y}_1 | a, \mathbf{R}) = \frac{1}{[\pi^N \det(\mathbf{R})]^{K+1}} \exp\left[-\operatorname{tr}\left(\mathbf{R}^{-1}\mathbf{T}_1\right) \right], \end{cases} \tag{5.3}$$

where

$$\begin{cases} \mathbf{T}_0 = \mathbf{Y}_1\mathbf{Y}_1^\dagger + (\mathbf{y} - \mathbf{w})(\mathbf{y} - \mathbf{w})^\dagger, \\ \mathbf{T}_1 = \mathbf{Y}_1\mathbf{Y}_1^\dagger + (\mathbf{y} - a\mathbf{p})(\mathbf{y} - a\mathbf{p})^\dagger, \end{cases} \tag{5.4}$$

with $\mathbf{w} = \mathbf{0}$ for problem (5.1) and $\mathbf{w} = \mathbf{p}_\perp$ for problem (5.2).

Under the above assumptions, we solve the conventional detection problem (5.1) by applying the Rao test design criterion, whereas the modified problem (5.2) is solved resorting to the GLRT design criterion.

5.2 Detector Design

As a preliminary step toward the derivation of the receivers, observing that exploiting the persymmetric properties, we have

$$\begin{cases} f_0(\mathbf{y}, \mathbf{Y}_1 | \mathbf{w}, \mathbf{R}) = \frac{1}{[\pi^N \det(\mathbf{R})]^{K+1}} \exp\left[-\operatorname{tr}\left(\mathbf{R}^{-1}\mathbf{T}_{p0}\right)\right], \\ f_1(\mathbf{y}, \mathbf{Y}_1 | \mathbf{a}, \mathbf{R}) = \frac{1}{[\pi^N \det(\mathbf{R})]^{K+1}} \exp\left[-\operatorname{tr}\left(\mathbf{R}^{-1}\mathbf{T}_{p1}\right)\right], \end{cases} \tag{5.5}$$

where

$$\begin{cases} \mathbf{T}_{p0} = \hat{\mathbf{R}} + (\mathbf{Y}_p - \mathbf{w}_p)(\mathbf{Y}_p - \mathbf{w}_p)^\dagger, \\ \mathbf{T}_{p1} = \hat{\mathbf{R}} + (\mathbf{Y}_p - \mathbf{p}\mathbf{a}_p^T)(\mathbf{Y} - \mathbf{p}\mathbf{a}_p^T)^\dagger, \end{cases} \tag{5.6}$$

$$\hat{\mathbf{R}} = \frac{1}{2}\left[\mathbf{Y}_1\mathbf{Y}_1^\dagger + \mathbf{J}\left(\mathbf{Y}_1\mathbf{Y}_1^\dagger\right)^*\mathbf{J}\right] \in \mathbb{C}^{N \times N}, \tag{5.7}$$

$$\mathbf{Y}_p = [\mathbf{y}_e, \mathbf{y}_o] \in \mathbb{C}^{N \times 2}, \tag{5.8}$$

$$\mathbf{a}_p = [a_e, a_o]^T \in \mathbb{C}^{2 \times 1}, \tag{5.9}$$

$$\mathbf{w}_p = [\mathbf{w}, \ \mathbf{0}] \in \mathbb{C}^{N \times 2}, \tag{5.10}$$

with

$$\begin{cases} \mathbf{y}_e = \frac{1}{2}\left(\mathbf{y} + \mathbf{J}\mathbf{y}^*\right), \\ \mathbf{y}_o = \frac{1}{2}\left(\mathbf{y} - \mathbf{J}\mathbf{y}^*\right), \end{cases} \tag{5.11}$$

and

$$\begin{cases} a_e = \frac{1}{2}(a + a^*) = \mathfrak{Re}(a), \\ a_o = \frac{1}{2}(a - a^*) = \jmath\,\mathfrak{Im}(a). \end{cases} \tag{5.12}$$

5.2.1 Persymmetric Rao Test

Denoting by $\boldsymbol{\theta}_r = [a_e, a_i]^T \in \mathbb{R}^{2 \times 1}$ the signal parameter vector, with $a_e = \mathfrak{Re}(a)$ and $a_i = \mathfrak{Im}(a)$, $\boldsymbol{\theta}_s = \mathbf{g}(\mathbf{R})$ the nuisance parameter vector, with $\mathbf{g}(\mathbf{R})$ an N^2 dimensional vector that contains in univocal way the elements of $\mathbf{R}$, and $\boldsymbol{\theta} = \left[\boldsymbol{\theta}_r^T, \boldsymbol{\theta}_s^T\right]^T$ the overall unknown parameters, $\mathbf{F}(\boldsymbol{\theta})$ is the FIM which can be formulated as

$$\mathbf{F}(\boldsymbol{\theta}) = \begin{bmatrix} \mathbf{F}_{rr}(\boldsymbol{\theta}) & \mathbf{F}_{rs}(\boldsymbol{\theta}) \\ \mathbf{F}_{sr}(\boldsymbol{\theta}) & \mathbf{F}_{ss}(\boldsymbol{\theta}) \end{bmatrix}, \tag{5.13}$$

where

$$\mathbf{F}_{ab}(\boldsymbol{\theta}) = -\mathrm{E}\left[\frac{\partial^2 \ln f_1(\mathbf{y}, \mathbf{Y}_1 | \boldsymbol{\theta})}{\partial \boldsymbol{\theta}_a \partial \boldsymbol{\theta}_b^T}\right]. \tag{5.14}$$

The inverse of $\mathbf{F}(\boldsymbol{\theta})$ can be written as

$$\mathbf{F}^{-1}(\boldsymbol{\theta}) = \begin{bmatrix} \mathbf{J}_{rr}(\boldsymbol{\theta}) & \mathbf{J}_{rs}(\boldsymbol{\theta}) \\ \mathbf{J}_{sr}(\boldsymbol{\theta}) & \mathbf{J}_{ss}(\boldsymbol{\theta}) \end{bmatrix}, \tag{5.15}$$

where

$$\mathbf{J}_{rr}(\boldsymbol{\theta}) = \left[\mathbf{F}_{rr}(\boldsymbol{\theta}) - \mathbf{F}_{rs}(\boldsymbol{\theta})\mathbf{F}_{ss}^{-1}(\boldsymbol{\theta})\mathbf{F}_{sr}(\boldsymbol{\theta})\right]^{-1} \qquad (5.16)$$

is the subblock of the inverse of the FIM formed by selecting its first two rows and the first two columns.

The Rao test for the problem at hand can be written as follows:

$$\Lambda_0 = \left[\frac{\partial \ln f_1(\mathbf{y}, \mathbf{Y}_1|\boldsymbol{\theta})}{\partial \boldsymbol{\theta}_r}\right]^T_{\boldsymbol{\theta}=\hat{\boldsymbol{\theta}}_0} \mathbf{J}_{rr}(\boldsymbol{\theta})\Big|_{\boldsymbol{\theta}=\hat{\boldsymbol{\theta}}_0} \left[\frac{\partial \ln f_1(\mathbf{y}, \mathbf{Y}_1|\boldsymbol{\theta})}{\partial \boldsymbol{\theta}_r}\right]_{\boldsymbol{\theta}=\hat{\boldsymbol{\theta}}_0} \mathop{\gtrless}^{H_1}_{H_0} \eta_0, \qquad (5.17)$$

where η_0 is the detection threshold to be set in order to ensure the preassigned PFA, and $\hat{\boldsymbol{\theta}}_0$ is the estimate of $\boldsymbol{\theta}$ under H_0.

It can be shown that [6, 10]

$$\mathbf{J}_{rr}(\boldsymbol{\theta}) = (2\mathbf{I}_2\mathbf{p}^\dagger\mathbf{R}^{-1}\mathbf{p})^{-1}, \qquad (5.18)$$

$$\frac{\partial \ln f_1(\mathbf{y}, \mathbf{Y}_1|\boldsymbol{\theta})}{\partial \boldsymbol{\theta}_r} = \left[\begin{array}{c} 2\Re\{\mathbf{p}^\dagger\mathbf{R}^{-1}(\mathbf{y} - a\mathbf{p})\} \\ 2\Im\{\mathbf{p}^\dagger\mathbf{R}^{-1}(\mathbf{y} - a\mathbf{p})\} \end{array}\right]. \qquad (5.19)$$

Gathering all the above results and exploiting persymmetry, the Rao test can be written as

$$\Lambda = \frac{\left|\mathbf{p}^\dagger(\hat{\mathbf{R}} + \mathbf{Y}_p\mathbf{Y}_p^\dagger)\mathbf{y}\right|^2}{\mathbf{p}^\dagger(\hat{\mathbf{R}} + \mathbf{Y}_p\mathbf{Y}_p^\dagger)\mathbf{p}} \mathop{\gtrless}^{H_1}_{H_0} \eta, \qquad (5.20)$$

which is referred to as persymmetric Rao test for homogeneous environment (P-RAO-HE).

5.2.2 Persymmetric GLRT

For the problem (5.2), the GLRT based on both the CUT and secondary data can be expressed as

$$\Xi_0 = \frac{\max_{\{a,\mathbf{R}\}} f_1(\mathbf{y}, \mathbf{Y}_1|a, \mathbf{R})}{\max_{\{\mathbf{p}_\perp, \mathbf{R}\}} f_0(\mathbf{y}, \mathbf{Y}_1|\mathbf{p}_\perp, \mathbf{R})} \mathop{\gtrless}^{H_1}_{H_0} \xi_0, \qquad (5.21)$$

where ξ_0 is the threshold to be set according to the desired PFA. In order to maximize $f_0(\mathbf{y}, \mathbf{Y}_1|\mathbf{p}_\perp, \mathbf{R})$ with respect to $\mathbf{p}_\perp$, we note that

$$\text{tr}\left[\mathbf{R}^{-1}(\mathbf{Y}_p - \mathbf{w}_p)(\mathbf{Y}_p - \mathbf{w}_p)^\dagger\right] = (\mathbf{y}_e - \mathbf{p}_\perp)^\dagger\mathbf{R}^{-1}(\mathbf{y}_e - \mathbf{p}_\perp) + \mathbf{y}_o^\dagger\mathbf{R}^{-1}\mathbf{y}_o. \qquad (5.22)$$

It follows that

$$\begin{aligned} \hat{\mathbf{p}}_\perp &= \arg\min_{\mathbf{p}_\perp}\left\{\text{tr}\left[\mathbf{R}^{-1}(\mathbf{Y}_p - \mathbf{w}_p)(\mathbf{Y}_p - \mathbf{w}_p)^\dagger\right]\right\} \\ &= \arg\min_{\mathbf{p}_\perp}\left[(\mathbf{y}_e - \mathbf{p}_\perp)^\dagger\mathbf{R}^{-1}(\mathbf{y}_e - \mathbf{p}_\perp)\right], \end{aligned} \qquad (5.23)$$

and

$$\begin{aligned} &\min_{\mathbf{p}_\perp}\left[(\mathbf{y}_e - \mathbf{p}_\perp)^\dagger\mathbf{R}^{-1}(\mathbf{y}_e - \mathbf{p}_\perp)\right] \\ &= \mathbf{y}_e^\dagger\mathbf{R}^{-1}\mathbf{y}_e - \mathbf{y}_e^\dagger\mathbf{R}^{-1/2}\mathbf{P}_{\mathbf{R}^{-1/2}\mathbf{p}}^\perp\mathbf{R}^{-1/2}\mathbf{y}_e, \end{aligned} \qquad (5.24)$$

where $\mathbf{P}^{\perp}_{\mathbf{R}^{-1/2}\mathbf{p}} = \mathbf{I}_N - \mathbf{P}_{\mathbf{R}^{-1/2}\mathbf{p}}$, with $\mathbf{P}_{\mathbf{R}^{-1/2}\mathbf{p}} = \mathbf{R}^{-1/2}\mathbf{p}(\mathbf{p}^{\dagger}\mathbf{R}^{-1}\mathbf{p})^{-1}\mathbf{p}^{\dagger}\mathbf{R}^{-1/2}$ a projection matrix onto the space spanned by $\mathbf{R}^{-1/2}\mathbf{p}$ and $\mathbf{I}_N$ the N-dimensional identity matrix. Using the above result yields

$$f_0(\mathbf{y}, \mathbf{Y}_1|\hat{\mathbf{p}}_{\perp}, \mathbf{R}) = \frac{1}{[\pi^N \det(\mathbf{R})]^{K+1}} \exp\left\{\mathrm{tr}\left[-\mathbf{R}^{-1}\hat{\mathbf{R}}_p\right]\right\}$$
$$\times \exp\left\{\mathbf{y}_e^{\dagger}\mathbf{R}^{-1/2}\mathbf{P}^{\perp}_{\mathbf{R}^{-1/2}\mathbf{p}}\mathbf{R}^{-1/2}\mathbf{y}_e\right\}, \qquad (5.25)$$

where $\hat{\mathbf{R}}_p = \hat{\mathbf{R}} + \mathbf{Y}_p\mathbf{Y}_p^{\dagger}$.

The maximization over $\mathbf{R}$ can be accomplished following the lead of [11]. More precisely, it is not difficult to show that setting to zero the derivative with respect to $\mathbf{R}$ of the natural logarithm of (5.25) leads to

$$(K+1)\mathbf{R} = \hat{\mathbf{R}}_p - \mathbf{R}^{1/2}\mathbf{P}^{\perp}_{\mathbf{p}_\omega}\mathbf{y}_{\omega e}\mathbf{y}_{\omega e}^{\dagger}\mathbf{P}^{\perp}_{\mathbf{p}_\omega}\mathbf{R}^{1/2}, \qquad (5.26)$$

which has a unique solution [11] given by

$$(K+1)\hat{\mathbf{R}}_0 = \hat{\mathbf{R}}_p - \hat{\mathbf{R}}_p^{1/2}\mathbf{P}^{\perp}_{\mathbf{p}_0}\mathbf{y}_{0e}\mathbf{y}_{0e}^{\dagger}\mathbf{P}^{\perp}_{\mathbf{p}_0}\hat{\mathbf{R}}_p^{1/2}, \qquad (5.27)$$

with $\mathbf{p}_\omega = \mathbf{R}^{-1/2}\mathbf{p}$, $\mathbf{y}_{\omega e} = \mathbf{R}^{-1/2}\mathbf{y}_e$, $\mathbf{p}_0 = \hat{\mathbf{R}}_p^{-1/2}\mathbf{p}$, and $\mathbf{y}_{0e} = \hat{\mathbf{R}}_p^{-1/2}\mathbf{y}_e$. Gathering all the results, the denominator of (5.21) can be written as

$$f_0(\mathbf{y}, \mathbf{Y}_1|\hat{\mathbf{p}}_{\perp}, \hat{\mathbf{R}}_0) = \left[(e\pi)^N \det\left(\hat{\mathbf{R}}_0\right)\right]^{-(K+1)}. \qquad (5.28)$$

The optimization problem under H_1 is equivalent to that solved in [8] assuming an extended target occupying $K_p = 2$ consecutive range cells, namely, $[\mathbf{y}_e\ \mathbf{y}_o]$. As a consequence, the compressed likelihood function under H_1 is given by

$$f_1(\mathbf{y}, \mathbf{Y}_1|\hat{a}, \hat{\mathbf{R}}_1) = \left[(e\pi)^N \det\left(\hat{\mathbf{R}}_1\right)\right]^{-(K+1)}, \qquad (5.29)$$

where

$$\hat{\mathbf{R}}_1 = \frac{1}{K+1}\left[\hat{\mathbf{R}} + \left(\mathbf{Y}_p - \mathbf{p}\frac{\mathbf{p}^{\dagger}\hat{\mathbf{R}}^{-1}\mathbf{Y}_p}{\mathbf{p}^{\dagger}\hat{\mathbf{R}}^{-1}\mathbf{p}}\right)\left(\mathbf{Y}_p - \mathbf{p}\frac{\mathbf{p}^{\dagger}\hat{\mathbf{R}}^{-1}\mathbf{Y}_p}{\mathbf{p}^{\dagger}\hat{\mathbf{R}}^{-1}\mathbf{p}}\right)^{\dagger}\right]. \qquad (5.30)$$

Finally, the GLRT can be written as

$$\Xi = \frac{\det\left(\hat{\mathbf{R}}_0\right)}{\det\left(\hat{\mathbf{R}}_1\right)} \underset{H_0}{\overset{H_1}{\gtrless}} \xi, \qquad (5.31)$$

which is referred to as persymmetric whitened adaptive beamformer orthogonal rejection test (P-W-ABORT) in the sequel. It is tedious but not difficult to show that the left-hand side of the above decision rule has the following expression (see the Appendix for further details)

$$
\frac{\det\left(\hat{\mathbf{R}}_0\right)}{\det\left(\hat{\mathbf{R}}_1\right)} = \left[\frac{\mathbf{e}_1^T \left(\mathbf{I}_2 + \mathbf{Y}_p^\dagger \hat{\mathbf{R}}^{-1} \mathbf{Y}_p\right)^{-1} \mathbf{Y}_p^\dagger \hat{\mathbf{R}}^{-1} \mathbf{p}}{\mathbf{p}^\dagger \hat{\mathbf{R}}^{-1} \mathbf{p} - \mathbf{p}^\dagger \hat{\mathbf{R}}^{-1} \mathbf{Y}_p \left(\mathbf{I}_2 + \mathbf{Y}_p^\dagger \hat{\mathbf{R}}^{-1} \mathbf{Y}_p\right)^{-1} \mathbf{Y}_p^\dagger \hat{\mathbf{R}}^{-1} \mathbf{p}} \right.
$$

$$
\times\, \mathbf{p}^\dagger \hat{\mathbf{R}}^{-1} \mathbf{Y}_p \left(\mathbf{I}_2 + \mathbf{Y}_p^\dagger \hat{\mathbf{R}}^{-1} \mathbf{Y}_p\right)^{-1} \mathbf{e}_1
$$

$$
\left. +\, \mathbf{e}_1^T \left(\mathbf{I}_2 + \mathbf{Y}_p^\dagger \hat{\mathbf{R}}^{-1} \mathbf{Y}_p\right)^{-1} \mathbf{e}_1 \right]
$$

$$
\times \left[\frac{1}{1 - \dfrac{\mathbf{p}^\dagger \hat{\mathbf{R}}^{-1} \mathbf{Y}_p \left(\mathbf{I}_2 + \mathbf{Y}_p^\dagger \hat{\mathbf{R}}^{-1} \mathbf{Y}_p\right)^{-1} \mathbf{Y}_p^\dagger \hat{\mathbf{R}}^{-1} \mathbf{p}}{\mathbf{p}^\dagger \hat{\mathbf{R}}^{-1} \mathbf{p}}} \right],
$$

where $\mathbf{e}_1 = [1, 0, \cdots, 0]^T \in \mathbb{R}^{N \times 1}$.

5.3   Numerical Examples

In this section, we analyze the performance of the P-RAO-HE and P-W-ABORT by means of numerical examples for both matched and mismatched signals. It is important to stress here that both of the detectors guarantee the CFAR property with respect to $\mathbf{R}$ (see [12] and references therein).

In the first part of this section, we compare the P-RAO-HE and the P-W-ABORT to the Rao test [6], the W-ABORT [8], Kelly's GLRT [13], which are designed without any assumption on the structure of the noise covariance matrix, and the PGLR introduced in [12]. Specifically, we evaluate the detection performance for matched signals. On the other hand, in the second part of this section, we focus on mismatched detection performance and compare the the P-RAO-HE and the P-W-ABORT to the PGLR. To this end, we resort to the so-called mesa plots [1]. The aim is to highlight in what extent proposed receivers can guarantee a trade-off between detection performance and rejection capabilities of unwanted signals. Since closed form expressions for both P_{fa} and P_d are not available, we resort to standard MC counting techniques. More precisely, in order to evaluate the threshold necessary to ensure a preassigned $P_{fa} = 0.001$ and the values of P_d, we resort to $100/P_{fa}$ and 10^4 independent trials, respectively. The SNR is defined as SNR $= |a|^2 \mathbf{v}_m(\theta_{az})^\dagger \mathbf{R}^{-1} \mathbf{v}_m(\theta_{az})$, where θ_{az} is the target azimuthal angle and

$$
\mathbf{v}_m(\theta_{az}) = \left[\exp\left(\jmath\pi \frac{d}{\lambda} (N-1)\cos\theta_{az} \right), \cdots, 1, \right.
$$
$$
\left. \cdots, \exp\left(-\jmath\pi \frac{d}{\lambda} (N-1)\cos\theta_{az} \right) \right] / \sqrt{N},
\tag{5.32}
$$

with d the element spacing, λ the working wavelength.

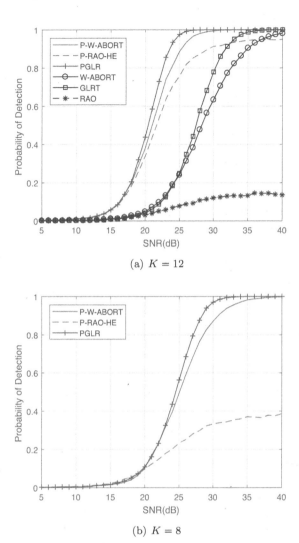

(a) $K = 12$

(b) $K = 8$

FIGURE 5.1
Detection probability versus SNR for $N = 10$.

We consider a clutter-dominated environment, and the (i, j)th element of the clutter covariance matrix of the primary data is chosen as $\mathbf{R}(i,j) = \sigma^2 0.9^{|i-j|}$, where σ^2 is the clutter power.

In Fig. 5.1, we plot P_d against SNR assuming $N = 10$ with different values of K. Inspection of Fig. 5.1(a) highlights that the PGLR achieves the best detection performance with a gain of about 1.6 dB at $P_d = 0.9$ with respect to the P-W-ABORT, which, in turn, outperforms the P-RAO-HE of about 4 dB (again at $P_d = 0.9$). The loss of the P-RAO-HE and P-W-ABORT is due to the fact that they are more selective than the PGLR as shown in the sequel. On the other hand, Kelly's detector and the W-ABORT exhibit a loss of about 8.2 and 10.6 dB, respectively, at $P_d = 0.9$ with respect to the P-W-ABORT. Observe that the performance of Rao test is very poor because of the lack of secondary data to estimate the noise covariance matrix. In Fig. 5.1(b), we plot P_d against SNR assuming $K < N$. In this case, we only consider three persymmetric detectors. It is seen that the P-W-ABORT ensures a good detection performance when $K < N$, but P-RAO-HE works poorly. Moreover, the best performance is still attained by the PGLR, whereas the P-W-ABORT experiences a loss of about 2.1 dB at $P_d = 0.9$.

The mismatched signal detection performance of the receivers is analyzed inspecting the contours of constant P_d represented as a function of $\cos^2 \phi$ plotted vertically, and the SNR plotted horizontally. Observe that reading the values of P_d on horizontal lines of a mesa plot returns the performance of the receiver in terms of P_d against SNR for a preassigned value of the mismatch angle ϕ. In Figs. 5.2 and 5.3, we compare the P-W-ABORT, the P-RAO-HE, and the PGLR for several values of N and K. From Fig. 5.2, it turns out that the P-RAO-HE is slightly more selective than the P-W-ABORT for SNR values greater than 30 dB ($K = 28$) or 36 dB ($K = 36$). The PGLR is less sensitive to mismatched signals than the other two detectors. Observe that the P-RAO-HE is asymptotically equivalent to the PGLR and, hence, when N (and/or K) increases the P-RAO-HE loses its selectivity (for a given SNR). This trend can be observed in Fig. 5.3, where the P-W-ABORT ensures the strongest performance in terms of rejection of mismatched signals for almost all of the illustrated values of SNR.

5.A   Derivations of (5.32)

Let $\mathbf{e}_1 = [1, 0, \cdots, 0]^T \in \mathbb{R}^{N \times 1}$. Plugging (5.27) and (5.30) into (5.31) yields

$$
\frac{\det\left(\hat{\mathbf{R}}_0\right)}{\det\left(\hat{\mathbf{R}}_1\right)} = \frac{\left[\det(\hat{\mathbf{R}})\right]^{-1} \det(\hat{\mathbf{R}}_p) \det\left(\mathbf{I}_N - \mathbf{P}_{\mathbf{p}_0}^{\perp} \mathbf{y}_{0e} \mathbf{y}_{0e}^{\dagger} \mathbf{P}_{\mathbf{p}_0}^{\perp}\right)}{\det\left[\mathbf{I}_N + \hat{\mathbf{R}}^{-1/2}\left(\mathbf{Y}_p - \mathbf{p}\frac{\mathbf{p}^{\dagger}\hat{\mathbf{R}}^{-1}\mathbf{Y}_p}{\mathbf{p}^{\dagger}\hat{\mathbf{R}}^{-1}\mathbf{p}}\right)\left(\mathbf{Y}_p - \mathbf{p}\frac{\mathbf{p}^{\dagger}\hat{\mathbf{R}}^{-1}\mathbf{Y}_p}{\mathbf{p}^{\dagger}\hat{\mathbf{R}}^{-1}\mathbf{p}}\right)^{\dagger}\hat{\mathbf{R}}^{-1/2}\right]}
$$

$$
= \frac{\det\left(\mathbf{I}_2 + \mathbf{Y}_p^{\dagger}\hat{\mathbf{R}}^{-1}\mathbf{Y}_p\right)\left(1 - \mathbf{y}_e^{\dagger}\hat{\mathbf{R}}_p^{-1/2}\mathbf{P}_{\mathbf{p}_0}^{\perp}\hat{\mathbf{R}}_p^{-1/2}\mathbf{y}_e\right)}{\det\left[\mathbf{I}_2 + \left(\mathbf{Y}_p - \mathbf{p}\frac{\mathbf{p}^{\dagger}\hat{\mathbf{R}}^{-1}\mathbf{Y}_p}{\mathbf{p}^{\dagger}\hat{\mathbf{R}}^{-1}\mathbf{p}}\right)^{\dagger}\hat{\mathbf{R}}^{-1}\left(\mathbf{Y}_p - \mathbf{p}\frac{\mathbf{p}^{\dagger}\hat{\mathbf{R}}^{-1}\mathbf{Y}_p}{\mathbf{p}^{\dagger}\hat{\mathbf{R}}^{-1}\mathbf{p}}\right)\right]}
$$

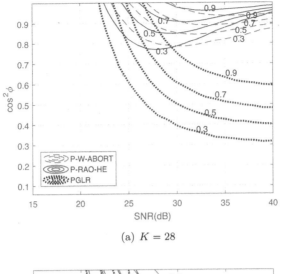

(a) $K = 28$

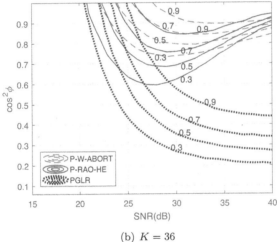

(b) $K = 36$

FIGURE 5.2
Contours for constant P_d, $N = 32$.

$$= \frac{\det\left(\mathbf{I}_2 + \mathbf{Y}_p^\dagger \hat{\mathbf{R}}^{-1} \mathbf{Y}_p\right)}{\det\left(\mathbf{I}_2 + \mathbf{Y}_p^\dagger \hat{\mathbf{R}}^{-1} \mathbf{Y}_p - \frac{\mathbf{Y}_p^\dagger \hat{\mathbf{R}}^{-1} \mathbf{p} \mathbf{p}^\dagger \hat{\mathbf{R}}^{-1} \mathbf{Y}_p}{\mathbf{p}^\dagger \hat{\mathbf{R}}^{-1} \mathbf{p}}\right)}$$
$$\times \left[1 + \mathbf{y}_e^\dagger \hat{\mathbf{R}}_p^{-1} \mathbf{p} \mathbf{p}^\dagger \hat{\mathbf{R}}_p^{-1} \mathbf{y}_e \left(\mathbf{p}^\dagger \hat{\mathbf{R}}_p^{-1} \mathbf{p}\right)^{-1} - \mathbf{y}_e^\dagger \hat{\mathbf{R}}_p^{-1} \mathbf{y}_e\right]$$

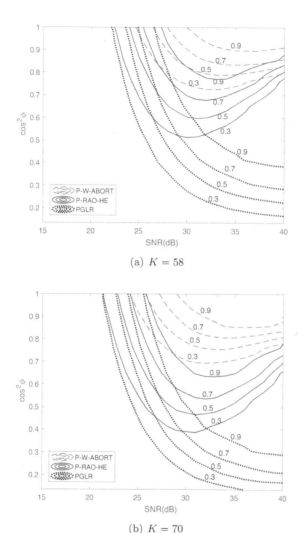

(a) $K = 58$

(b) $K = 70$

FIGURE 5.3
Contours for constant P_d, $N = 64$.

$$= \frac{\mathbf{e}_1^T \left[\mathbf{I}_2 - \mathbf{Y}_p^\dagger \hat{\mathbf{R}}_p^{-1} \mathbf{Y}_p + \dfrac{\mathbf{Y}_p^\dagger \hat{\mathbf{R}}_p^{-1} \mathbf{p} \mathbf{p}^\dagger \hat{\mathbf{R}}_p^{-1} \mathbf{Y}_p}{\mathbf{p}^\dagger \hat{\mathbf{R}}_p^{-1} \mathbf{p}} \right] \mathbf{e}_1}{1 - \dfrac{\mathbf{p}^\dagger \hat{\mathbf{R}}^{-1} \mathbf{Y}_p \left(\mathbf{I}_2 + \mathbf{Y}_p^\dagger \hat{\mathbf{R}}^{-1} \mathbf{Y}_p \right)^{-1} \mathbf{Y}_p^\dagger \hat{\mathbf{R}}^{-1} \mathbf{p}}{\mathbf{p}^\dagger \hat{\mathbf{R}}^{-1} \mathbf{p}}}$$

$$= \frac{\mathbf{e}_1^T \left[\mathbf{A} + \dfrac{\mathbf{B} \mathbf{Y}_p^\dagger \hat{\mathbf{R}}^{-1} \mathbf{p} \mathbf{p}^\dagger \hat{\mathbf{R}}^{-1} \mathbf{Y}_p \mathbf{B}^\dagger}{\mathbf{p}^\dagger \hat{\mathbf{R}}^{-1} \mathbf{p} - \mathbf{p}^\dagger \hat{\mathbf{R}}^{-1} \mathbf{Y}_p \left(\mathbf{I}_2 + \mathbf{Y}_p^\dagger \hat{\mathbf{R}}^{-1} \mathbf{Y}_p \right)^{-1} \mathbf{Y}_p^\dagger \hat{\mathbf{R}}^{-1} \mathbf{p}} \right] \mathbf{e}_1}{1 - \dfrac{\mathbf{p}^\dagger \hat{\mathbf{R}}^{-1} \mathbf{Y}_p \left(\mathbf{I}_2 + \mathbf{Y}_p^\dagger \hat{\mathbf{R}}^{-1} \mathbf{Y}_p \right)^{-1} \mathbf{Y}_p^\dagger \hat{\mathbf{R}}^{-1} \mathbf{p}}{\mathbf{p}^\dagger \hat{\mathbf{R}}^{-1} \mathbf{p}}} \, ,$$

where we have used the following identity

$$\det\left(\mathbf{I}_M + \mathbf{AB}^\dagger\right) = \det\left(\mathbf{I}_N + \mathbf{B}^\dagger\mathbf{A}\right), \quad \mathbf{A}, \mathbf{B} \in \mathbb{C}^{M\times N}, \qquad (5.A.1)$$

and

$$\begin{cases} \mathbf{A} = \mathbf{I}_2 - \mathbf{Y}_p^\dagger\hat{\mathbf{R}}^{-1}\mathbf{Y}_p + \mathbf{Y}_p^\dagger\hat{\mathbf{R}}^{-1}\mathbf{Y}_p \left(\mathbf{I}_2 + \mathbf{Y}_p^\dagger\hat{\mathbf{R}}^{-1}\mathbf{Y}_p\right)^{-1} \mathbf{Y}_p^\dagger\hat{\mathbf{R}}^{-1}\mathbf{Y}_p, \\ \mathbf{B} = \mathbf{I}_2 - \mathbf{Y}_p^\dagger\hat{\mathbf{R}}^{-1}\mathbf{Y}_p \left(\mathbf{I}_2 + \mathbf{Y}_p^\dagger\hat{\mathbf{R}}^{-1}\mathbf{Y}_p\right)^{-1}. \end{cases}$$

$$(5.A.2)$$

Now observe that the eigendecomposition of $\mathbf{Y}_p^\dagger\hat{\mathbf{R}}^{-1}\mathbf{Y}_p$ is

$$\mathbf{Y}_p^\dagger\hat{\mathbf{R}}^{-1}\mathbf{Y}_p = \mathbf{U}\mathbf{\Lambda}\mathbf{U}^\dagger, \qquad (5.A.3)$$

where $\mathbf{\Lambda} \in \mathbb{C}^{2\times 2}$ is a diagnol matrix and $\mathbf{U} \in \mathbb{C}^{2\times 2}$ is a unitary matrix. It follows that

$$\begin{aligned} \mathbf{A} &= \mathbf{I}_2 - \mathbf{U}\mathbf{\Lambda}\mathbf{U}^\dagger + \mathbf{U}\mathbf{\Lambda}\mathbf{U}^\dagger\left(\mathbf{I}_2 + \mathbf{U}\mathbf{\Lambda}\mathbf{U}^\dagger\right)^{-1}\mathbf{U}\mathbf{\Lambda}\mathbf{U}^\dagger \\ &= \mathbf{U}\left[\mathbf{I}_2 - \mathbf{\Lambda} + \mathbf{\Lambda}\left(\mathbf{I}_2 + \mathbf{\Lambda}\right)^{-1}\mathbf{\Lambda}\right]\mathbf{U}^\dagger \\ &= \mathbf{U}\left[\left(\mathbf{I}_2 + \mathbf{\Lambda}\right)^{-1}\right]\mathbf{U}^\dagger \\ &= \left(\mathbf{I}_2 + \mathbf{Y}_p^\dagger\hat{\mathbf{R}}^{-1}\mathbf{Y}_p\right)^{-1}, \end{aligned} \qquad (5.A.4)$$

and

$$\begin{aligned} \mathbf{B} &= \mathbf{I}_2 - \mathbf{U}\mathbf{\Lambda}\mathbf{U}^\dagger\left(\mathbf{I}_2 + \mathbf{U}\mathbf{\Lambda}\mathbf{U}^\dagger\right)^{-1} \\ &= \mathbf{U}\left[\mathbf{I}_2 - \mathbf{\Lambda}\left(\mathbf{I}_2 + \mathbf{\Lambda}\right)^{-1}\right]\mathbf{U}^\dagger \\ &= \mathbf{U}\left[\left(\mathbf{I}_2 + \mathbf{\Lambda}\right)^{-1}\right]\mathbf{U}^\dagger \\ &= \left(\mathbf{I}_2 + \mathbf{Y}_p^\dagger\hat{\mathbf{R}}^{-1}\mathbf{Y}_p\right)^{-1}. \end{aligned} \qquad (5.A.5)$$

Substituting (5.A.4) and (5.A.5) into (5.A.1), it is easy to show that the GLRT is given by (5.32).

Bibliography

[1] N. B. Pulsone and C. M. Rader, "Adaptive beamformer orthogonal rejection test," *IEEE Transactions on Signal Processing*, vol. 49, no. 3, pp. 521–529, March 2001.

[2] C. Hao, D. Orlando, X. Ma, S. Yan, and C. Hou, "Persymmetric detectors with enhanced rejection capabilities," *IET Radar, Sonar and Navigation*, vol. 8, no. 5, pp. 557–563, 2014.

[3] F. Bandiera, D. Orlando, and G. Ricci, *Advanced Radar Detection Schemes under Mismatched Signal Models in Synthesis Lectures on Signal Processing.* San Rafael, CA, USA Morgan & Claypool, 2009.

[4] F. Gini, A. Farina, and M. Greco, "Radar detection and pre-classification based on multiple hypotheses testing," *IEEE Transactions on Aerospace and Electronic Systems*, vol. 40, no. 3, pp. 1046–1059, 2004.

[5] M. Greco, F. Gini, and M. Diani, "Robust CFAR detection of random signals in compound-Gaussian clutter plus thermal noise," *IEE Proceedings on Radar, Sonar and Navigation*, vol. 148, no. 4, pp. 227–232, August 2001.

[6] A. D. Maio, "Rao test for adaptive detection in Gaussian interference with unknown covariance matrix," *IEEE Transactions on Signal Processing*, vol. 55, no. 7, pp. 3577–3584, July 2007.

[7] D. Orlando and G. Ricci, "A Rao test with enhanced selectivity properties in homogeneous scenarios," *IEEE Transactions on Signal Processing*, vol. 58, no. 10, pp. 5385–5390, October 2010.

[8] F. Bandiera, O. Besson, D. Orlando, and G. Ricci, "An ABORT-like detector with improved mismatched signals rejection capabilities," *IEEE Transactions on Signal Processing*, vol. 56, no. 1, pp. 14–25, January 2008.

[9] C. Hao, J. Yang, X. Ma, C. Hou, and D. Orlando, "Adaptive detection of distributed targets with orthogonal rejection," *IET Radar, Sonar and Navigation*, vol. 6, no. 6, pp. 483–493, 2012.

[10] D. Orlando and G. Ricci, "A Rao test with enhanced selectivity properties in homogeneous scenarios," *IEEE Transactions on Signal Processing*, vol. 58, no. 10, pp. 5385–5390, October 2010.

[11] F. Bandiera, O. Besson, D. Orlando, and G. Ricci, "Derivation and analysis of an adaptive detector with enhanced mismatched signals rejection capabilities," in *Proc. 41st Asilomar Conf. Signals, Systems and Computers*, Pacific Grove, CA, USA, 2007.

[12] L. Cai and H. Wang, "A persymmetric multiband GLR algorithm," *IEEE Transactions on Aerospace and Electronic Systems*, vol. 28, no. 3, pp. 806–816, July 1992.

[13] E. J. Kelly, "An adaptive detection algorithm," *IEEE Transactions on Aerospace and Electronic Systems*, vol. 22, no. 1, pp. 115–127, March 1986.

6

Distributed Target Detection in Homogeneous Environments

In this chapter, we consider a multichannel adaptive detection problem with a widely used data model [1, 2]:

$$H_0 : \begin{cases} \mathbf{y}_j = \mathbf{n}_j, \ j = 1, \ldots, J, \\ \mathbf{y}_k = \mathbf{n}_k, \ k = J+1, J+2, \ldots, J+K, \end{cases} \tag{6.1a}$$

$$H_1 : \begin{cases} \mathbf{y}_j = a_j \, \mathbf{p} + \mathbf{n}_j, \ j = 1, \ldots, J, \\ \mathbf{y}_k = \mathbf{n}_k, \ k = J+1, J+2, \ldots, J+K, \end{cases} \tag{6.1b}$$

where $\mathbf{Y} = [\mathbf{y}_1, \mathbf{y}_2, \ldots, \mathbf{y}_J] \in \mathbb{C}^{N \times J}$ denotes the test data (also called primary data), while $\mathbf{Y}_1 = [\mathbf{y}_{J+1}, \mathbf{y}_{J+2}, \ldots, \mathbf{y}_{J+K}] \in \mathbb{C}^{N \times K}$ denotes the training data (also called secondary data), which are often collected from the cells adjacent to the cells under test. The numbers of test and training data vectors are denoted by J and K, respectively, and the dimension of the data is N. The complex-valued scalar a_j is an unknown quantity accounting for the target reflectivity and channel propagation effects, $\mathbf{p}$, is the target signal steering vector, and $\mathbf{n}_j$ denotes the clutter in the jth cell. Assume that the clutter vectors $\{\mathbf{n}_j\}$ are independent of each other, and each has a circularly symmetric, complex Gaussian distribution with zero-mean and unknown covariance matrix $\mathbf{R}$, i.e., $\mathbf{n}_j \sim \mathcal{CN}_N(\mathbf{0}, \mathbf{R})$ for $j = 1, 2, \ldots, J+K$.

When the number of test data is more than one (i.e., $J > 1$), the model in (6.1) can correspond to adaptive detection problems for three cases. The first case is a distributed target detection problem, where the target echoes occupy multiple range cells. Based on the principles of the one-step GLRT and two-step GLRT, several detectors were designed in [3, 4] for the distributed target detection problem. The second case is the target detection problem in multi-band radar. In such a case, the test data are collected from multiple bands. The GLRT and SMI algorithms were proposed in [5] and [6], respectively. The third case is the adaptive detection problem where the test data are collected from multiple CPIs or dwells. In [7], Raghavan derived the maximal invariants for this case and analyzed several invariant tests including the one-step GLRT and two-step GLRT.

The focus of this chapter is to design two adaptive detectors with persymmetry by using the principles of one-step and two-step GLRTs. Analytical

DOI: 10.1201/9781003340232-6

113

performance of the two detectors is provided, which is verified via MC simulations.

6.1 Persymmetric One-Step GLRT

6.1.1 Detector Design

For the detection problem (6.1), we assume that the covariance matrix $\mathbf{R}$ and the steering vector $\mathbf{p}$ have persymmetric structures. This assumption is valid when the receiver uses a symmetrically spaced linear array and/or symmetrically spaced pulse trains. The persymmetry of $\mathbf{R}$ means $\mathbf{R} = \mathbf{J}\mathbf{R}^*\mathbf{J}$, and the persymmetry in the steering vector $\mathbf{p}$ indicates $\mathbf{p} = \mathbf{J}\mathbf{p}^*$. Take a pulsed Doppler radar with symmetrically spaced pulse trains as an example. The persymmetric steering vector $\mathbf{p}$ can be expressed as

$$
\begin{cases}
\mathbf{p} = \left[e^{-\jmath 2\pi \bar{f}\frac{(N-1)}{2}}, \ldots, e^{-\jmath 2\pi \bar{f}\frac{1}{2}}, \ e^{\jmath 2\pi \bar{f}\frac{1}{2}}, \ldots, e^{\jmath 2\pi \bar{f}\frac{(N-1)}{2}} \right]^T, & \text{for even } N, \\
\mathbf{p} = \left[e^{-\jmath 2\pi \bar{f}\frac{(N-1)}{2}}, \ldots, e^{-\jmath 2\pi \bar{f}}, \ 1, \ e^{\jmath 2\pi \bar{f}}, \ldots, e^{\jmath 2\pi \bar{f}\frac{(N-1)}{2}} \right]^T, & \text{for odd } N,
\end{cases}
\tag{6.2}
$$

where $\bar{f}$ is a normalized target Doppler frequency.

The persymmetric one-step GLRT can be given by

$$
\Xi_0 = \frac{\max_{\{\mathbf{a},\mathbf{R}\}} f_1}{\max_{\{\mathbf{R}\}} f_0} \underset{H_0}{\overset{H_1}{\gtrless}} \xi_0,
\tag{6.3}
$$

where ξ_0 is the detection threshold, $\mathbf{a} = [a_1, a_2, \ldots, a_J]^T$, f_1 and f_0 denote the PDFs of the secondary data $\mathbf{Y}_1 = [\mathbf{y}_1, \mathbf{y}_2, \ldots, \mathbf{y}_J]$ under H_1 and H_0, respectively.

As the statistical independence among the input vectors is assumed, we have

$$
\begin{cases}
f_0 = \left\{ \frac{1}{\pi^N \det(\mathbf{R})} \exp\left[-\operatorname{tr}\left(\mathbf{R}^{-1}\mathbf{T}_0 \right) \right] \right\}^{K+J}, \\
f_1 = \left\{ \frac{1}{\pi^N \det(\mathbf{R})} \exp\left[-\operatorname{tr}\left(\mathbf{R}^{-1}\mathbf{T}_1 \right) \right] \right\}^{K+J},
\end{cases}
\tag{6.4}
$$

where

$$
\begin{cases}
\mathbf{T}_0 = \frac{1}{K+J}(\mathbf{Y}_1\mathbf{Y}_1^\dagger + \mathbf{Y}\mathbf{Y}^\dagger) \in \mathbb{C}^{N\times N}, \\
\mathbf{T}_1 = \frac{1}{K+J}\left[\mathbf{Y}_1\mathbf{Y}_1^\dagger + (\mathbf{Y} - \mathbf{p}\mathbf{a}^T)(\mathbf{Y} - \mathbf{p}\mathbf{a}^T)^\dagger \right] \in \mathbb{C}^{N\times N}.
\end{cases}
\tag{6.5}
$$

Since $\mathbf{R}$ is persymmetric, (6.4) can be rewritten in a persymmetric form, i.e.,

$$
\begin{cases}
f_0 = \left\{ \frac{1}{\pi^N \det(\mathbf{R})} \exp\left[-\operatorname{tr}\left(\mathbf{R}^{-1}\mathbf{T}_{p0} \right) \right] \right\}^{K+J}, \\
f_1 = \left\{ \frac{1}{\pi^N \det(\mathbf{R})} \exp\left[-\operatorname{tr}\left(\mathbf{R}^{-1}\mathbf{T}_{p1} \right) \right] \right\}^{K+J},
\end{cases}
\tag{6.6}
$$

where

$$\begin{cases} \mathbf{T}_{p0} = \frac{1}{2} \left(\mathbf{T}_0 + \mathbf{J} \mathbf{T}_0^* \mathbf{J} \right) \in \mathbb{C}^{N \times N}, \\ \mathbf{T}_{p1} = \frac{1}{2} \left(\mathbf{T}_1 + \mathbf{J} \mathbf{T}_1^* \mathbf{J} \right) \in \mathbb{C}^{N \times N}. \end{cases} \tag{6.7}$$

Define

$$\hat{\mathbf{R}} = \frac{1}{2} \left[\mathbf{Y}_1 \mathbf{Y}_1^\dagger + \mathbf{J} \left(\mathbf{Y}_1 \mathbf{Y}_1^\dagger \right)^* \mathbf{J} \right] \in \mathbb{C}^{N \times N}, \tag{6.8}$$

we can use (6.5) to obtain

$$\begin{cases} \mathbf{T}_{p0} = \frac{1}{K+J} \left[\hat{\mathbf{R}} + \mathbf{Y}_p \mathbf{Y}_p^H \right], \\ \mathbf{T}_{p1} = \frac{1}{K+J} \left[\hat{\mathbf{R}} + \left(\mathbf{Y}_p - \mathbf{p} \mathbf{a}_p^T \right) \left(\mathbf{Y}_p - \mathbf{p} \mathbf{a}_p^T \right)^\dagger \right], \end{cases} \tag{6.9}$$

where

$$\mathbf{Y}_p = [\mathbf{y}_{e1}, \dots \mathbf{y}_{eJ}, \mathbf{y}_{o1}, \dots, \mathbf{y}_{oJ}] \in \mathbb{C}^{N \times 2J}, \tag{6.10}$$

and

$$\mathbf{a}_p = [a_{e1}, \dots, a_{eJ}, a_{o1}, \dots, a_{oJ}]^T \in \mathbb{C}^{2J \times 1}, \tag{6.11}$$

with

$$\begin{cases} \mathbf{y}_{ej} = \frac{1}{2} \left(\mathbf{y}_j + \mathbf{J} \mathbf{y}_j^* \right), \\ \mathbf{y}_{oj} = \frac{1}{2} \left(\mathbf{y}_j - \mathbf{J} \mathbf{y}_j^* \right), \end{cases} \tag{6.12}$$

and

$$\begin{cases} a_{ej} = \frac{1}{2}(a_j + a_j^*) = \mathfrak{Re}(a_j), \\ a_{oj} = \frac{1}{2}(a_j - a_j^*) = j \, \mathfrak{Im}(a_j). \end{cases} \tag{6.13}$$

As shown in [8], the maximum of f_0 (or f_1) can be achieved at $\mathbf{R} = \mathbf{T}_{p0}$ (or $\mathbf{R} = \mathbf{T}_{p1}$) for fixed $\mathbf{a}$, and

$$\begin{cases} \max_{\{\mathbf{R}\}} f_0 = \left\{ \frac{1}{(e\pi)^N \det(\mathbf{T}_{p0})} \right\}^{K+J}, \\ \max_{\{\mathbf{R}\}} f_1 = \left\{ \frac{1}{(e\pi)^N \det(\mathbf{T}_{p1})} \right\}^{K+J}. \end{cases} \tag{6.14}$$

The test statistic Ξ_0 in (6.3) can be equivalent to

$$\Xi_1 = \frac{\det\left(\mathbf{T}_{p0} \right)}{\min_{\{\mathbf{a}\}} \left[\det\left(\mathbf{T}_{p1} \right) \right]} \underset{H_0}{\overset{H_1}{\gtrless}} \xi_1. \tag{6.15}$$

Using (6.9), we have

$$\det\left(\mathbf{T}_{p0} \right) = (K+J)^{-N} \det\left(\hat{\mathbf{R}} \right) \det\left(\mathbf{I}_{2J} + \mathbf{Y}_p^\dagger \hat{\mathbf{R}}^{-1} \mathbf{Y}_p \right), \tag{6.16}$$

and

$$\det\left(\mathbf{T}_{p1} \right) = (K+J)^{-N} \det\left(\hat{\mathbf{R}} \right)$$
$$\times \det\left[\mathbf{I}_{2J} + \left(\mathbf{Y}_p - \mathbf{p} \mathbf{a}_p^T \right)^\dagger \hat{\mathbf{R}}_p^{-1} \left(\mathbf{Y}_p - \mathbf{p} \mathbf{a}_p^T \right) \right]. \tag{6.17}$$

Expanding (6.17) and completing square on those terms containing $\mathbf{a}_p$, one can easily verify

$$\min_{\mathbf{a}} \{\det(\mathbf{T}_{p1})\} = (K+J)^{-N} \det\left(\hat{\mathbf{R}}\right)$$

$$\times \det\left\{\mathbf{I}_{2J} + \mathbf{Y}_p^\dagger \hat{\mathbf{R}}^{-1} \mathbf{Y}_p - \frac{\mathbf{Y}_p^\dagger \hat{\mathbf{R}}^{-1} \mathbf{p} \mathbf{p}^\dagger \hat{\mathbf{R}}^{-1} \mathbf{Y}_p}{\mathbf{p}^\dagger \hat{\mathbf{R}}^{-1} \mathbf{p}}\right\}. \quad (6.18)$$

It then follows that Ξ_1 can be equivalently written as

$$\Xi_2 = \left\{\det\left[\mathbf{I}_{2J} - \left[\mathbf{I}_{2J} + \mathbf{Y}_p^\dagger \hat{\mathbf{R}}^{-1} \mathbf{Y}_p\right]^{-1/2} \mathbf{Y}_p^\dagger \hat{\mathbf{R}}^{-1} \mathbf{p}\right.\right.$$

$$\left.\left.\times \frac{\mathbf{p}^\dagger \hat{\mathbf{R}}^{-1} \mathbf{Y}_p \left[\mathbf{I}_{2J} + \mathbf{Y}_p^\dagger \hat{\mathbf{R}}^{-1} \mathbf{Y}_p\right]^{-1/2}}{\mathbf{p}^\dagger \hat{\mathbf{R}}^{-1} \mathbf{p}}\right]\right\}^{-1}$$

$$= \left\{1 - \frac{\mathbf{p}^\dagger \hat{\mathbf{R}}^{-1} \mathbf{Y}_p \left[\mathbf{I}_{2J} + \mathbf{Y}_p^\dagger \hat{\mathbf{R}}^{-1} \mathbf{Y}_p\right]^{-1} \mathbf{Y}_p^\dagger \hat{\mathbf{R}}^{-1} \mathbf{p}}{\mathbf{p}^\dagger \hat{\mathbf{R}}^{-1} \mathbf{p}}\right\}^{-1}. \quad (6.19)$$

Finally the decision rule is equivalent to

$$\Xi = \frac{\mathbf{p}^\dagger \hat{\mathbf{R}}^{-1} \mathbf{Y}_p \left(\mathbf{I}_{2J} + \mathbf{Y}_p^\dagger \hat{\mathbf{R}}^{-1} \mathbf{Y}_p\right)^{-1} \mathbf{Y}_p^\dagger \hat{\mathbf{R}}^{-1} \mathbf{p}}{\mathbf{p}^\dagger \hat{\mathbf{R}}^{-1} \mathbf{p}} \underset{H_0}{\overset{H_1}{\gtrless}} \xi, \quad (6.20)$$

where ξ is the detection threshold. This detector is referred to as persymmetric GLRT for distributed targets (PGLRT-D), for ease of reference.

6.1.2   Analytical Performance

In this part, we derive analytical expressions for the detection probability of the PGLRT-D detector in the presence of steering vector mismatch.

6.1.2.1   Transformation from Complex Domain to Real Domain

Define a unitary matrix

$$\mathbf{D} = \frac{1}{2}[(\mathbf{I}_N + \mathbf{J}) + \jmath(\mathbf{I}_N - \mathbf{J})] \in \mathbb{C}^{N \times N}. \quad (6.21)$$

Using (6.10), (6.12), and (6.21), we obtain

$$\mathbf{D}\mathbf{Y}_p = [\mathbf{y}_{er1}, \ldots, \mathbf{y}_{erJ}, \jmath\mathbf{y}_{or1}, \ldots, \jmath\mathbf{y}_{orJ}], \quad (6.22)$$

where

$$\mathbf{y}_{erj} = \mathbf{D}\mathbf{y}_{ej} = \frac{1}{2}[(\mathbf{I}_N + \mathbf{J})\Re\mathfrak{e}(\mathbf{y}_j) - (\mathbf{I}_N - \mathbf{J})\Im\mathfrak{m}(\mathbf{y}_j)] \in \mathbb{R}^{N \times 1}, \quad (6.23)$$

and

$$\mathbf{y}_{orj} = -{}_J\mathbf{D}\mathbf{y}_{oj} = \frac{1}{2}\left[(\mathbf{I}_N - \mathbf{J})\mathfrak{Re}(\mathbf{y}_j) + (\mathbf{I}_N + \mathbf{J})\mathfrak{Im}(\mathbf{y}_j)\right] \in \mathbb{R}^{N \times 1}, \quad (6.24)$$

for $j = 1, 2, \ldots, J$. Note that $\mathbf{y}_{erj}$ and $\mathbf{y}_{orj}$ are N-dimensional real-valued vectors.

Define another unitary matrix

$$\mathbf{V} = \begin{bmatrix} \mathbf{I}_J & \mathbf{0} \\ \mathbf{0} & -{}_J\mathbf{I}_J \end{bmatrix} \in \mathbb{C}^{2J \times 2J}. \quad (6.25)$$

We use the two unitary matrices $\mathbf{D}$ and $\mathbf{V}$ to transform all complex-valued quantities in (6.20) to be real-valued ones, i.e.,

$$\hat{\mathbf{R}}_r = \mathbf{D}\hat{\mathbf{R}}\mathbf{D}^\dagger = \mathfrak{Re}(\hat{\mathbf{R}}) + \mathbf{J}\mathfrak{Im}(\hat{\mathbf{R}}) \in \mathbb{R}^{N \times N}, \quad (6.26)$$

$$\mathbf{p}_r = \mathbf{D}\mathbf{p} = \mathfrak{Re}(\mathbf{p}) - \mathfrak{Im}(\mathbf{p}) \in \mathbb{R}^{N \times 1}, \quad (6.27)$$

and

$$\mathbf{Y}_r = \mathbf{D}\mathbf{Y}_p\mathbf{V} = [\mathbf{y}_{er1}, \ldots, \mathbf{y}_{erJ}, \mathbf{y}_{or1}, \ldots, \mathbf{y}_{orJ}] \in \mathbb{R}^{N \times 2J}. \quad (6.28)$$

Using these definitions, the PGLRT-D detector in (6.20) can be rewritten as

$$\Xi = \frac{\mathbf{p}_r^T\hat{\mathbf{R}}_r^{-1}\mathbf{Y}_r \left(\mathbf{I}_{2J} + \mathbf{Y}_r^T\hat{\mathbf{R}}_r^{-1}\mathbf{Y}_r\right)^{-1} \mathbf{Y}_r^T\hat{\mathbf{R}}_r^{-1}\mathbf{p}_r}{\mathbf{p}_r^T\hat{\mathbf{R}}_r^{-1}\mathbf{p}_r} \underset{H_0}{\overset{H_1}{\gtrless}} \xi. \quad (6.29)$$

It is worth noting that all quantities in (6.29) are real-valued. Next, we perform statistical analysis on the PGLRT-D detector in the real-valued domain.

6.1.2.2 Statistical Properties

Denote by $\mathbf{q}$ the true target steering vector which is assumed to be persymmetric (i.e., $\mathbf{q} = \mathbf{J}\mathbf{q}^*$). In practical scenarios, there usually exist mismatches between the assumed target steering vector $\mathbf{p}$ and the true steering vector $\mathbf{q}$ (e.g., due to pointing errors[1]) [9, 10]. The mismatch angle ϕ between the true steering vector $\mathbf{q}$ and the assumed $\mathbf{p}$ is defined by

$$\cos^2 \phi = \frac{|\mathbf{p}^\dagger\mathbf{R}^{-1}\mathbf{q}|^2}{(\mathbf{p}^\dagger\mathbf{R}^{-1}\mathbf{p})(\mathbf{q}^\dagger\mathbf{R}^{-1}\mathbf{q})}. \quad (6.30)$$

In the matched case, $\mathbf{q} = \mathbf{p}$ and $\cos^2 \phi = 1$. Otherwise, $\mathbf{q} \neq \mathbf{p}$ and $\cos^2 \phi < 1$. In the following, we derive the statistical properties of the PGLRT-D in the mismatched case.

---

[1]Here, we assume that the true steering vector $\mathbf{q}$ is persymmetric. This assumption is valid when the mismatch comes from pointing errors. It should be pointed out that the true steering vector $\mathbf{q}$ may not be persymmetric, when the mismatches are due to calibration errors or distortions in array geometry.

As derived in Appendix 6.A, the PGLRT-D detector in (6.29) is equivalent to the following form:

$$m_{11}\mathbf{v}^T\mathbf{v} \underset{H_0}{\overset{H_1}{\gtrless}} \frac{\xi}{1-\xi}, \tag{6.31}$$

where the random variables m_{11} and $\mathbf{v}$, defined in (6.A.21) and (6.A.34), respectively, are conditionally independent. Let

$$\nu = \mathbf{v}^T\mathbf{v}. \tag{6.32}$$

It follows from (6.A.35) that,

$$\nu \sim \begin{cases} \chi_{2J}^2, & \text{under } H_0, \\ \chi_{2J}'^2(\rho_1\delta), & \text{under } H_1, \end{cases} \tag{6.33}$$

where

$$\delta = \mathbf{a}_r^T\mathbf{a}_r\mathbf{q}_r^T\mathbf{R}_r^{-1}\mathbf{q}_r\cos^2\phi, \tag{6.34}$$

and

$$\rho_1 = \frac{\mathbf{a}_r^T\left(\mathbf{I}_{2J}+\boldsymbol{\Sigma}\right)^{-1}\mathbf{a}_r}{\mathbf{a}_r^T\mathbf{a}_r}, \tag{6.35}$$

with $\boldsymbol{\Sigma}$ and $\mathbf{a}_r$ defined in (6.A.29) and (6.A.14), respectively. Let

$$\tau = m_{11}^{-1}. \tag{6.36}$$

Then, the distribution of τ is [11, Theorem 3.2.10]

$$\begin{aligned} \tau &\sim \mathcal{W}_1(2K - N + 1, 1) \\ &= \chi_{2K-N+1}^2, \end{aligned} \tag{6.37}$$

where the second equality is obtained from [11, p. 87]. In addition, the random quantities ν and τ are independent given ρ_1. Using (6.32) and (6.36), we can write (6.31) as

$$\frac{\nu}{\tau} \underset{H_0}{\overset{H_1}{\gtrless}} \frac{\xi}{1-\xi}. \tag{6.38}$$

Note that the noncentral parameter of the distribution of ν includes the random variable ρ_1. As derived in Appendix 6.B, the distribution of ρ_1 in the mismatched case is given by

$$\rho_1 \overset{\text{d}}{=} \begin{cases} \left(1 + \dfrac{\chi_{N-1}^2}{\chi_{2K+2J-N+1}^2}\right)^{-1}, & \text{under } H_0, \\[3mm] \left(1 + \dfrac{\chi_{N-1}'^2(c)}{\chi_{2K+2J-N+1}^2}\right)^{-1}, & \text{under } H_1, \end{cases} \tag{6.39}$$

where

$$c = \mathbf{a}_r^T\mathbf{a}_r\mathbf{q}_r^T\mathbf{R}_r^{-1}\mathbf{q}_r\sin^2\phi. \tag{6.40}$$

Note that the statistical characterizations under H_0 have already been obtained in [12], which are consistent with the results above. However, the statistical results under H_1 are new, which have never been obtained in the open literature.

6.1.2.3   Detection Probability

The random variable ρ_1 is fixed temporarily. We first derive the detection probability conditioned on ρ_1 and then obtain the unconditional detection probability by averaging over the random variable ρ_1. We derive closed-form expressions for the detection probability in two cases: 1) N is an odd integer and 2) N is an even integer.

N is an odd integer

According to [13, Appendix 2], the PDF of ρ_1 in (6.39) with odd N under H_1 can be derived as

$$
\begin{aligned}
f^o_{\rho_1}(\rho_1|N) = \exp\left(-\frac{c\rho_1}{2}\right) \rho_1^{K+J-\frac{N}{2}-\frac{1}{2}} \sum_{k=0}^{\frac{2K+2J-N+1}{2}} C^k_{\frac{2K+2J-N+1}{2}} \\
\times \frac{\Gamma(K+J)\left(\frac{c}{2}\right)^k (1-\rho_1)^{k+\frac{N}{2}-\frac{3}{2}}}{\Gamma(K+J-\frac{N}{2}+\frac{1}{2})\Gamma(k+\frac{N}{2}-\frac{1}{2})}, \quad 0 < \rho_1 < 1.
\end{aligned}
\tag{6.41}
$$

It is worth noting that the above expression is valid only for odd N. To emphasize this, we explicitly write the PDF as a function of N, i.e., $f^o_{\rho_1}(\rho_1|N)$ in (6.41).

Define

$$
D = \frac{2K-N+1}{2}.
\tag{6.42}
$$

Obviously, D is an integer when N is odd. Letting $\tau_1 = \frac{1}{2}\tau$ and $\nu_1 = \frac{1}{2}\nu$, we can obtain that under H_1,

$$
\begin{cases}
\tau_1 \sim \frac{1}{2}\chi^2_{2D}, \\
\nu_1 \sim \frac{1}{2}\chi'^2_{2J}(\rho_1\delta).
\end{cases}
\tag{6.43}
$$

Denoting by $f_{\nu_1|H_1}(\nu_1)$ and $f_{\tau_1}(\tau_1)$ the PDFs of ν_1 and τ_1 under H_1, respectively, we can derive the detection probability conditioned on ρ_1 as

$$
\begin{aligned}
P^o_{\mathrm{D}|\rho_1}(N) &= \int_0^\infty \int_{\frac{\xi\tau_1}{1-\xi}}^\infty f_{\nu_1|H_1}(\nu_1)\mathrm{d}\nu_1 f_{\tau_1}(\tau_1)\mathrm{d}\tau_1 \\
&= 1 - (1-\xi)^{\frac{2K+2J-N-1}{2}} \sum_{j=1}^{\frac{2K-N+1}{2}} C^{J+j-1}_{\frac{2K+2J-N-1}{2}} \left(\frac{\xi}{1-\xi}\right)^{J+j-1} \\
&\quad \times \exp\left[-\frac{\delta\rho_1(1-\xi)}{2}\right] \sum_{m=0}^{j-1} \frac{1}{m!}\left[\frac{\delta\rho_1(1-\xi)}{2}\right]^m,
\end{aligned}
\tag{6.44}
$$

where the second equality is obtained using the results in [14]. Further, the unconditional PD can be obtained by averaging over ρ_1 in (6.41), namely,

$$P_D^o(N) = \int_0^1 P_{D|\rho_1}^o(N) f_{\rho_1}^o(\rho_1|N) \, d\rho_1$$

$$= 1 - (1-\xi)^{\frac{2K+2J-N-1}{2}} \sum_{k=0}^{\frac{2K+2J-N+1}{2}} C_{\frac{2K+2J-N+1}{2}}^k \frac{\Gamma(K+J)\left(\frac{c}{2}\right)^k}{\Gamma\left(K+J-\frac{N}{2}+\frac{1}{2}\right)}$$

$$\times \sum_{j=1}^{\frac{2K-N+1}{2}} C_{\frac{2K+2J-N-1}{2}}^{J+j-1} \left(\frac{\xi}{1-\xi}\right)^{J+j-1}$$

$$\times \sum_{m=0}^{j-1} \frac{1}{m!} \left[\frac{\delta(1-\xi)}{2}\right]^m \frac{\Gamma\left(K+J+m-\frac{N}{2}+\frac{1}{2}\right)}{\Gamma(K+J+m+k)}$$

$$\times {}_1F_1\left(K+J+m-\frac{N}{2}+\frac{1}{2}; K+J+m+k; -\frac{c+\delta(1-\xi)}{2}\right),$$

$$\tag{6.45}$$

the second equality is obtained using [15, pp. 347, eq. (3.383.1)].

N is an even integer

When N is even, we cannot use the method in Section 6.1.2.3. Alternatively, we separately consider the matched and mismatched cases. In the matched case, an infinite-sum expression for the detection probability is derived, whereas in the mismatched case, an approximate expression for the detection probability is obtained.

First we consider the matched case where $\mathbf{p} = \mathbf{q}$. For fixed ρ_1, the CCDF of ν under H_1 is [16, eq. (29.2)]

$$G(\nu) = 1 - \exp(-\rho_1\delta/2) \sum_{j=0}^{\infty} \frac{(\rho_1\delta/2)^j}{2^{J+j}\Gamma(J+j)\,j!} \int_0^{\nu} y^{J+j-1} \exp\left(-\frac{y}{2}\right) dy$$

$$= 1 - \exp(-\rho_1\delta/2) \sum_{j=0}^{\infty} \frac{(\rho_1\delta/2)^j}{\Gamma(J+j)j!} \gamma\left(J+j, \frac{\nu}{2}\right),$$

$$\tag{6.46}$$

where the second equality is obtained from [15, eq. (3.381.1)], and γ is the incomplete Gamma function defined in [15, eq. (8.350.1)]. Further, the detection probability conditioned on ρ_1 in the case of even N can be expressed as

$$P_{D|\rho_1}^e = \int_0^{+\infty} G(\eta\tau) f_\tau(\tau) d\tau$$

$$= 1 - \sum_{j=0}^{\infty} \frac{(\rho_1\delta/2)^j \exp(-\rho_1\delta/2)}{2^{\frac{2K-N+1}{2}}\Gamma(J+j)\Gamma\left(\frac{2K-N+1}{2}\right)j!}$$

$$\times \underbrace{\int_0^{\infty} \gamma\left(J+j, \frac{\eta\tau}{2}\right) \tau^{\frac{2K-N-1}{2}} \exp\left(-\frac{\tau}{2}\right) d\tau}_{\triangleq V_1},$$

$$\tag{6.47}$$

where

$$\eta = \frac{\xi}{1 - \xi}. \tag{6.48}$$

According to [15, eq. (6.455.2)], we have

$$V_1 = \frac{\left(\frac{\eta}{2}\right)^{J+j} \Gamma\left(\frac{2K+2J+2j-N+1}{2}\right)}{(J+j)\left(\frac{1+\eta}{2}\right)^{\frac{2K+2J+2j-N+1}{2}}}$$
$$\times {}_2F_1\left(1, \frac{2K+2J+2j-N+1}{2}; J+j+1; \frac{\eta}{1+\eta}\right). \tag{6.49}$$

It follows from (6.39) that the PDF of ρ_1 with even N in the matched case is

$$f_{\rho_1}(\rho_1) = \frac{\Gamma(K+J)\rho_1^{\frac{2K+2J-N-1}{2}}(1-\rho_1)^{\frac{N-3}{2}}}{\Gamma\left(\frac{2K+2J-N+1}{2}\right)\Gamma\left(\frac{N-1}{2}\right)}, \qquad 0 < \rho_1 < 1. \tag{6.50}$$

Further, the detection probability of the PGLRT-D detector is obtained by averaging over ρ_1, i.e.,

$$P_{\mathrm{D}} = \int_0^1 P_{\mathrm{D}|\rho_1} f_{\rho_1}(\rho_1)\,\mathrm{d}\rho_1,$$

$$= 1 - \sum_{j=0}^{\infty} \frac{(\delta/2)^j V_1 \Gamma(K+J)}{2^{\frac{2K-N+1}{2}} \Gamma(J+j) \Gamma\left(\frac{2K-N+1}{2}\right) j! \Gamma\left(\frac{2K+2J-N+1}{2}\right) \Gamma\left(\frac{N-1}{2}\right)}$$
$$\times \underbrace{\int_0^1 \exp(-\rho_1 \delta/2)\rho_1^{\frac{2K+2J+2j-N-1}{2}}(1-\rho_1)^{\frac{N-3}{2}}\,\mathrm{d}\rho_1}_{\triangleq V_2}. \tag{6.51}$$

Due to [15, eq. (3.383.1)], we can obtain

$$V_2 = \frac{\Gamma\left(\frac{N-1}{2}\right)\Gamma\left(\frac{2K+2J+2j-N+1}{2}\right)}{\Gamma(K+J+j)}$$
$$\times {}_1F_1\left(\frac{2K+2J+2j-N+1}{2}; K+J+j; \frac{-\delta}{2}\right). \tag{6.52}$$

Substituting (6.49) and (6.52) into (6.51) produces

$$
\begin{aligned}
P_{\mathrm{D}} = 1 &- \frac{\Gamma\left(K+J\right)}{2^{\frac{2K-N+1}{2}}\Gamma\left(\frac{2K-N+1}{2}\right)\Gamma\left(\frac{2K+2J-N+1}{2}\right)} \\
&\times \sum_{j=0}^{\infty} \frac{\left(\frac{\delta}{2}\right)^{j}\left(\frac{\eta}{2}\right)^{J+j}\Gamma^{2}\left(\frac{2K+2J+2j-N+1}{2}\right)}{j!\,(J+j)!\left(\frac{1+\eta}{2}\right)^{\frac{2K+2J+2j-N+1}{2}}\Gamma\left(K+J+j\right)} \\
&\times {}_{2}F_{1}\left(1,\frac{2K+2J+2j-N+1}{2};J+j+1;\frac{\eta}{1+\eta}\right) \\
&\times {}_{1}F_{1}\left(\frac{2K+2J+2j-N+1}{2};K+J+j;\frac{-\delta}{2}\right),
\end{aligned}
\tag{6.53}
$$

where η and δ are defined in (6.48) and (6.34), respectively.

In fact, (6.53) also holds true for odd N in the matched case. This is to say, (6.53) equals to (6.45) with odd N in the matched case, even though they appear in completely different forms. Obviously, (6.45) is preferred when N is odd, since its finite-sum form is easy to calculate.

In the mismatched case where $\mathbf{q} \neq \mathbf{p}$, the PDF of ρ_1 with even N is infinite-sum, and the resulting expression for the detection probability is in doubly infinite-sum form. Obviously, the doubly infinite-sum expression is not computationally efficient to calculate in practice. As an alternative, we obtain an approximate but finite-sum expression for the detection probability with even N in an intuitive way, i.e.,

$$
P_{\mathrm{D}}^{\mathrm{e}} = \frac{P_{\mathrm{D}}^{\mathrm{o}}(N-1) + P_{\mathrm{D}}^{\mathrm{o}}(N+1)}{2},
\tag{6.54}
$$

where $P_{\mathrm{D}}^{\mathrm{o}}(\cdot)$ is defined in (6.45). As will be shown in next section, the accuracy of this approximate expression for the detection probability is acceptable in the case of even N.

6.1.2.4 Probability of False Alarm

According to (6.38), the PFA can be represented as

$$
\begin{aligned}
P_{\mathrm{FA}} &= \int_{0}^{\infty}\int_{\frac{\xi\tau}{1-\xi}}^{\infty} f_{\nu|H_0}(\nu)\mathrm{d}\nu f_{\tau}(\tau)\mathrm{d}\tau \\
&= (1-\xi)^{\frac{2K-N+1}{2}}\sum_{j=1}^{J}\frac{\Gamma(K+J-j-\frac{N}{2}+\frac{1}{2})}{\Gamma(K-\frac{N}{2}+\frac{1}{2})\Gamma(J-j+1)}\xi^{J-j},
\end{aligned}
\tag{6.55}
$$

which is consistent with the result in [12]. It follows that the PGLRT-D detector exhibits the desirable CFAR property against the clutter covariance matrix, since the PFA in (6.55) is independent of $\mathbf{R}$.

6.2 Persymmetric Two-Step GLRT

This section is to design a persymmetric adaptive detector according to the criterion of two-step GLRT.

6.2.1 Detector Design

In the first step, the covariance matrix $\mathbf{R}$ is assumed known. Then the detector is given by

$$\Lambda_0 = \frac{\max_{\{\mathbf{a}\}} f\left(\mathbf{Y}|H_1\right)}{f\left(\mathbf{Y}|H_0\right)} \underset{H_0}{\overset{H_1}{\gtrless}} \lambda_0, \tag{6.56}$$

where λ_0 is the detection threshold, $\mathbf{a} = [a_1, a_2, \ldots, a_J]^T$, $f\left(\mathbf{Y}|H_1\right)$ and $f\left(\mathbf{Y}|H_0\right)$ denote the PDFs of the primary data $\mathbf{Y} = [\mathbf{y}_1, \mathbf{y}_2, \ldots, \mathbf{y}_J]$ under H_1 and H_0, respectively.

Due to the independence among the primary data, the PDF of $\mathbf{Y}$ under H_q $(q = 0, 1)$ can be written as

$$f(\mathbf{Y}|H_q) = \frac{1}{\pi^{NJ} \det(\mathbf{R})^J} \exp\left[-\operatorname{tr}(\mathbf{R}^{-1}\mathbf{F}_q)\right], \quad q = 0, 1, \tag{6.57}$$

where

$$\mathbf{F}_q = \left(\mathbf{Y} - q\mathbf{p}\mathbf{a}^T\right)\left(\mathbf{Y} - q\mathbf{p}\mathbf{a}^T\right)^\dagger \in \mathbb{C}^{N \times N}. \tag{6.58}$$

Using the persymmetry structure in the covariance matrix $\mathbf{R}$, we have

$$\begin{aligned}
\operatorname{tr}(\mathbf{R}^{-1}\mathbf{F}_q) &= \operatorname{tr}[\mathbf{J}(\mathbf{R}^*)^{-1}\mathbf{J}\mathbf{F}_q] \\
&= \operatorname{tr}[(\mathbf{R}^*)^{-1}\mathbf{J}\mathbf{F}_q\mathbf{J}] \\
&= \operatorname{tr}[\mathbf{R}^{-1}\mathbf{J}\mathbf{F}_q^*\mathbf{J}], \quad q = 0, 1.
\end{aligned} \tag{6.59}$$

As a result, (6.57) can be rewritten as

$$f(\mathbf{Y}|H_q) = \frac{1}{\pi^{NJ} \det(\mathbf{R})^J} \exp\left[-\operatorname{tr}(\mathbf{R}^{-1}\mathbf{H}_q)\right], \quad q = 0, 1, \tag{6.60}$$

where

$$\mathbf{H}_q = \frac{1}{2}\left(\mathbf{F}_q + \mathbf{J}\mathbf{F}_q^*\mathbf{J}\right) \in \mathbb{C}^{N \times N}. \tag{6.61}$$

Substituting (6.58) into (6.61), after some algebra, leads to

$$\mathbf{H}_q = \left(\mathbf{Y}_p - q\mathbf{p}\mathbf{a}_p^T\right)\left(\mathbf{Y}_p - q\mathbf{p}\mathbf{a}_p^T\right)^\dagger, \quad q = 0, 1, \tag{6.62}$$

where $\mathbf{Y}_p$ and $\mathbf{a}_p$ are defined in (6.10) and (6.11), respectively.

The maximization of $f(\mathbf{Y}|H_1)$ with respect to $\mathbf{a}_p$ can be achieved at

$$\hat{\mathbf{a}}_p = \frac{\mathbf{p}^\dagger \mathbf{R}^{-1} \mathbf{Y}_p}{\mathbf{p}^\dagger \mathbf{R}^{-1} \mathbf{p}}. \tag{6.63}$$

As a result, we have

$$\max_{\{\mathbf{a}_p\}} f\left(\mathbf{Y}|H_1\right) = \frac{1}{\pi^{NJ} \det(\mathbf{R})^J} \exp\left[-\operatorname{tr}(\mathbf{R}^{-1} \mathbf{Y}_p \mathbf{Y}_p^\dagger)\right]$$
$$\times \exp\left[\operatorname{tr}\left(\frac{\mathbf{R}^{-1} \mathbf{Y}_p \mathbf{Y}_p^\dagger \mathbf{R}^{-1} \mathbf{p}\mathbf{p}^\dagger}{\mathbf{p}^\dagger \mathbf{R}^{-1} \mathbf{p}}\right)\right]. \tag{6.64}$$

It is obvious that

$$f\left(\mathbf{Y}|H_0\right) = \frac{1}{\pi^{NJ} \det(\mathbf{R})^J} \exp\left[-\operatorname{tr}(\mathbf{R}^{-1} \mathbf{Y}_p \mathbf{Y}_p^\dagger)\right]. \tag{6.65}$$

Substituting (6.64) and (6.65) into (6.56), after some equivalent transformations, we can obtain the detector with known covariance matrix as

$$\Lambda_1 = \frac{\mathbf{p}^\dagger \mathbf{R}^{-1} \mathbf{Y}_p \mathbf{Y}_p^\dagger \mathbf{R}^{-1} \mathbf{p}}{\mathbf{p}^\dagger \mathbf{R}^{-1} \mathbf{p}} \underset{H_0}{\overset{H_1}{\gtrless}} \lambda_1, \tag{6.66}$$

where λ_1 is the detection threshold. This detector cannot be used, since the covariance matrix $\mathbf{R}$ is unknown in practice.

In the second step, we need to obtain an estimate of the covariance matrix to take the place of $\mathbf{R}$ in (6.66). Based on the training data, the covariance matrix estimate $\hat{\mathbf{R}}$ is given in (6.8). Using $\hat{\mathbf{R}}$ to replace $\mathbf{R}$ in (6.66) leads to

$$\Lambda = \frac{\mathbf{p}^\dagger \hat{\mathbf{R}}^{-1} \mathbf{Y}_p \mathbf{Y}_p^\dagger \hat{\mathbf{R}}^{-1} \mathbf{p}}{\mathbf{p}^\dagger \hat{\mathbf{R}}^{-1} \mathbf{p}} \underset{H_0}{\overset{H_1}{\gtrless}} \lambda, \tag{6.67}$$

where λ is the detection threshold. This detector is referred to as persymmetric AMF for distributed targets (PAMF-D), because it is designed using the two-step method as in [17].

We now make the following observations on computational complexity. Note that the PGLRT-D requires $\mathcal{O}(Q_1 N^2) + \mathcal{O}(Q_2 J^2)$ floating-point operations (flops), where $Q_1 = \max(K, N, J)$ and $Q_2 = \max(N, J)$, whereas the PAMF-D involves $\mathcal{O}(Q_1 N^2)$ flops. Compared with the PGLRT-D detector in (6.20), the PAMF-D detector in (6.67) is computationally efficient to implement, since the PAMF-D detector does not need the online inversion of the matrix $\mathbf{I}_{2J} + \mathbf{Y}_p^\dagger \hat{\mathbf{R}}^{-1} \mathbf{Y}_p$.

6.2.2 Analytical Performance

In this section, analytical expressions for the detection probability and the PFA of the PAMF-D detector in (6.67) are derived for the matched case where the true target steering vector $\mathbf{q}$ is aligned with the nominal one $\mathbf{p}$. Analytical performance of the PAMF-D in the mismatched case is still an open problem.

Similar to the derivations in Section 6.1.2.1, the PAMF-D detector in (6.67) can be rewritten as

$$\Lambda = \frac{\mathbf{p}_r^T \hat{\mathbf{R}}_r^{-1} \mathbf{Y}_r \mathbf{Y}_r^T \hat{\mathbf{R}}_r^{-1} \mathbf{p}_r}{\mathbf{p}_r^T \hat{\mathbf{R}}_r^{-1} \mathbf{p}_r} \underset{H_0}{\overset{H_1}{\gtrless}} \lambda. \tag{6.68}$$

It is worth pointing out that all quantities in (6.68) are real-valued. Next, we will perform statistical analysis on the PAMF-D detector in the real-valued domain.

6.2.2.1 Statistical Properties

As derived in Appendix 6.C, the PAMF-D detector in (6.68) is statistically equivalent to the following form:

$$m_{11} \mathbf{v}^T \mathbf{v} \underset{H_0}{\overset{H_1}{\gtrless}} \lambda \rho, \tag{6.69}$$

where the random variables m_{11}, $\mathbf{v}$, and ρ are defined in (6.A.30), (6.C.9), and (6.C.7), respectively. Note that m_{11} is independent of $\mathbf{v}$ given ρ.

Let

$$\nu = \mathbf{v}^T \mathbf{v}. \tag{6.70}$$

It follows from (6.C.10) that conditioned on ρ,

$$\nu \sim \begin{cases} \chi^2_{2J}, & \text{under } H_0, \\ \chi'^2_{2J}(\rho\delta), & \text{under } H_1, \end{cases} \tag{6.71}$$

with

$$\delta = \mathbf{a}_r^T \mathbf{a}_r \mathbf{p}_r^T \mathbf{R}_r^{-1} \mathbf{p}_r. \tag{6.72}$$

Note that the performance of the PAMF-D detector is affected by the target scatterers through $\mathbf{a}_r^T \mathbf{a}_r$ which is determined by the total energy of the scatterers (instead of their distribution). Define

$$\tau = m_{11}^{-1}. \tag{6.73}$$

Then, τ is distributed as [11, Theorem 3.2.10]

$$\begin{aligned} \tau &\sim \mathcal{W}_1(2K - N + 1, 1) \\ &= \chi^2_{2K-N+1}, \end{aligned} \tag{6.74}$$

where the second equality is obtained from [11, p. 87]. In addition, the random quantities ν and τ are independent given ρ. Using (6.70) and (6.73), (6.69) can be expressed as

$$\frac{\nu}{\tau} \underset{H_0}{\overset{H_1}{\gtrless}} \lambda \rho. \tag{6.75}$$

In (6.75), only the distribution of ρ has not been derived so far. We now proceed with introducing a theorem before obtaining the distribution of ρ defined in (6.C.7).

Theorem 6.2.1. *Let* $\mathbf{W} \sim \mathcal{N}_{n \times m}(\mathbf{0}, \mathbf{I}_n \otimes \mathbf{I}_m)$ *where* $m \geqslant n$, *and*

$$\mathbf{G} = \mathbf{W}\mathbf{W}^T = \begin{bmatrix} g_{11} & \mathbf{g}_{21}^T \\ \mathbf{g}_{21} & \mathbf{G}_{22} \end{bmatrix} \in \mathbb{R}^{n \times n}, \tag{6.76}$$

where g_{11} *is a scalar,* $\mathbf{g}_{21} \in \mathbb{R}^{(n-1) \times 1}$, *and* $\mathbf{G}_{22} \in \mathbb{R}^{(n-1) \times (n-1)}$. *Then,*

$$(1 + \mathbf{g}_{21}^T \mathbf{G}_{22}^{-2} \mathbf{g}_{21})^{-1} \sim \beta \left(\frac{m-n+2}{2}, \frac{n-1}{2} \right). \tag{6.77}$$

Proof. See Appendix 6.D. □

Recall that ρ is defined in (6.C.7). According to Theorem 6.2.1, we have

$$\rho \sim \beta \left(\frac{2K-N+2}{2}, \frac{N-1}{2} \right). \tag{6.78}$$

So far, we have derived all the distributions of the random quantities τ, ν, and ρ in (6.75). Based on these results, we derive analytical expressions for the PFA and PD of the PAMF-D detector in the following.

6.2.2.2 Probability of False Alarm

Under H_0, the PDF of ν is

$$f_{\nu|H_0}(\nu) = \frac{\nu^{J-1}}{2^J \Gamma(J)} \exp\left(-\frac{\nu}{2}\right), \quad \nu > 0. \tag{6.79}$$

From (6.74), we obtain that the PDF of τ is

$$f_\tau(\tau) = \frac{1}{2^{\frac{2K-N+1}{2}} \Gamma\left(\frac{2K-N+1}{2}\right)} \tau^{\frac{2K-N-1}{2}} \exp\left(-\frac{\tau}{2}\right), \quad \tau > 0. \tag{6.80}$$

According to (6.75), we can obtain the PFA conditioned on ρ as

$$\begin{aligned} P_{\text{FA}|\rho} &= \int_0^\infty \int_{\lambda\rho\tau}^\infty f_{\nu|H_0}(\nu) \mathrm{d}\nu f_\tau(\tau) \mathrm{d}\tau \\ &= \sum_{j=1}^J \frac{\Gamma(K+J-j-\frac{N}{2}+\frac{1}{2})}{\Gamma(K-\frac{N}{2}+\frac{1}{2})\Gamma(J-j+1)} (\lambda\rho)^{J-j} (1+\lambda\rho)^{-(K+J-j-\frac{N}{2}+\frac{1}{2})}. \end{aligned} \tag{6.81}$$

As indicated in (6.78), the PDF of ρ is given by

$$f_\rho(\rho) = \frac{\Gamma\left(\frac{2K+1}{2}\right) \rho^{\frac{2K-N}{2}} (1-\rho)^{\frac{N-3}{2}}}{\Gamma\left(\frac{2K-N+2}{2}\right) \Gamma\left(\frac{N-1}{2}\right)}, \quad 0 < \rho < 1. \tag{6.82}$$

As a result, the PFA can be obtained by averaging over ρ, i.e.,

$$
\begin{aligned}
P_{\text{FA}} &= \int_0^1 P_{\text{FA}|\rho} f_\rho(\rho) \, d\rho \\
&= \frac{\Gamma(K + \frac{1}{2})}{\Gamma(K - \frac{N}{2} + 1)} \sum_{j=1}^J \frac{\Gamma(J + K - j - \frac{N}{2} + \frac{1}{2})}{\Gamma(J + K - j + \frac{1}{2})} \frac{\Gamma(J + K - j - \frac{N}{2} + 1) \lambda^{J-j}}{\Gamma(K - \frac{N}{2} + \frac{1}{2})\Gamma(J - j + 1)} \\
&\quad \times {}_2F_1\left(J + K - j - \frac{N}{2} + \frac{1}{2}, J + K - j - \frac{N}{2} + 1; \right. \\
&\qquad\qquad \left. J + K - j + \frac{1}{2}; -\lambda \right).
\end{aligned}
$$

$$(6.83)$$

It follows that the detector exhibits the desirable CFAR property against the clutter covariance matrix, since the PFA in (6.83) is irrelevant to $\mathbf{R}$.

6.2.2.3 Detection Probability

The random variable ρ is fixed temporarily. We first derive the detection probability conditioned on ρ and then obtain the unconditional detection probability by averaging over the random variable ρ. Closed-form expressions are derived for the detection probability in two cases: 1) N is an odd integer and 2) N is an even integer. In the former case, a finite-sum expression is derived for the detection probability. In the second case, an infinite-sum expression is derived for the detection probability.

Odd integer for N:

Define

$$
D = \frac{2K - N + 1}{2}.
$$

$$(6.84)$$

Obviously, D is an integer when N is odd. Letting

$$
\tau_1 = \frac{1}{2}\tau, \quad \nu_1 = \frac{1}{2}\nu,
$$

$$(6.85)$$

we can obtain that under H_1

$$
\begin{cases}
\tau_1 \sim \frac{1}{2}\chi_{2D}^2, \\
\nu_1 \sim \frac{1}{2}\chi_{2J}'^2(\rho\delta).
\end{cases}
$$

$$(6.86)$$

Denote by $f_{\nu_1|H_1}(\nu_1)$ and $f_{\tau_1}(\tau_1)$ the PDFs of ν_1 and τ_1 under H_1, respectively. Based on the results in [14], we can obtain the detection probability conditioned on ρ as

$$P_{D|\rho} = \int_0^\infty \int_{\lambda\rho\tau_1}^\infty f_{\nu_1|H_1}(\nu_1)\mathrm{d}\nu_1 f_{\tau_1}(\tau_1)\mathrm{d}\tau_1$$

$$=1 - (\lambda\rho)^{J-1}(1+\lambda\rho)^{-(D+J-1)}\sum_{j=1}^{D} C_{D+J-1}^{J+j-1}(\lambda\rho)^j \qquad (6.87)$$

$$\times \exp\left[-\frac{\delta\rho}{2(1+\lambda\rho)}\right]\sum_{m=0}^{j-1}\frac{1}{m!}\left[\frac{\delta\rho}{2(1+\lambda\rho)}\right]^m.$$

It is worth noting that the above expression is in force only for the case where D is an integer (i.e., N has to be an odd integer). Consequently, the unconditional PD can be obtained by averaging over ρ, namely,

$$P_D = \int_0^1 P_{D|\rho}f_\rho(\rho)\,\mathrm{d}\rho, \qquad (6.88)$$

where $f_\rho(\rho)$ is given by (6.82).

Even integer for N:

For the case of even N, (6.87) is no longer valid, since D is not an integer. As an alternative, we seek an infinite-sum expression for the detection probability in the case of even N.

When ρ is fixed, the CCDF of ν under H_1 is [16, p. 435, eq. (29.2)]

$$G(\nu) =1 - \exp\left(-\frac{\rho\delta}{2}\right)\sum_{j=0}^{\infty}\frac{\left(\frac{\rho\delta}{2}\right)^j \int_0^\nu y^{J+j-1}\exp\left(-\frac{y}{2}\right)\mathrm{d}y}{2^{J+j}\Gamma(J+j)j!}$$

$$= \exp\left(-\frac{\nu+\rho\delta}{2}\right)\sum_{j=0}^{\infty}\frac{(\rho\delta)^j}{j!2^j}\sum_{m=0}^{J+j-1}\frac{\nu^m}{m!2^m}, \qquad (6.89)$$

where we have used [15, p. 346, eq. (3.381.1)] and [15, p. 899, eq. (8.352.1)]. As a consequence, the detection probability conditioned on ρ can be given as

$$P_{D|\rho} = \int_0^{+\infty} G(\lambda\rho\tau) f_\tau(\tau)\mathrm{d}\tau$$

$$= \frac{1}{2^{\frac{2K-N+1}{2}}\Gamma\left(\frac{2K-N+1}{2}\right)}\sum_{j=0}^{\infty}\frac{\delta^j}{j!2^j}\sum_{m=0}^{J+j-1}\frac{\lambda^m}{m!2^m} \qquad (6.90)$$

$$\times \int_0^\infty \exp\left(-\frac{\lambda\rho\tau+\tau+\rho\delta}{2}\right)\tau^{\frac{2K+2m-N-1}{2}}\rho^{m+j}\mathrm{d}\tau.$$

Further, the detection probability of the PAMF-D detector is obtained by averaging over ρ, i.e.,

$$
\begin{aligned}
P_{\mathrm{D}} &= \int_0^1 P_{\mathrm{D}|\rho} f_\rho(\rho)\,\mathrm{d}\rho \\
&= \frac{1}{2^{\frac{2K-N+1}{2}}\Gamma\left(\frac{2K-N+1}{2}\right)} \sum_{j=0}^{\infty} \frac{\delta^j}{j!2^j} \sum_{m=0}^{J+j-1} \frac{\lambda^m \Gamma\left(\frac{2K+1}{2}\right) L}{m!2^m \Gamma\left(\frac{2K-N+2}{2}\right)\Gamma\left(\frac{N-1}{2}\right)},
\end{aligned}
$$

$$(6.91)$$

where

$$
L = \int_0^1 g(\rho)\rho^{K+j+m-\frac{N}{2}}(1-\rho)^{\frac{N-3}{2}}\exp(-\rho\delta/2)\mathrm{d}\rho \tag{6.92}
$$

with

$$
g(\rho) = \int_0^\infty \exp\left[-\left(\frac{1+\lambda\rho}{2}\right)\tau\right]\tau^{\frac{2K+2m-N-1}{2}}\mathrm{d}\tau. \tag{6.93}
$$

According to [15, p. 346, eq. (3.381.4)], we obtain

$$
g(\rho) = \left(\frac{1+\lambda\rho}{2}\right)^{-\frac{2K+2m-N+1}{2}}\Gamma\left(\frac{2K+2m-N+1}{2}\right). \tag{6.94}
$$

Applying (6.94) to (6.92) results in

$$
\begin{aligned}
L = &\; 2^{\frac{2K+2m-N+1}{2}}\Gamma\left(\frac{2K+2m-N+1}{2}\right) \\
&\times \int_0^1 \frac{\rho^{K+j+m-\frac{N}{2}}(1-\rho)^{\frac{N-3}{2}}\exp(-\rho\delta/2)}{(1+\lambda\rho)^{\frac{2K+2m-N+1}{2}}}\mathrm{d}\rho.
\end{aligned}
$$

$$(6.95)$$

Taking (6.95) back into (6.91), after some algebra, yields

$$
\begin{aligned}
P_{\mathrm{D}} = &\; \frac{\Gamma\left(\frac{2K+1}{2}\right)}{\Gamma\left(\frac{2K-N+1}{2}\right)\Gamma\left(\frac{2K-N+2}{2}\right)\Gamma\left(\frac{N-1}{2}\right)} \\
&\times \sum_{j=0}^{\infty} \frac{\delta^j}{j!2^j} \sum_{m=0}^{J+j-1} \frac{\lambda^m}{m!}\Gamma\left(\frac{2K+2m-N+1}{2}\right) \\
&\times \int_0^1 \frac{\rho^{K+j+m-\frac{N}{2}}(1-\rho)^{\frac{N-3}{2}}\exp(-\rho\delta/2)}{(1+\lambda\rho)^{\frac{2K+2m-N+1}{2}}}\mathrm{d}\rho.
\end{aligned}
$$

$$(6.96)$$

It has to be emphasized here that the analytical expression in (6.96) for the detection probability holds true for any positive integer N. When the data dimension N takes on odd integers, (6.96) is equal to (6.88), although they have completely different forms.

6.3　Numerical Examples

In this section, numerical simulations are provided to verify the above theoretical results. Assume that a pulsed Doppler radar is used, which transmits symmetrically spaced pulse trains. The number of pulses in a coherent processing interval is given by N. The (i,j)th element of the clutter covariance matrix is chosen as $\mathbf{R}(i,j) = \sigma^2 0.9^{|i-j|}$, where σ^2 is the clutter power. The SCR in decibel is defined as

$$\text{SCR} = 10\log_{10} \frac{\frac{1}{J}\sum_{j=1}^{J}|a_j|^2}{\sigma^2}. \tag{6.97}$$

For comparison purposes, conventional GLRT and AMF for distributed target detection are considered. Note that these conventional adaptive detectors bear the CFAR property against the clutter covariance matrix. The GLRT and AMF are given in eqs. (27) and (28) in [3], respectively, i.e.,

$$\Xi_{\text{GLRT}} = \frac{\mathbf{p}^{\dagger}(\mathbf{Y}_1\mathbf{Y}_1^{\dagger})^{-1}\mathbf{Y}\left(\mathbf{I}_J + \mathbf{Y}^{\dagger}(\mathbf{Y}_1\mathbf{Y}_1^{\dagger})^{-1}\mathbf{Y}\right)^{-1}\mathbf{Y}^{\dagger}(\mathbf{Y}_1\mathbf{Y}_1^{\dagger})^{-1}\mathbf{p}}{\mathbf{p}^{\dagger}(\mathbf{Y}_1\mathbf{Y}_1^{\dagger})^{-1}\mathbf{p}} \underset{H_0}{\overset{H_1}{\gtrless}} \xi_{\text{GLRT}}, \tag{6.98}$$

and

$$\Lambda_{\text{AMF}} = \frac{\mathbf{p}^{\dagger}(\mathbf{Y}_1\mathbf{Y}_1^{\dagger})^{-1}\mathbf{Y}\mathbf{Y}^{\dagger}(\mathbf{Y}_1\mathbf{Y}_1^{\dagger})^{-1}\mathbf{p}}{\mathbf{p}^{\dagger}(\mathbf{Y}_1\mathbf{Y}_1^{\dagger})^{-1}\mathbf{p}} \underset{H_0}{\overset{H_1}{\gtrless}} \lambda_{\text{AMF}}, \tag{6.99}$$

where ξ_{GLRT} and λ_{AMF} are the detection thresholds.

First, we consider the matched case where the nominal steering vector is aligned with the true one. The ROC curves of all detectors considered here are plotted for different K in Fig. 6.1, where SCR = 8 dB, $N = 8$, and $J = 2$. We can observe from Fig. 6.1 that the PAMF-D and PGLRT-D detectors outperform their counterparts, especially in the case of insufficient training data. Note that we plot the ROC curves of the PAMF-D and PGLRT-D detectors with a small number of training data in Fig. 6.1(c), where the training data size $K = 6$ is less than the dimension of the received data. For this sample-starved case, the conventional adaptive detectors cannot work. Hence, their performance is not presented in Fig. 6.1(c).

In Fig. 6.2, we plot the ROC curves of the considered detectors for the case of odd $N = 9$. The parameters are set as $J = 2$ and SCR = 8 dB. Results in Fig. 6.2 highlight that the PAMF-D and PGLRT-D detectors achieve performance gains over their competitors. The performance gain becomes large as the number of training data decreases. It has to be emphasized again that the PAMF-D and PGLRT-D detectors can work in the limited training data case where the conventional adaptive detectors cannot work.

In Fig. 6.3, we depict the detection probability versus the number of training data K, where SCR = 8 dB, $N = 9$, $J = 2$, and $P_{\text{FA}} = 10^{-3}$. It highlights that the PAMF-D and PGLRT-D detectors always outperform the

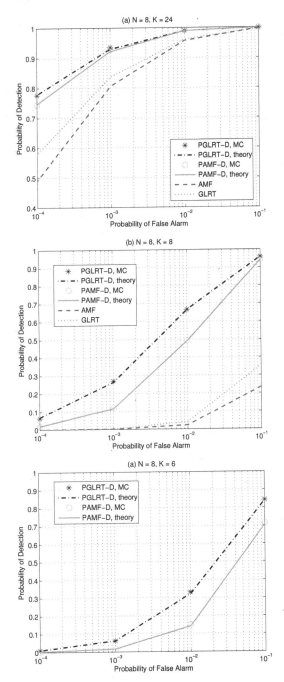

FIGURE 6.1
Receiver operating characteristic curves with SCR = 8 dB for $N = 8$ and $J = 2$.

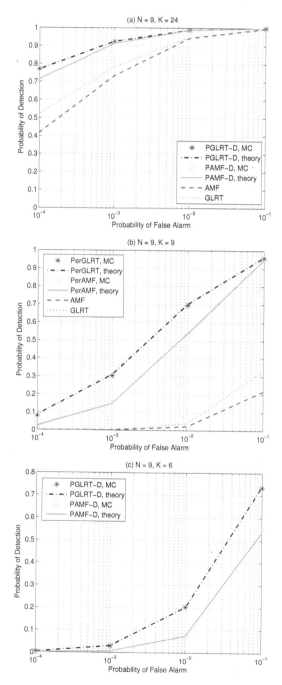

FIGURE 6.2

Receiver operating characteristic curves with SCR $= 8$ dB for $N = 9$ and $J = 2$.

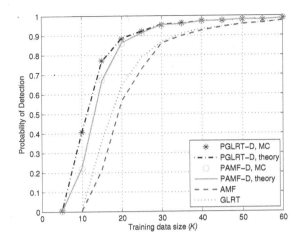

FIGURE 6.3
Probability of detection versus the training data size K for $N = 9$, $J = 2$ and $P_{FA} = 10^{-3}$.

conventional adaptive detectors, and the performance gains are obvious for the case of limited training data. When the training data are sufficient, all the detectors perform almost the same. This is to say, the PAMF-D and PGLRT-D detectors are more suitably applied to the case where the number of training data is insufficient.

It should be noted that the PGLRT-D detector generally outperforms the PAMF-D detector in the parameter setting in Figs. 6.1–6.3, when there is no mismatch in the steering vector. Nevertheless, different behaviors can be observed when there exists a mismatch in the steering vector, as shown next.

In Fig. 6.4, we depict the detection probability with respect to the SCR with odd $N = 9$ for different $\cos^2 \phi$ in the case where $J = 2$, $K = 12$ and $P_{FA} = 10^{-3}$. When no mismatch occurs in the target steering vector (i.e., $\cos^2 \phi = 1$), the PGLRT-D detector outperforms its counterparts. When mismatch happens in the target steering vector (e.g., the nominal Doppler frequency is 0.12 and the resulting $\cos^2 \phi = 0.9092$ in Fig. 6.4(b)), we can observe that the performance of the PGLRT-D and GLRT detectors is obviously degraded, and the performance of the PAMF-D and AMF detectors is slightly degraded. As the mismatch becomes more severe (e.g., the nominal Doppler frequency is 0.14 and the resulting $\cos^2 \phi = 0.6806$ in Fig. 6.4(c)), we can see an interesting phenomenon that the PAMF-D detector performs the best, and the PGLRT-D detector even performs worse than the AMF detector in the high SCR region. These results in Fig. 6.4 highlight that the PGLRT-D detector is sensitive to the target steering vector mismatch, whereas the PAMF-D detector is robust to it.

In Fig. 6.5, the detection probability as a function of SCR is plotted for the case of even $N = 8$. Here, we choose $J = 2$, $K = 12$, and $P_{FA} = 10^{-3}$. Note

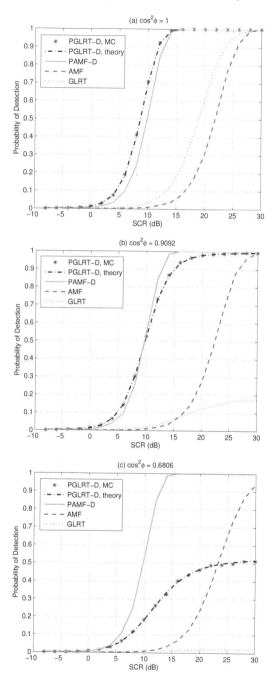

FIGURE 6.4
Detection probability versus SCR for $J = 2$, $N = 9$, $K = 12$, and $P_{\text{FA}} = 10^{-3}$.

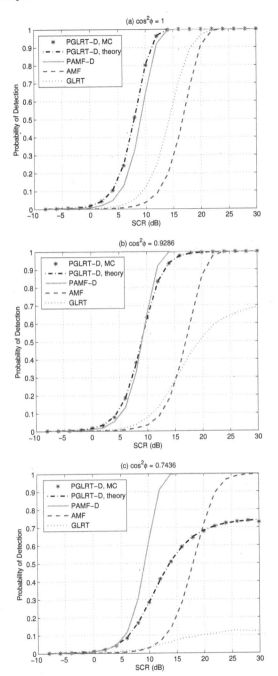

FIGURE 6.5
Detection probability versus SCR for $J = 2$, $N = 8$, $K = 12$, and $P_{\mathrm{FA}} = 10^{-3}$.

that when the true Doppler frequencies are 0.12 and 0.14, the resulting $\cos^2 \phi$ turns out to be 0.9286 and 0.7436 in this case, respectively. The phenomena observed in Fig. 6.5 are similar to those in Fig. 6.4, which can be explained in the similar way.

6.A   Derivations of (6.31)

Similar to the derivations in [12, eq. (B11)], we can obtain

$$\hat{\mathbf{R}}_r \sim \mathcal{W}_N(2K, \mathbf{R}_r), \tag{6.A.1}$$

and

$$\begin{cases} \mathbf{y}_{erj} \sim \mathcal{N}_N(\mathbf{0}, \mathbf{R}_r), \\ \mathbf{y}_{orj} \sim \mathcal{N}_N(\mathbf{0}, \mathbf{R}_r), \end{cases} \tag{6.A.2}$$

where $j = 1, 2, \ldots, J$, $\mathbf{y}_{erj}$ and $\mathbf{y}_{orj}$ are independent of each other, and

$$\mathbf{R}_r = \frac{1}{2}\mathbf{D}\mathbf{R}\mathbf{D}^\dagger = \frac{1}{2}\left[\mathfrak{Re}(\mathbf{R}) + \mathbf{J}\mathfrak{Im}(\mathbf{R})\right] \in \mathbb{R}^{N \times N}. \tag{6.A.3}$$

By defining

$$\mathbf{q}_r \triangleq \mathbf{D}\mathbf{q} = \mathfrak{Re}(\mathbf{q}) - \mathfrak{Im}(\mathbf{q}) \in \mathbb{R}^{N \times 1}, \tag{6.A.4}$$

(6.30) can be rewritten as

$$\cos^2 \phi = \frac{(\mathbf{p}_r^\dagger \mathbf{R}_r^{-1} \mathbf{q}_r)^2}{(\mathbf{p}_r^\dagger \mathbf{R}_r^{-1} \mathbf{p}_r)(\mathbf{q}_r^\dagger \mathbf{R}_r^{-1} \mathbf{q}_r)}, \tag{6.A.5}$$

where $\mathbf{p}_r$ is defined in (6.27).

Obviously, there exists an orthogonal matrix $\mathbf{U} \in \mathbb{R}^{N \times N}$ such that

$$\frac{\mathbf{U}^T \mathbf{R}_r^{-1/2} \mathbf{p}_r}{(\mathbf{p}_r^T \mathbf{R}_r^{-1} \mathbf{p}_r)^{1/2}} = \mathbf{e} \triangleq [1, 0, \ldots, 0]^T \in \mathbb{R}^{N \times 1}. \tag{6.A.6}$$

Employing (6.A.5) and (6.A.6), we can obtain

$$\frac{\mathbf{U}^T \mathbf{R}_r^{-1/2} \mathbf{q}_r}{(\mathbf{q}_r^T \mathbf{R}_r^{-1} \mathbf{q}_r)^{1/2}} = \cos \phi \mathbf{e} + \sin \phi \mathbf{g}, \tag{6.A.7}$$

where

$$\mathbf{g} = [0, \mathbf{g}_1^T]^T \in \mathbb{R}^{N \times 1}, \tag{6.A.8}$$

with $\mathbf{g}_1$ being a unit vector of dimension $(N-1) \times 1$. Note that

$$\begin{cases} \mathbf{e}^T \mathbf{g} = 0, \\ \mathbf{g}_1^T \mathbf{g}_1 = 1. \end{cases} \tag{6.A.9}$$

Using the orthogonal matrix $\mathbf{U}$, we define

$$\mathbf{X} = \mathbf{U}^T \mathbf{R}_r^{-1/2} \mathbf{Y}_r \in \mathbb{R}^{N \times 2J}, \tag{6.A.10}$$

and

$$\mathbf{Q} = \mathbf{U}^T \mathbf{R}_r^{-1/2} \hat{\mathbf{R}}_r \mathbf{R}_r^{-1/2} \mathbf{U} \in \mathbb{R}^{N \times N}. \tag{6.A.11}$$

Using (6.A.6), (6.A.10), and (6.A.11), we can recast (6.29) to

$$\Xi = \frac{\mathbf{e}^T \mathbf{Q}^{-1} \mathbf{X} \left(\mathbf{I}_{2J} + \mathbf{X}^T \mathbf{Q}^{-1} \mathbf{X}\right)^{-1} \mathbf{X}^T \mathbf{Q}^{-1} \mathbf{e}}{\mathbf{e}^T \mathbf{Q}^{-1} \mathbf{e}} \underset{H_0}{\overset{H_1}{\gtrless}} \xi. \tag{6.A.12}$$

Based on (6.28), (6.A.2), and (6.A.10), we can obtain that

$$\mathbf{X} \sim \begin{cases} \mathcal{N}_{N \times 2J}(\mathbf{0}, \mathbf{I}_N \otimes \mathbf{I}_{2J}), & \text{under } H_0, \\ \mathcal{N}_{N \times 2J}(\mathbf{U}^T \mathbf{R}_r^{-1/2} \mathbf{q}_r \mathbf{a}_r^T, \mathbf{I}_N \otimes \mathbf{I}_{2J}), & \text{under } H_1, \end{cases} \tag{6.A.13}$$

where

$$\mathbf{a}_r = [\mathfrak{Re}(a_1), \dots, \mathfrak{Re}(a_J), \mathfrak{Im}(a_1), \dots, \mathfrak{Im}(a_J)]^T \in \mathbb{R}^{2J \times 1}. \tag{6.A.14}$$

Combining (6.A.1) with (6.A.11), we have

$$\mathbf{Q} \sim \mathcal{W}_N(2K, \mathbf{I}_N). \tag{6.A.15}$$

As a result, $\mathbf{Q}$ is statistically equivalent to

$$\mathbf{Q} \overset{\mathrm{d}}{=} \sum_{k=1}^{2K} \mathbf{z}_k \mathbf{z}_k^T, \tag{6.A.16}$$

where $\mathbf{z}_k \sim \mathcal{N}_N(\mathbf{0}, \mathbf{I}_N)$, and these vectors $\mathbf{z}_k$'s are independent of each other. Write

$$\mathbf{z}_k = \begin{bmatrix} z_{k1} \\ \mathbf{z}_{k2} \end{bmatrix}, \tag{6.A.17}$$

where z_{k1} is a scalar, and $\mathbf{z}_{k2} \in \mathbb{R}^{(N-1) \times 1}$ is a column vector. Then, we can express $\mathbf{Q}$ as

$$\mathbf{Q} = \begin{bmatrix} q_{11} & \mathbf{q}_{21}^T \\ \mathbf{q}_{21} & \mathbf{Q}_{22} \end{bmatrix}, \tag{6.A.18}$$

where q_{11} is a scalar, $\mathbf{q}_{21}$ is a column vector of dimension $N - 1$, and

$$\mathbf{Q}_{22} = \sum_{k=1}^{2K} \mathbf{z}_{k2} \mathbf{z}_{k2}^T \in \mathbb{R}^{(N-1) \times (N-1)}. \tag{6.A.19}$$

Define

$$\mathbf{M} = \mathbf{Q}^{-1} = \begin{bmatrix} m_{11} & \mathbf{m}_{21}^T \\ \mathbf{m}_{21} & \mathbf{M}_{22} \end{bmatrix} \in \mathbb{R}^{N \times N}. \tag{6.A.20}$$

According to the partitioned matrix inversion theorem [13], we have

$$m_{11} = (q_{11} - \mathbf{q}_{21}^T \mathbf{Q}_{22}^{-1} \mathbf{q}_{21})^{-1}, \tag{6.A.21}$$

$$\mathbf{m}_{21} = -m_{11} \mathbf{Q}_{22}^{-1} \mathbf{q}_{21} \in \mathbb{R}^{(N-1) \times 1}, \tag{6.A.22}$$

and

$$\mathbf{M}_{22} = (\mathbf{Q}_{22} - q_{11}^{-1} \mathbf{q}_{21} \mathbf{q}_{21}^T)^{-1} \in \mathbb{R}^{(N-1) \times (N-1)}. \tag{6.A.23}$$

For convenience, we partition $\mathbf{X}$ as

$$\mathbf{X} = \begin{bmatrix} \mathbf{x}_1^T \\ \mathbf{X}_2^T \end{bmatrix}, \tag{6.A.24}$$

where $\mathbf{x}_1 \in \mathbb{R}^{2J \times 1}$ and $\mathbf{X}_2 \in \mathbb{R}^{2J \times (N-1)}$. According to (6.A.13), we have

$$\mathbf{x}_1 \sim \begin{cases} \mathcal{N}_{2J}(\mathbf{0}, \mathbf{I}_{2J}), & \text{under } H_0, \\ \mathcal{N}_{2J}((\mathbf{q}_r^T \mathbf{R}_r^{-1} \mathbf{q}_r)^{1/2} \cos\phi \mathbf{a}_r, \mathbf{I}_{2J}), & \text{under } H_1, \end{cases} \tag{6.A.25}$$

and

$$\mathbf{X}_2 \sim \begin{cases} \mathcal{N}_{2J \times (N-1)}(\mathbf{0}, \mathbf{I}_{2J} \otimes \mathbf{I}_{N-1}), & \text{under } H_0, \\ \mathcal{N}_{2J \times (N-1)}((\mathbf{q}_r^T \mathbf{R}_r^{-1} \mathbf{q}_r)^{1/2} \sin\phi \mathbf{a}_r \mathbf{g}_1^T, \mathbf{I}_{2J} \otimes \mathbf{I}_{N-1}), & \text{under } H_1. \end{cases} \tag{6.A.26}$$

Define

$$\mathbf{w} = \mathbf{x}_1 - \mathbf{X}_2 \mathbf{Q}_{22}^{-1} \mathbf{q}_{21} \in \mathbb{R}^{2J \times 1}. \tag{6.A.27}$$

We proceed by fixing $\mathbf{X}_2$ and $\mathbf{z}_{k2}$ temporarily. It follows that

$$\mathbf{w} \sim \begin{cases} \mathcal{N}_{2J}(\mathbf{0}, \mathbf{I}_{2J} + \boldsymbol{\Sigma}), & \text{under } H_0, \\ \mathcal{N}_{2J}((\mathbf{q}_r^T \mathbf{R}_r^{-1} \mathbf{q}_r)^{1/2} \cos\phi \mathbf{a}_r, \mathbf{I}_{2J} + \boldsymbol{\Sigma}), & \text{under } H_1, \end{cases} \tag{6.A.28}$$

where

$$\boldsymbol{\Sigma} = \mathbf{X}_2 \mathbf{Q}_{22}^{-1} \mathbf{X}_2^T \in \mathbb{R}^{2J \times 2J}. \tag{6.A.29}$$

It is straightforward that

$$\mathbf{e}^T \mathbf{Q}^{-1} \mathbf{e} = m_{11}, \tag{6.A.30}$$

$$\mathbf{e}^T \mathbf{Q}^{-1} \mathbf{X} = m_{11} \mathbf{w}^T, \tag{6.A.31}$$

and

$$\mathbf{X}^T \mathbf{Q}^{-1} \mathbf{X} = m_{11} \mathbf{w}\mathbf{w}^T + \boldsymbol{\Sigma}. \tag{6.A.32}$$

Substituting (6.A.30), (6.A.31), and (6.A.29) into (6.A.12), and performing some equivalent transformations, yield

$$m_{11} \mathbf{w}^T (\mathbf{I}_{2J} + \boldsymbol{\Sigma})^{-1} \mathbf{w} \underset{H_0}{\overset{H_1}{\gtrless}} \frac{\xi}{1 - \xi}. \tag{6.A.33}$$

Similar to the derivations in [19, eq. (B31)], we can obtain that conditioned on $\mathbf{X}_2$ and $\mathbf{z}_{k2}$, the random vector $\mathbf{w}$ is independent of the random quantity m_{11}. Define

$$\mathbf{v} = (\mathbf{I}_{2J} + \boldsymbol{\Sigma})^{-1/2}\,\mathbf{w} \in \mathbb{R}^{2J \times 1}. \tag{6.A.34}$$

According to (6.A.28) and (6.A.34), we have

$$\mathbf{v} \sim \begin{cases} \mathcal{N}_{2J}(\mathbf{0}, \mathbf{I}_{2J}), & \text{under } H_0, \\ \mathcal{N}_{2J}(\boldsymbol{\mu}, \mathbf{I}_{2J}), & \text{under } H_1, \end{cases} \tag{6.A.35}$$

where

$$\boldsymbol{\mu} = (\mathbf{q}_r^T \mathbf{R}_r^{-1} \mathbf{q}_r)^{1/2} \cos\phi\,(\mathbf{I}_{2J} + \boldsymbol{\Sigma})^{-1/2}\,\mathbf{a}_r \in \mathbb{R}^{2J \times 1}. \tag{6.A.36}$$

Using (6.A.35), we can rewrite (6.A.33) as (6.31).

6.B   Derivations of (6.39)

In this Appendix, we derive the distribution of ρ defined in (6.35) in the mismatched case. Obviously, there always exists an orthogonal matrix $\mathbf{U}_a \in \mathbb{R}^{2J \times 2J}$ such that

$$\frac{\mathbf{U}_a \mathbf{a}_r}{(\mathbf{a}_r^T \mathbf{a}_r)^{1/2}} = \mathbf{e}_{2J} \triangleq [1, 0, \dots, 0]^T \in \mathbb{R}^{2J \times 1}, \tag{6.B.1}$$

where $\mathbf{a}_r$ is defined in (6.A.14). Then, ρ in (6.35) can be recast to

$$\rho = \mathbf{e}_{2J}^T \left(\mathbf{I}_{2J} + \tilde{\mathbf{X}} \mathbf{Q}_{22}^{-1} \tilde{\mathbf{X}}^T \right)^{-1} \mathbf{e}_{2J}, \tag{6.B.2}$$

where

$$\tilde{\mathbf{X}} = \mathbf{U}_a \mathbf{X}_2 \in \mathbb{R}^{2J \times (N-1)}. \tag{6.B.3}$$

According to the matrix inversion lemma [20, p. 1348], we have

$$\left(\mathbf{I}_{2J} + \tilde{\mathbf{X}} \mathbf{Q}_{22}^{-1} \tilde{\mathbf{X}}^T \right)^{-1} = \mathbf{I}_{2J} - \tilde{\mathbf{X}} \left(\mathbf{Q}_{22} + \tilde{\mathbf{X}}^T \tilde{\mathbf{X}} \right)^{-1} \tilde{\mathbf{X}}^T. \tag{6.B.4}$$

Inserting (6.B.4) into (6.B.2) leads to

$$\rho = 1 - \mathbf{e}_{2J}^T \tilde{\mathbf{X}} \left(\mathbf{Q}_{22} + \tilde{\mathbf{X}}^T \tilde{\mathbf{X}} \right)^{-1} \tilde{\mathbf{X}}^T \mathbf{e}_{2J}. \tag{6.B.5}$$

We partition $\tilde{\mathbf{X}}$ as

$$\tilde{\mathbf{X}} = \begin{bmatrix} \tilde{\mathbf{x}}_1^T \\ \tilde{\mathbf{X}}_2^T \end{bmatrix}, \tag{6.B.6}$$

where $\tilde{\mathbf{x}}_1 \in \mathbb{R}^{(N-1)\times 1}$ and $\tilde{\mathbf{X}}_2 \in \mathbb{R}^{(N-1)\times(2J-1)}$. Then, we have

$$\mathbf{e}_{2J}^T \tilde{\mathbf{X}} = \tilde{\mathbf{x}}_1^T, \tag{6.B.7}$$

and

$$\tilde{\mathbf{X}}^T \tilde{\mathbf{X}} = \tilde{\mathbf{x}}_1 \tilde{\mathbf{x}}_1^T + \tilde{\mathbf{X}}_2 \tilde{\mathbf{X}}_2^T. \tag{6.B.8}$$

Substituting (6.B.7) and (6.B.8) into (6.B.5) produces

$$\begin{aligned}
\rho &= 1 - \tilde{\mathbf{x}}_1^T \left(\mathbf{Q}_{22} + \tilde{\mathbf{x}}_1 \tilde{\mathbf{x}}_1^T + \tilde{\mathbf{X}}_2 \tilde{\mathbf{X}}_2^T \right)^{-1} \tilde{\mathbf{x}}_1 \\
&= \left[1 + \tilde{\mathbf{x}}_1^T \left(\mathbf{Q}_{22} + \tilde{\mathbf{X}}_2 \tilde{\mathbf{X}}_2^T \right)^{-1} \tilde{\mathbf{x}}_1 \right]^{-1},
\end{aligned} \tag{6.B.9}$$

where the second equality is obtained using the matrix inversion lemma [20, p. 1348]. Due to (6.A.26) and (6.B.3), we have

$$\tilde{\mathbf{X}} \sim \begin{cases} \mathcal{N}_{2J\times(N-1)}(\mathbf{0}, \mathbf{I}_{2J} \otimes \mathbf{I}_{N-1}), & \text{under } H_0, \\ \mathcal{N}_{2J\times(N-1)}((\mathbf{a}_r^T \mathbf{a}_r \mathbf{q}_r^T \mathbf{R}_r^{-1} \mathbf{q}_r)^{1/2} \sin\phi \mathbf{e}_{2J} \mathbf{g}_1^T, \mathbf{I}_{2J} \otimes \mathbf{I}_{N-1}), & \text{under } H_1. \end{cases} \tag{6.B.10}$$

Based on (6.B.6) and (6.B.10), we have

$$\tilde{\mathbf{x}}_1 \sim \begin{cases} \mathcal{N}_{N-1}(\mathbf{0}, \mathbf{I}_{N-1}), & \text{under } H_0, \\ \mathcal{N}_{N-1}((\mathbf{a}_r^T \mathbf{a}_r \mathbf{q}_r^T \mathbf{R}_r^{-1} \mathbf{q}_r)^{1/2} \sin\phi \mathbf{g}_1, \mathbf{I}_{N-1}), & \text{under } H_1, \end{cases} \tag{6.B.11}$$

and

$$\tilde{\mathbf{X}}_2 \sim \mathcal{N}_{(N-1)\times(2J-1)}(\mathbf{0}, \mathbf{I}_{N-1} \otimes \mathbf{I}_{2J-1}). \tag{6.B.12}$$

It has to be emphasized that the distribution of $\tilde{\mathbf{X}}_2$ is the same under both H_0 and H_1. Using (6.A.19) and (6.B.12), we can obtain

$$\mathbf{Q}_{22} + \tilde{\mathbf{X}}_2 \tilde{\mathbf{X}}_2^T \sim \mathcal{W}_{N-1}(2K + 2J - 1, \mathbf{I}_{N-1}). \tag{6.B.13}$$

Substituting (6.B.13) and (6.B.11) into (6.B.9), and using Theorem 5.2.2 in [21], we can derive the distribution of ρ, as shown in (6.39).

6.C   Derivations of (6.69)

We consider the case where the true and nonimal steering vectors are consistent (i.e., $\mathbf{p} = \mathbf{q}$ or $\mathbf{p}_r = \mathbf{q}_r$).

Similar to (6.A.12), (6.68) can be recast to

$$\Lambda = \frac{\mathbf{e}^T \mathbf{Q}^{-1} \mathbf{X} \mathbf{X}^T \mathbf{Q}^{-1} \mathbf{e}}{\mathbf{e}^T \mathbf{Q}^{-1} \mathbf{e}} \overset{H_1}{\underset{H_0}{\gtrless}} \lambda, \tag{6.C.1}$$

where

$$\mathbf{X} \sim \begin{cases} \mathcal{N}_{N \times 2J}(\mathbf{0}, \mathbf{I}_N \otimes \mathbf{I}_{2J}), & \text{under } H_0, \\ \mathcal{N}_{N \times 2J}(\mathbf{U}^T \mathbf{R}_r^{-1/2} \mathbf{p}_r \mathbf{a}_r^T, \mathbf{I}_N \otimes \mathbf{I}_{2J}), & \text{under } H_1. \end{cases} \tag{6.C.2}$$

Substituting (6.A.30) and (6.A.31) into (6.C.1) yields

$$\Lambda = m_{11} \mathbf{w}^T \mathbf{w} \underset{H_0}{\overset{H_1}{\gtrless}} \lambda. \tag{6.C.3}$$

Note that the following derivations for the PAMF-D will be different to those for the PGLRT-D.

We now proceed by fixing $\mathbf{q}_{21}$ and $\mathbf{z}_{k2}$ temporarily. It is obvious that conditioned on $\mathbf{q}_{21}$ and $\mathbf{z}_{k2}$, the random vector $\mathbf{w}$ is independent of the random quantity m_{11}. Based on the results in [eq. (B35)] [12], we can obtain the conditional mean of $\mathbf{w}$ as

$$\mathrm{E}(\mathbf{w}) = \begin{cases} \mathbf{0}, & \text{under } H_0, \\ (\mathbf{p}_r^T \mathbf{R}_r^{-1} \mathbf{p}_r)^{1/2} \mathbf{a}_r, & \text{under } H_1. \end{cases} \tag{6.C.4}$$

Denote by w_j the jth element of the column vector $\mathbf{w}$. Then, we have

$$w_j = x_{1j} - \mathbf{x}_{2j}^T \mathbf{Q}_{22}^{-1} \mathbf{q}_{21}, \tag{6.C.5}$$

where x_{1j} is the jth element of the column vector $\mathbf{x}_1$, and $\mathbf{x}_{2j}^T$ is the jth row vector of the matrix $\mathbf{X}_2$. Due to (6.C.2), it is easy to obtain that conditioned on $\mathbf{q}_{21}$ and $\mathbf{z}_{k2}$,

$$\mathrm{E}(w_j w_k) = \begin{cases} \rho^{-1}, & \text{for } k = j, \\ 0, & \text{for } k \neq j, \end{cases} \tag{6.C.6}$$

where

$$\rho = (1 + \mathbf{q}_{21}^T \mathbf{Q}_{22}^{-2} \mathbf{q}_{21})^{-1}. \tag{6.C.7}$$

It means that

$$\mathbf{w} \sim \begin{cases} \mathcal{N}_{2J}(\mathbf{0}, \rho^{-1} \mathbf{I}_{2J}), & \text{under } H_0, \\ \mathcal{N}_{2J}((\mathbf{p}_r^T \mathbf{R}_r^{-1} \mathbf{p}_r)^{1/2} \mathbf{a}_r, \rho^{-1} \mathbf{I}_{2J}), & \text{under } H_1. \end{cases} \tag{6.C.8}$$

Defining

$$\mathbf{v} = \rho^{1/2} \mathbf{w}, \tag{6.C.9}$$

we have

$$\mathbf{v} \sim \begin{cases} \mathcal{N}_{2J}(\mathbf{0}, \mathbf{I}_{2J}), & \text{under } H_0, \\ \mathcal{N}_{2J}((\rho \mathbf{p}_r^T \mathbf{R}_r^{-1} \mathbf{p}_r)^{1/2} \mathbf{a}_r, \mathbf{I}_{2J}), & \text{under } H_1. \end{cases} \tag{6.C.10}$$

Using (6.C.9), we can rewrite (6.C.3) as (6.69).

6.D   Proof of Theorem 6.2.1

Let us start by defining

$$\mathbf{W} = [\mathbf{w}_1^T \ \mathbf{W}_2^T]^T \in \mathbb{R}^{m \times n}, \tag{6.D.1}$$

where $\mathbf{w}_1 \in \mathbb{R}^{m \times 1}$ and $\mathbf{W}_2 \in \mathbb{R}^{m \times (n-1)}$. Then,

$$\begin{cases} \mathbf{w}_1 \sim \mathcal{N}_m(\mathbf{0}, \mathbf{I}_m), \\ \mathbf{W}_2 \sim \mathcal{N}_{m \times (n-1)}(\mathbf{0}_{m \times (n-1)}, \mathbf{I}_m \otimes \mathbf{I}_{n-1}). \end{cases} \tag{6.D.2}$$

Further, we have

$$g_{11} = \mathbf{w}_1^T \mathbf{w}_1, \qquad \mathbf{g}_{21} = \mathbf{W}_2^T \mathbf{w}_1, \tag{6.D.3}$$

and

$$\mathbf{G}_{22} = \mathbf{W}_2^T \mathbf{W}_2 \sim \mathcal{W}_{n-1}(m, \mathbf{I}_{n-1}). \tag{6.D.4}$$

Define

$$\mathbf{t} = \mathbf{G}_{22}^{-1/2} \mathbf{g}_{21} \in \mathbb{R}^{(n-1) \times 1}. \tag{6.D.5}$$

Applying (6.D.3) and (6.D.4) to (6.D.5) leads to

$$\mathbf{t} = (\mathbf{W}_2^T \mathbf{W}_2)^{-1/2} \mathbf{W}_2^T \mathbf{w}_1. \tag{6.D.6}$$

Conditioning on $\mathbf{W}_2$, we obtain

$$\mathbf{t} \sim \mathcal{N}_{n-1}(\mathbf{0}, \mathbf{I}_{n-1}). \tag{6.D.7}$$

That is to say, $\mathbf{t}$ is independent of $\mathbf{W}_2$. Hence, $\mathbf{t}$ is independent of $\mathbf{G}_{22}$.
Using (6.D.5), we have

$$(1 + \mathbf{g}_{21}^T \mathbf{G}_{22}^{-2} \mathbf{g}_{21})^{-1} = (1 + \mathbf{t}^T \mathbf{G}_{22}^{-1} \mathbf{t})^{-1}. \tag{6.D.8}$$

Based on (6.D.4), (6.D.7), and (6.D.8), we can derive that [21, p. 177]

$$(1 + \mathbf{g}_{21}^T \mathbf{G}_{22}^{-2} \mathbf{g}_{21})^{-1} \sim \beta \left(\frac{m-n+2}{2}, \frac{n-1}{2} \right). \tag{6.D.9}$$

The proof is completed.

Bibliography

[1] J. Liu and J. Li, "Mismatched signal rejection performance of the per-symmetric GLRT detector," *IEEE Transactions on Signal Processing*, vol. 67, no. 6, pp. 1610–1619, March 15 2019.

[2] J. Liu, W. Liu, B. Tang, J. Zheng, and S. Xu, "Distributed target detection exploiting persymmetry in Gaussian clutter," *IEEE Transactions on Signal Processing*, vol. 67, no. 4, pp. 1022–1033, February 2019.

[3] E. Conte, A. De Maio, and G. Ricci, "GLRT-based adaptive detection algorithms for range-spread targets," *IEEE Transactions on Signal Processing*, vol. 49, no. 7, pp. 1336–1348, July 2001.

[4] X. Shuai, L. Kong, and J. Yang, "Adaptive detection for distributed targets in Gaussian noise with Rao and Wald tests," *Science China Information Sciences*, vol. 55, no. 6, pp. 1290–1300, 2012.

[5] H. Wang and L. Cai, "On adaptive multiband signal detection with GLR algorithm," *IEEE Transactions on Aerospace and Electronic Systems*, vol. 27, no. 2, pp. 225–233, March 1991.

[6] ——, "On adaptive multiband signal detection with SMI algorithm," *IEEE Transactions on Aerospace and Electronic Systems*, vol. 26, no. 5, pp. 768–773, September 1990.

[7] R. S. Raghavan, "Maximal invariants and performance of some invariant hypothesis tests for an adaptive detection problem," *IEEE Transactions on Signal Processing*, vol. 61, no. 14, pp. 3607–3619, July 15 2013.

[8] R. Nitzberg, "Application of maximum likelihood estimation of persymmetric covariance matrices to adaptive processing," *IEEE Transactions on Aerospace and Electronic Systems*, vol. AES-16, no. 1, pp. 124–127, January 1980.

[9] F. Bandiera, D. Orlando, and G. Ricci, *Advanced Radar Detection Schemes under Mismatched Signal Models in Synthesis Lectures on Signal Processing.* San Rafael, CA, USA, Morgan & Claypool, 2009.

[10] J. Li and P. Stoica, *Robust Adaptive Beamforming.* Hoboken, New Jersey, John Wiley & Sons, Inc., 2005.

[11] R. J. Muirhead, *Aspects of Multivariate Statistical Theory.* New York: Wiley, 1982.

[12] L. Cai and H. Wang, "A persymmetric multiband GLR algorithm," *IEEE Transactions on Aerospace and Electronic Systems*, vol. 28, no. 3, pp. 806–816, July 1992.

[13] E. J. Kelly and K. Forsythe, "Adaptive detection and parameter estimation for multidimensional signal models," Lincoln Laboratory, MIT, Technical Report 848, 1989.

[14] E. J. Kelly, "Finite-sum expression for signal detection probabilities," Lincoln Laboratory, MIT, Technical Report 566, 1981.

[15] I. S. Gradshteyn and I. M. Ryzhik, *Table of Integrals, Series, and Products*, 7th ed. San Diego: Academic Press, 2007.

[16] N. L. Johnson, S. Kotz, and N. Balakrishnan, *Continuous Univariate Distributions (Volumn 2)*, 2nd ed. Hoboken, New Jersey, John Wiley & Sons, Inc., 1995.

[17] F. C. Robey, D. R. Fuhrmann, E. J. Kelly, and R. Nitzberg, "A CFAR adaptive matched filter detector," *IEEE Transactions on Aerospace and Electronic Systems*, vol. 28, no. 1, pp. 208–216, January 1992.

[18] J. Liu, W. Liu, B. Chen, H. Liu, and H. Li, "Detection probability of a CFAR matched filter with signal steering vector errors," *IEEE Signal Processing Letters*, vol. 22, no. 12, pp. 2474–2478, December 2015.

[19] J. Liu, Z.-J. Zhang, Y. Yang, and H. Liu, "A CFAR adaptive subspace detector for first-order or second-order Gaussian signals based on a single observation," *IEEE Transactions on Signal Processing*, vol. 59, no. 11, pp. 5126–5140, November 2011.

[20] H. L. Van Trees, *Optimum Array Processing, Part IV of Detection, Estimation, and Modulation Theory.* New Yok, Wiley-Interscience, 2002.

[21] T. Anderson, *An Introduction to Multivariate Statistical Analysis*, 3rd ed. New York, USA: Wiley, 2003.

7

Robust Detection in Homogeneous Environments

Chapter 6 does not consider the uncertainty of the steering vector into the procedures of detector design. This chapter discusses the target detection problem in the presence of steering vector uncertainties. In order to take into consideration mismatches of the target steering vector, we adopt a subspace model where the target steering vector is assumed to lie in a subspace spanned by the column vectors of a known matrix with unknown target coordinates [1–3]. By exploiting persymmetric structures, several adaptive detectors are designed according to *ad hoc* modifications of GLRT, Wald test, and Rao test in this chapter.

7.1 Problem Formulation

Assume that the received data are collected from N (temporal, spatial, or space-time) channels. We consider the problem of detecting the presence of target signals with J observations. These observations, called primary (test) data, are denoted by $\mathbf{y}_j = a_j^* \mathbf{q} + \mathbf{n}_j$, $j = 1, \ldots, J$, where a_j^* is an unknown complex scalar, $\mathbf{q}$ is the true target signal steering vector, and $\mathbf{n}_j$ denotes wide-sense noise possibly including interference, clutter, and thermal noise. Assume that the noise $\mathbf{n}_j \sim \mathcal{CN}_N(\mathbf{0}, \mathbf{R})$ with unknown covariance matrix $\mathbf{R}$ for $j = 1, 2, \ldots, J$.

In what follows, we assume that $\mathbf{q}$ and $\mathbf{R}$ are persymmetric, namely, $\mathbf{R} = \mathbf{J}\mathbf{R}^*\mathbf{J}$ and $\mathbf{q} = \mathbf{J}\mathbf{q}^*$, where $\mathbf{J} \in \mathbb{R}^{N \times N}$ is the permutation matrix defined in (1.2). To estimate the covariance matrix, a set of homogeneous secondary (training) data $\{\mathbf{y}_k\}_{k=J+1}^{J+K}$ (K is the number of training data) is assumed available, which are free of the target echoes, i.e., $\mathbf{y}_k = \mathbf{n}_k \sim \mathcal{CN}_N(\mathbf{0}, \mathbf{R})$, for $k = J + 1, J + 2, \ldots, J + K$. Suppose further that the noise $\mathbf{n}_k$s for $k = 1, 2, \ldots, J + K$ are IID. Due to the exploitation of persymmetry, we impose a constraint $K \geqslant \lceil \frac{N}{2} \rceil$ [4] in the following. This constraint is much less restrictive than the one (i.e., $K \geqslant N$) required in [5].

In practice, target detection is performed on a grid of spatial and/or Doppler frequencies, while the true target frequency may be located between the grid. Here, we consider such a mismatched case where the true frequency is not aligned with the nominal one during the grid search stage. To account for the mismatch, in the stage of detector design we assume that

$$\mathbf{q} = \mathbf{Sb}, \tag{7.1}$$

where $\mathbf{S} \in \mathbb{C}^{N \times q}$ is a known full-column-rank matrix, and $\mathbf{b}$ is an unknown q-dimensional coordinate vector. In such a subspace model, $\mathbf{S}$ contains the nominal steering vector and other directions which account for the possible mismatches. The aim of the subspace is to collect as much energy as possible, and this is accomplished through the additional directions. We would like to emphasize that the true steering vector $\mathbf{q}$ is not guaranteed to belong to the chosen subspace, since $\mathbf{q}$ is unknown in practice. Strictly speaking, the true steering vector $\mathbf{q}$ is just approximated by using the linear combination of the column vectors of the carefully designed matrix $\mathbf{S}$ in the subspace model (7.1).

We make two assumptions in the subspace model: 1) each column vector of $\mathbf{S}$ is persymmetric and 2) $\mathbf{b}$ is a real-valued vector of dimension q. The two assumptions are required to make $\mathbf{Sb}$ persymmetric, namely, $\mathbf{Sb} = \mathbf{J}(\mathbf{Sb})^* = \mathbf{JS}^*\mathbf{b}^*$, which can be satisfied when $\mathbf{S} = \mathbf{JS}^*$ and $\mathbf{b} = \mathbf{b}^*$. Both assumptions play a vital role in the derivations of robust detectors in the next section. In summary, we can formulate the detection problem[1] as

$$H_0 : \begin{cases} \mathbf{y}_j \sim \mathcal{CN}_N(\mathbf{0}, \mathbf{R}), \ j = 1, \ldots, J, \\ \mathbf{y}_k \sim \mathcal{CN}_N(\mathbf{0}, \mathbf{R}), \ k = J + 1, J + 2, \ldots, J + K, \end{cases} \tag{7.2a}$$

$$H_1 : \begin{cases} \mathbf{y}_j \sim \mathcal{CN}_N(a_j^* \mathbf{Sb}, \mathbf{R}), \ j = 1, \ldots, J, \\ \mathbf{y}_k \sim \mathcal{CN}_N(\mathbf{0}, \mathbf{R}), \ k = J + 1, J + 2, \ldots, J + K, \end{cases} \tag{7.2b}$$

where a_j, $\mathbf{b}$, and $\mathbf{R}$ are all unknown.

7.2   Detection Design

In this section, we design several decision schemes for (7.2) exploiting the one-step GLRT, Wald test, and Rao test, as well as their two-step variations.

Define the test data matrix $\mathbf{Y} = [\mathbf{y}_1, \mathbf{y}_2, \ldots, \mathbf{y}_J] \in \mathbb{C}^{N \times J}$ and the secondary data matrix $\mathbf{Y}_s = [\mathbf{y}_{J+1}, \mathbf{y}_{J+2}, \ldots, \mathbf{y}_{J+K}] \in \mathbb{C}^{N \times K}$. Using the persymmetry in $\mathbf{S}$ and $\mathbf{R}$, we can write the joint PDF of $\mathbf{Y}$ and $\mathbf{Y}_s$ as

$$f(\mathbf{Y}, \mathbf{Y}_s | H_i) = \frac{1}{\pi^{N(J+K)} \det(\mathbf{R})^{J+K}} \exp\left[-\operatorname{tr}(\mathbf{R}^{-1}\mathbf{H}_i)\right], \tag{7.3}$$

---

[1]Note that the subspace model in (7.1) is also used for Doppler distributed targets in [6–8]. Hence, the problem of detecting Doppler distributed targets can also be formulated as the binary hypothesis test in (7.2).

where
$$\mathbf{H}_i = \hat{\mathbf{R}} + \left(\mathbf{Y}_p - i\mathbf{Sba}_p^\dagger\right)\left(\mathbf{Y}_p - i\mathbf{Sba}_p^\dagger\right)^\dagger, \quad i = 0, 1, \tag{7.4}$$

with
$$\hat{\mathbf{R}} = \frac{1}{2}\left[\mathbf{Y}_s\mathbf{Y}_s^\dagger + \mathbf{J}\left(\mathbf{Y}_s\mathbf{Y}_s^\dagger\right)^*\mathbf{J}\right] \in \mathbb{C}^{N\times N}, \tag{7.5}$$

$$\mathbf{Y}_p = [\mathbf{y}_{e1}, \ldots \mathbf{y}_{eJ}, \mathbf{y}_{o1}, \ldots, \mathbf{y}_{oJ}] \in \mathbb{C}^{N\times 2J}, \tag{7.6}$$

and
$$\mathbf{a}_p = [a_{e1}, \ldots, a_{eJ}, a_{o1}, \ldots, a_{oJ}]^T \in \mathbb{C}^{2J\times 1}. \tag{7.7}$$

In (7.6) and (7.7),
$$\begin{cases} \mathbf{y}_{ej} = \frac{1}{2}\left(\mathbf{y}_j + \mathbf{J}\mathbf{y}_j^*\right) \in \mathbb{C}^{N\times 1}, \\ \mathbf{y}_{oj} = \frac{1}{2}\left(\mathbf{y}_j - \mathbf{J}\mathbf{y}_j^*\right) \in \mathbb{C}^{N\times 1}, \end{cases} \tag{7.8}$$

and
$$\begin{cases} a_{ej} = \frac{1}{2}(a_j + a_j^*) = \mathfrak{Re}(a_j), \\ a_{oj} = \frac{1}{2}(a_j - a_j^*) = \jmath\,\mathfrak{Im}(a_j). \end{cases} \tag{7.9}$$

Define two unitary matrices:
$$\mathbf{D} = \frac{1}{2}[(\mathbf{I}_N + \mathbf{J}) + \jmath\,(\mathbf{I}_N - \mathbf{J})] \in \mathbb{C}^{N\times N}, \tag{7.10}$$

and
$$\mathbf{V} = \begin{bmatrix} \mathbf{I}_J & \mathbf{0} \\ \mathbf{0} & -\jmath\mathbf{I}_J \end{bmatrix} \in \mathbb{C}^{2J\times 2J}. \tag{7.11}$$

Letting
$$\begin{aligned} \mathbf{a}_r &\triangleq \mathbf{V}^\dagger\mathbf{a}_p \\ &= [\mathfrak{Re}(a_1), \ldots, \mathfrak{Re}(a_J), -\mathfrak{Im}(a_1), \ldots, -\mathfrak{Im}(a_J)]^T \in \mathbb{R}^{2J\times 1}, \end{aligned} \tag{7.12}$$

and
$$\begin{cases} \mathbf{Y}_r \triangleq \mathbf{D}\mathbf{Y}_p\mathbf{V} = [\mathbf{y}_{er1}, \ldots, \mathbf{y}_{erJ}, \mathbf{y}_{or1}, \ldots, \mathbf{y}_{orJ}] \in \mathbb{R}^{N\times 2J}, \\ \hat{\mathbf{R}}_r \triangleq \mathbf{D}\hat{\mathbf{R}}\mathbf{D}^\dagger = \mathfrak{Re}(\hat{\mathbf{R}}) + \mathbf{J}\mathfrak{Im}(\hat{\mathbf{R}}) \in \mathbb{R}^{N\times N}, \\ \mathbf{S}_r \triangleq \mathbf{D}\mathbf{S} = \mathfrak{Re}(\mathbf{S}) - \mathfrak{Im}(\mathbf{S}) \in \mathbb{R}^{N\times q}, \end{cases} \tag{7.13}$$

we can rewrite $f(\mathbf{Y}, \mathbf{Y}_s|H_i)$ in (7.3) as
$$f(\mathbf{Y}, \mathbf{Y}_s|H_i) = \frac{\exp\left\{-\operatorname{tr}\left[\mathbf{R}_r^{-1}\tilde{\mathbf{H}}_i\right]\right\}}{\pi^{N(J+K)}\det(\mathbf{R}_r)^{J+K}}, \tag{7.14}$$

where $i = 0$ or 1,
$$\mathbf{R}_r = \mathbf{D}\mathbf{R}\mathbf{D}^\dagger = [\mathfrak{Re}(\mathbf{R}) + \mathbf{J}\mathfrak{Im}(\mathbf{R})] \in \mathbb{R}^{N\times N}, \tag{7.15}$$

and
$$\tilde{\mathbf{H}}_i = \hat{\mathbf{R}}_r + \left(\mathbf{Y}_r - i\mathbf{S}_r\mathbf{ba}_r^T\right)\left(\mathbf{Y}_r - i\mathbf{S}_r\mathbf{ba}_r^T\right)^T. \tag{7.16}$$

Note that all the quantities in (7.14) are real-valued.

7.2.1 GLRT Criterion

7.2.1.1 One-Step GLRT

The one-step GLRT detector is given by

$$\frac{\max\limits_{\{\mathbf{a}_r,\mathbf{b},\mathbf{R}_r\}} f\left(\mathbf{Y},\mathbf{Y}_s|H_1\right)}{\max\limits_{\{\mathbf{R}_r\}} f\left(\mathbf{Y},\mathbf{Y}_s|H_0\right)} \underset{H_0}{\overset{H_1}{\gtrless}} \xi_0, \tag{7.17}$$

where ξ_0 is a detection threshold. It is straightforward to see that

$$\begin{cases} \max\limits_{\mathbf{R}_r} f(\mathbf{Y},\mathbf{Y}_s|H_1) = \dfrac{\det(\tilde{\mathbf{H}}_1)^{-(J+K)}}{(e\pi)^{N(J+K)}}, \\ \max\limits_{\mathbf{R}_r} f(\mathbf{Y},\mathbf{Y}_s|H_0) = \dfrac{\det(\tilde{\mathbf{H}}_0)^{-(J+K)}}{(e\pi)^{N(J+K)}}. \end{cases} \tag{7.18}$$

Then, (7.17) can be rewritten as

$$\frac{\det\left(\mathbf{I}_{2J} + \tilde{\mathbf{Y}}^T\tilde{\mathbf{Y}}\right)}{\min\limits_{\{\mathbf{a}_r,\mathbf{b}\}} \det\left(\mathbf{I}_{2J} + \left(\tilde{\mathbf{Y}} - \tilde{\mathbf{S}}\mathbf{b}\mathbf{a}_r^T\right)^T \left(\tilde{\mathbf{Y}} - \tilde{\mathbf{S}}\mathbf{b}\mathbf{a}_r^T\right)\right)} \underset{H_0}{\overset{H_1}{\gtrless}} \xi_1, \tag{7.19}$$

where ξ_1 is a detection threshold, and

$$\begin{cases} \tilde{\mathbf{Y}} = \hat{\mathbf{R}}_r^{-1/2}\mathbf{Y}_r \in \mathbb{R}^{N\times 2J}, \\ \tilde{\mathbf{S}} = \hat{\mathbf{R}}_r^{-1/2}\mathbf{S}_r \in \mathbb{R}^{N\times q}. \end{cases} \tag{7.20}$$

Exploiting the procedure of [9, eq. (16)], the detector can be expressed as

$$\Xi_{\text{1S-PGLRT-H}} = \lambda_{\max}\left\{\mathbf{P}_{\tilde{\mathbf{S}}}\tilde{\mathbf{Y}}\left(\mathbf{I}_{2J} + \tilde{\mathbf{Y}}^T\tilde{\mathbf{Y}}\right)^{-1}\tilde{\mathbf{Y}}^T\right\} \underset{H_0}{\overset{H_1}{\gtrless}} \xi_{\text{1S-PGLRT-H}}, \tag{7.21}$$

where $\xi_{\text{1S-PGLRT-H}}$ is a detection threshold, and $\mathbf{P}_{\tilde{\mathbf{S}}} = \tilde{\mathbf{S}}\left(\tilde{\mathbf{S}}^T\tilde{\mathbf{S}}\right)^{-1}\tilde{\mathbf{S}}^T$. This detector is referred to as one-step persymmetric GLRT in homogeneous environments (1S-PGLRT-H), for ease of reference.

7.2.1.2 Two-Step GLRT

In the first step, the covariance matrix $\mathbf{R}$ is assumed known. Under this assumption, the GLRT detector is given by

$$\frac{\max\limits_{\{\mathbf{a},\mathbf{b}\}} f\left(\mathbf{Y}|H_1\right)}{f\left(\mathbf{Y}|H_0\right)} \underset{H_0}{\overset{H_1}{\gtrless}} \xi_2, \tag{7.22}$$

where ξ_2 is a detection threshold, $f\left(\mathbf{Y}|H_i\right)$, $i = 0, 1$, denote the PDFs of the primary data $\mathbf{Y}$ under H_i, i.e.,

$$f(\mathbf{Y}|H_i) = \frac{1}{\pi^{NJ}\det(\mathbf{R})^J} \exp\left[-\text{tr}(\mathbf{R}^{-1}\bar{\mathbf{H}}_i)\right], \tag{7.23}$$

where

$$\bar{\mathbf{H}}_i = \left(\mathbf{Y}_p - i\mathbf{Sba}_p^\dagger\right)\left(\mathbf{Y}_p - i\mathbf{Sba}_p^\dagger\right)^\dagger, \tag{7.24}$$

with $\mathbf{Y}_p$ and $\mathbf{a}_p$ given in (7.6) and (7.7), respectively.

The maximization of $f(\mathbf{Y}|H_1)$ with respect to $\mathbf{a}_p$ can be achieved at

$$\hat{\mathbf{a}}_p = \frac{\mathbf{Y}_p^\dagger \mathbf{R}^{-1}\mathbf{Sb}}{\mathbf{b}^\dagger \mathbf{S}^\dagger \mathbf{R}^{-1}\mathbf{Sb}}. \tag{7.25}$$

As a result,

$$
\begin{aligned}
\max_{\{\mathbf{a}\}} f\left(\mathbf{Y}|H_1\right) &= \max_{\{\mathbf{a}_p\}} f\left(\mathbf{Y}|H_1\right) \\
&= \frac{1}{\pi^{NJ} \det(\mathbf{R})^J} \exp\left[-\operatorname{tr}(\mathbf{R}^{-1}\mathbf{Y}_p\mathbf{Y}_p^\dagger)\right] \\
&\quad \times \exp\left[\operatorname{tr}\left(\frac{\mathbf{R}^{-1}\mathbf{Y}_p\mathbf{Y}_p^\dagger \mathbf{R}^{-1}\mathbf{Sbb}^\dagger \mathbf{S}^\dagger}{\mathbf{b}^\dagger \mathbf{S}^\dagger \mathbf{R}^{-1}\mathbf{Sb}}\right)\right].
\end{aligned}
\tag{7.26}
$$

Substituting (7.26) into (7.22), it is not difficult to show that

$$\Xi_1 = \lambda_{\max}\left\{\left(\mathbf{S}_r^T\mathbf{R}_r^{-1}\mathbf{S}_r\right)^{-1}\mathbf{S}_r^T\mathbf{R}_r^{-1}\mathbf{Y}_r\mathbf{Y}_r^T\mathbf{R}_r^{-1}\mathbf{S}_r\right\}. \tag{7.27}$$

Note that all the quantities in (7.27) are real-valued.

In the second step, replacing $\mathbf{R}$ in (7.27) with $\hat{\mathbf{R}}$, we can obtain the detector as

$$\Xi_{\text{2S-PGLRT-H}} = \lambda_{\max}\left\{\tilde{\mathbf{Y}}^T\mathbf{P}_{\tilde{\mathbf{S}}}\tilde{\mathbf{Y}}\right\} \underset{H_0}{\overset{H_1}{\gtrless}} \xi_{\text{2S-PGLRT-H}}, \tag{7.28}$$

where $\xi_{\text{2S-PGLRT-H}}$ is a detection threshold. This detector is referred to as two-step persymmetric GLRT in homogeneous environments (2S-PGLRT-H), for ease of reference.

7.2.2 Wald Criterion

7.2.2.1 One-Step Wald Test

Let $\boldsymbol{\Theta}$ be the parameter vector partitioned as $\boldsymbol{\Theta} = [\boldsymbol{\Theta}_r^T, \boldsymbol{\Theta}_s^T]^T$, where the relevant parameter vector $\boldsymbol{\Theta}_r = [\mathbf{b}^T, \mathbf{a}_r^T]^T$, and the nuisance parameter vector is given by $\boldsymbol{\Theta}_s = \Upsilon(\mathbf{R}_r)$ with $\Upsilon(\cdot)$ a suitable function that selects the relevant entries of $\mathbf{R}_r$ according to its structure. In fact, any constraint on $\mathbf{R}_r$ can be addressed by suitably choosing the nuisance parameter vector, and hence we can apply the conventional unconstrained approach to compute the FIM [10]. The FIM $\mathbf{F}(\boldsymbol{\Theta})$ associated with $f(\mathbf{Y}, \mathbf{Y}_s|H_1)$ can be expressed as

$$\mathbf{F}(\boldsymbol{\Theta}) \triangleq \begin{bmatrix} \mathbf{F}_{\boldsymbol{\Theta}_r, \boldsymbol{\Theta}_r} & \mathbf{F}_{\boldsymbol{\Theta}_r, \boldsymbol{\Theta}_s} \\ \mathbf{F}_{\boldsymbol{\Theta}_s, \boldsymbol{\Theta}_r} & \mathbf{F}_{\boldsymbol{\Theta}_s, \boldsymbol{\Theta}_s} \end{bmatrix}, \tag{7.29}$$

where

$$\mathbf{F}_{\mathbf{\Theta}_r,\mathbf{\Theta}_r} = \mathrm{E}\left\{\frac{\partial \ln f(\mathbf{Y},\mathbf{Y}_s|H_1)}{\partial \mathbf{\Theta}_r}\frac{\partial \ln f(\mathbf{Y},\mathbf{Y}_s|H_1)}{\partial \mathbf{\Theta}_r^T}\right\}, \tag{7.30}$$

$$\mathbf{F}_{\mathbf{\Theta}_r,\mathbf{\Theta}_s} = \mathrm{E}\left\{\frac{\partial \ln f(\mathbf{Y},\mathbf{Y}_s|H_1)}{\partial \mathbf{\Theta}_r}\frac{\partial \ln f(\mathbf{Y},\mathbf{Y}_s|H_1)}{\partial \mathbf{\Theta}_s^T}\right\}, \tag{7.31}$$

$$\mathbf{F}_{\mathbf{\Theta}_s,\mathbf{\Theta}_s} = \mathrm{E}\left\{\frac{\partial \ln f(\mathbf{Y},\mathbf{Y}_s|H_1)}{\partial \mathbf{\Theta}_s}\frac{\partial \ln f(\mathbf{Y},\mathbf{Y}_s|H_1)}{\partial \mathbf{\Theta}_s^T}\right\}, \tag{7.32}$$

$$\mathbf{F}_{\mathbf{\Theta}_s,\mathbf{\Theta}_r} = \mathrm{E}\left\{\frac{\partial \ln f(\mathbf{Y},\mathbf{Y}_s|H_1)}{\partial \mathbf{\Theta}_s}\frac{\partial \ln f(\mathbf{Y},\mathbf{Y}_s|H_1)}{\partial \mathbf{\Theta}_r^T}\right\}. \tag{7.33}$$

According to [11, p. 188], the Wald test in the real-valued domain is given by

$$\left(\hat{\mathbf{\Theta}}_{r_1} - \hat{\mathbf{\Theta}}_{r_0}\right)^T\left\{\left[\mathbf{F}^{-1}(\hat{\mathbf{\Theta}}_1)\right]_{\mathbf{\Theta}_r,\mathbf{\Theta}_r}\right\}^{-1}\left(\hat{\mathbf{\Theta}}_{r_1} - \hat{\mathbf{\Theta}}_{r_0}\right) \underset{H_0}{\overset{H_1}{\gtrless}} \xi_4, \tag{7.34}$$

where ξ_4 is a detection threshold, $\hat{\mathbf{\Theta}}_{r_1}$ and $\hat{\mathbf{\Theta}}_{r_0}$ are the MLE and the value of $\mathbf{\Theta}_r$ under H_1 and H_0, respectively, and

$$\left\{\left[\mathbf{F}^{-1}(\hat{\mathbf{\Theta}}_1)\right]_{\mathbf{\Theta}_r,\mathbf{\Theta}_r}\right\}^{-1} = \mathbf{F}_{\mathbf{\Theta}_r,\mathbf{\Theta}_r} - \mathbf{F}_{\mathbf{\Theta}_r,\mathbf{\Theta}_s}\mathbf{F}_{\mathbf{\Theta}_s,\mathbf{\Theta}_s}^{-1}\mathbf{F}_{\mathbf{\Theta}_s,\mathbf{\Theta}_r} \tag{7.35}$$

is evaluated at the MLE $\hat{\mathbf{\Theta}}_1$ of $\mathbf{\Theta}$ under H_1.

As derived in Appendix 7.A, we have

$$\mathbf{F}_{\mathbf{\Theta}_r,\mathbf{\Theta}_r} = 4\left[\begin{array}{cc} \mathbf{a}_r^T\mathbf{a}_r\mathbf{S}_r^T\mathbf{R}_r^{-1}\mathbf{S}_r & \mathbf{S}_r^T\mathbf{R}_r^{-1}\mathbf{S}_r\mathbf{b}\mathbf{a}_r^T \\ (\mathbf{S}_r^T\mathbf{R}_r^{-1}\mathbf{S}_r\mathbf{b}\mathbf{a}_r^T)^T & \mathbf{b}^T\mathbf{S}_r^T\mathbf{R}_r^{-1}\mathbf{S}_r\mathbf{b}\mathbf{I}_{2J} \end{array}\right]. \tag{7.36}$$

Note that $\mathbf{F}_{\mathbf{\Theta}_r,\mathbf{\Theta}_s}$ is a null matrix. Then, we apply (7.36) to (7.34) to produce the Wald test with known parameters as

$$16\mathbf{a}_r^T\mathbf{a}_r\mathbf{b}^T\mathbf{S}_r^T\mathbf{R}_r^{-1}\mathbf{S}_r\mathbf{b} \underset{H_0}{\overset{H_1}{\gtrless}} \xi_4, \tag{7.37}$$

where ξ_4 is a detection threshold. The MLE of $\mathbf{R}_r$ under H_1 is given by

$$\bar{\mathbf{R}}_r = \frac{1}{K+J}\left[\hat{\mathbf{R}}_r + \left(\mathbf{Y}_r - \mathbf{S}_r\mathbf{b}\mathbf{a}_r^T\right)\left(\mathbf{Y}_r - \mathbf{S}_r\mathbf{b}\mathbf{a}_r^T\right)^T\right]. \tag{7.38}$$

The MLE of $\mathbf{a}_r$ can be rewritten as

$$\hat{\mathbf{a}}_r = \frac{\tilde{\mathbf{Y}}^T\mathbf{g}}{\mathbf{g}^T\mathbf{g}}, \tag{7.39}$$

where $\mathbf{g} = \tilde{\mathbf{S}}\mathbf{b}$. Using the MLEs in (7.38) and (7.39) to replace the corresponding parameters in (7.37), with the constant dropped, produces the Wald test as

$$\Xi_2 = \frac{\mathbf{g}^T\tilde{\mathbf{Y}}\tilde{\mathbf{Y}}^T\mathbf{g}}{(\mathbf{g}^T\mathbf{g})^2}\mathbf{g}^T\mathbf{K}\mathbf{g} \underset{H_0}{\overset{H_1}{\gtrless}} \xi_5, \tag{7.40}$$

where ξ_5 is a detection threshold, and

$$\mathbf{K} = \left[\mathbf{I}_N + \mathbf{P_g^\perp}\tilde{\mathbf{Y}}\tilde{\mathbf{Y}}^T\mathbf{P_g^\perp}\right]^{-1}. \qquad (7.41)$$

According to the matrix inversion lemma [12, p.99], we have

$$\mathbf{K} = \mathbf{I}_N - \mathbf{P_g^\perp}\tilde{\mathbf{Y}}\left(\mathbf{I}_{2J} + \tilde{\mathbf{Y}}^T\mathbf{P_g^\perp}\tilde{\mathbf{Y}}\right)^{-1}\tilde{\mathbf{Y}}^T\mathbf{P_g^\perp}. \qquad (7.42)$$

Taking (7.42) back into (7.40) produces

$$\Xi_2 = \frac{\mathbf{g}^T\tilde{\mathbf{Y}}\tilde{\mathbf{Y}}^T\mathbf{g}}{\mathbf{g}^T\mathbf{g}} = \frac{\mathbf{b}^T\tilde{\mathbf{S}}^T\tilde{\mathbf{Y}}\tilde{\mathbf{Y}}^T\tilde{\mathbf{S}}\mathbf{b}}{\mathbf{b}^T\tilde{\mathbf{S}}^T\tilde{\mathbf{S}}\mathbf{b}}. \qquad (7.43)$$

The MLE of $\mathbf{b}$ under H_1 is given as

$$\hat{\mathbf{b}} = \mathcal{P}_e\left\{\left(\tilde{\mathbf{S}}^T\tilde{\mathbf{S}}\right)^{-1}\tilde{\mathbf{S}}^T\tilde{\mathbf{Y}}\left(\mathbf{I}_{2J} + \tilde{\mathbf{Y}}^T\tilde{\mathbf{Y}}\right)^{-1}\tilde{\mathbf{Y}}^T\tilde{\mathbf{S}}\right\}. \qquad (7.44)$$

Finally, the Wald test can be derived as

$$\Xi_{\text{1S-PWald-H}} = \frac{\hat{\mathbf{b}}^T\tilde{\mathbf{S}}^T\tilde{\mathbf{Y}}\tilde{\mathbf{Y}}^T\tilde{\mathbf{S}}\hat{\mathbf{b}}}{\hat{\mathbf{b}}^T\tilde{\mathbf{S}}^T\tilde{\mathbf{S}}\hat{\mathbf{b}}} \underset{H_0}{\overset{H_1}{\gtrless}} \xi_{\text{1S-PWald-H}}, \qquad (7.45)$$

where $\xi_{\text{1S-PWald-H}}$ is a detection threshold. This detector is called one-step persymmetric Wald test in homogeneous environments (1S-PWald-H), for ease of reference.

7.2.2.2 Two-Step Wald Test

In the first step, the covariance matrix $\mathbf{R}$ is assumed known. It means that there is no nuisance parameter. In such a case, $\mathbf{\Theta} = \mathbf{\Theta_r} = [\mathbf{b}^T, \ \mathbf{a}_r^T]^T$, namely, $\mathbf{\Theta}$ only contains signal parameters. The FIM $\mathbf{F}(\mathbf{\Theta})$ associated with $f(\mathbf{Y}|H_1)$ can be expressed as

$$\mathbf{F}(\mathbf{\Theta_r}) = \mathrm{E}\left\{\frac{\partial \ln f(\mathbf{Y}|H_1)}{\partial \mathbf{\Theta_r}}\frac{\partial \ln f(\mathbf{Y}|H_1)}{\partial \mathbf{\Theta}_r^T}\right\}, \qquad (7.46)$$

where

$$f(\mathbf{Y}|H_1) = \frac{\exp\left\{-\mathrm{tr}\left[\mathbf{R}_r^{-1}\left(\mathbf{Y}_r - \mathbf{S}_r\mathbf{b}\mathbf{a}_r^T\right)\left(\mathbf{Y}_r - \mathbf{S}_r\mathbf{b}\mathbf{a}_r^T\right)^T\right]\right\}}{\pi^{NJ}\det(\mathbf{R}_r)^J}. \qquad (7.47)$$

Proceeding the same line as that for the derivation of (7.37), we can obtain the Wald test with known $\mathbf{b}$ and $\mathbf{a}_r$ as

$$\Xi_3 = \mathbf{a}_r^T\mathbf{a}_r\mathbf{b}^T\mathbf{S}_r^T\mathbf{R}_r^{-1}\mathbf{S}_r\mathbf{b}. \qquad (7.48)$$

Similar to (7.25), the MLE of $\mathbf{a}_r$ is given by

$$\bar{\mathbf{a}}_r = \frac{\mathbf{Y}_r^T \mathbf{R}_r^{-1} \mathbf{S}_r \mathbf{b}}{\mathbf{b}^T \mathbf{S}_r^T \mathbf{R}_r^{-1} \mathbf{S}_r \mathbf{b}}. \tag{7.49}$$

Substituting (7.49) into (7.48), we can rewrite the Wald test as

$$\Xi_4 = \frac{\mathbf{b}^T \mathbf{S}_r^T \mathbf{R}_r^{-1} \mathbf{Y}_r \mathbf{Y}_r^T \mathbf{R}_r^{-1} \mathbf{S}_r \mathbf{b}}{\mathbf{b}^T \mathbf{S}_r^T \mathbf{R}_r^{-1} \mathbf{S}_r \mathbf{b}}. \tag{7.50}$$

It is straightforward that the MLE of $\mathbf{b}$ is given by

$$\bar{\mathbf{b}} = \mathcal{P}_e \left\{ \left(\mathbf{S}_r^T \mathbf{R}_r^{-1} \mathbf{S}_r \right)^{-1} \mathbf{S}_r^T \mathbf{R}_r^{-1} \mathbf{Y}_r \mathbf{Y}_r^T \mathbf{R}_r^{-1} \mathbf{S}_r \right\}. \tag{7.51}$$

So the two-step Wald test with known $\mathbf{R}$ is

$$
\begin{aligned}
\Xi_5 &= \frac{\bar{\mathbf{b}}^T \mathbf{S}_r^T \mathbf{R}_r^{-1} \mathbf{Y}_r \mathbf{Y}_r^T \mathbf{R}_r^{-1} \mathbf{S}_r \bar{\mathbf{b}}}{\bar{\mathbf{b}}^T \mathbf{S}_r^T \mathbf{R}_r^{-1} \mathbf{S}_r \bar{\mathbf{b}}} \\
&= \lambda_{\max} \left\{ \left(\mathbf{S}_r^T \mathbf{R}_r^{-1} \mathbf{S}_r \right)^{-1} \mathbf{S}_r^T \mathbf{R}_r^{-1} \mathbf{Y}_r \mathbf{Y}_r^T \mathbf{R}_r^{-1} \mathbf{S}_r \right\} \\
&\underset{H_0}{\overset{H_1}{\gtrless}} \xi_7,
\end{aligned}
\tag{7.52}
$$

where ξ_7 is a detection threshold. Note that the MLE of $\mathbf{R}$ based on the training data is given in (7.5). In the second step, we replace $\mathbf{R}$ with $\hat{\mathbf{R}}$ in Ξ_5 to derive the two-step Wald test, which has the same form as the 2S-PGLRT-H in (7.28).

7.2.3   Rao Criterion

According to [11, p. 189], the Rao test in the real-valued domain is given by

$$\Xi_6 = \left. \frac{\partial \ln f(\mathbf{Y}|H_1)}{\partial \mathbf{\Theta}_r} \right|_{\mathbf{\Theta}=\hat{\mathbf{\Theta}}_0}^T \left[\mathbf{F}^{-1}(\hat{\mathbf{\Theta}}_0) \right]_{\mathbf{\Theta}_r,\mathbf{\Theta}_r} \left. \frac{\partial \ln (\mathbf{Y}|H_1)}{\partial \mathbf{\Theta}_r} \right|_{\mathbf{\Theta}=\hat{\mathbf{\Theta}}_0}, \tag{7.53}$$

where $\hat{\mathbf{\Theta}}_0$ is the MLE of $\mathbf{\Theta}$ under H_0, and $\left[\mathbf{F}^{-1}(\hat{\mathbf{\Theta}}_0) \right]_{\mathbf{\Theta}_r,\mathbf{\Theta}_r}$ is $\left[\mathbf{F}^{-1}(\mathbf{\Theta}) \right]_{\mathbf{\Theta}_r,\mathbf{\Theta}_r}$ evaluated at $\hat{\mathbf{\Theta}}_0$. It is easy to check that $\left[\mathbf{F}^{-1}(\hat{\mathbf{\Theta}}_0) \right]_{\mathbf{\Theta}_r,\mathbf{\Theta}_r}$ does not exist for both one-step and two-step Rao tests. That is to say, the Rao test does not exist.

Now we provide some remarks on the CFAR properties and computational complexity of those detectors. Following the lead of [13], we can show that $\tilde{\mathbf{Y}}^T \tilde{\mathbf{Y}}$ and $\tilde{\mathbf{Y}}^T \mathbf{P}_{\tilde{\mathbf{S}}} \tilde{\mathbf{Y}}$ are independent of $\mathbf{R}$ under H_0. Hence, the detectors exhibit the CFAR properties. As for the computational complexities of those detectors, the relationship is $\Xi_{\text{2S-PGLRT-H}} < \Xi_{\text{1S-PGLRT-H}} < \Xi_{\text{1S-PWald-H}}$.

7.3   Numerical Examples

In this section, numerical simulations are provided to evaluate the detection performance of those detectors. Assume that a pulsed Doppler radar is used, which transmits symmetrically spaced pulse trains. The number of pulses in a coherent processing interval is $N = 9$. The (i, j)th element of the noise covariance matrix is chosen as $\mathbf{R}(i, j) = \sigma^2 \rho^{|i-j|}$, where σ^2 is the noise power, and $\rho = 0.9$ is the noise correlation coefficient. We choose that the true Doppler frequency is 0.15, and the number of range cells is $J = 4$. The PFA is set to be 10^{-3}. We select that $q = 3$, and the subspace matrix $\mathbf{S}$ is created using q steering vectors at $\tilde{f}_d - 0.03$, $\tilde{f}_d$, and $\tilde{f}_d + 0.03$, where $\tilde{f}_d$ is the nominal Doppler frequency. It means that the subspace is spanned by the nominal steering vector and two additional steering vectors adjacent to the nominal one. A similar approach to choose the subspace matrix can be found in [14, 15]. Note that when $\tilde{f}_d = 0.15$, it corresponds to the matched case. Denote by $\mathbf{p}$ the nominal target steering vector. The mismatch angle ϕ between the true steering vector $\mathbf{q}$ and the assumed one $\mathbf{p}$ is defined by

$$\cos^2 \phi = \frac{|\mathbf{p}^\dagger \mathbf{R}^{-1} \mathbf{q}|^2}{(\mathbf{p}^\dagger \mathbf{R}^{-1} \mathbf{p})(\mathbf{q}^\dagger \mathbf{R}^{-1} \mathbf{q})}. \tag{7.54}$$

In the matched case, $\mathbf{q} = \mathbf{p}$ and $\cos^2 \phi = 1$. In the mismatched case, $\mathbf{q} \neq \mathbf{p}$ and $\cos^2 \phi < 1$. In the following simulations, we change the value of $\tilde{f}_d$ to alter the mismatch degree. The SNR in decibel is defined by

$$\text{SNR} = 10 \log_{10} \left(\mathbf{a}^\dagger \mathbf{a} \mathbf{q}^\dagger \mathbf{R}^{-1} \mathbf{q} \right). \tag{7.55}$$

For comparison purposes, the PGLRT-D, PAMF-D in Chapter 6, GLRT [5, eq. (27)], and AMF [5, eq. (28)] are considered.

In Fig. 7.1, we depict the detection probability as a function of SNR for different K. Here, we choose $\tilde{f}_d = 0.10$. In such a case, we have $\cos^2 \phi = 0.5427$. It can be observed that the three detectors are more robust than their counterparts. Interestingly, the 2S-PGLRT-H detector has the same performance as the 1S-PWald-H detectors. It has to be pointed out that when $K < N$, the conventional GLRT and AMF in [5] do not work, and hence their performance is not presented in Fig. 7.1(c).

In Fig. 7.2, we consider the case where $\tilde{f}_d = 0.12$ (i.e., $\cos^2 \phi = 0.8126$). The other parameters are chosen to be the same as those in Fig. 7.1. In this case, the mismatch in the steering vector reduces. We can see that the detectors designed in this chapter are still more robust than their counterparts, and the 1S-PGLRT-H detector is the most robust in these parameters.

In the matched case where the nominal steering vector is aligned with the true steering vector (i.e., $\cos^2 \phi = 1$), we present the detection probability as a function of SNR in Fig. 7.3. It is shown that the designed detectors exhibit

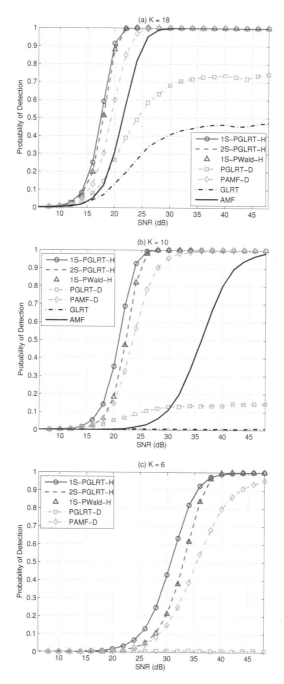

FIGURE 7.1
Detection probability versus SNR for $\cos^2 \phi = 0.5427$ in the mismatched case.

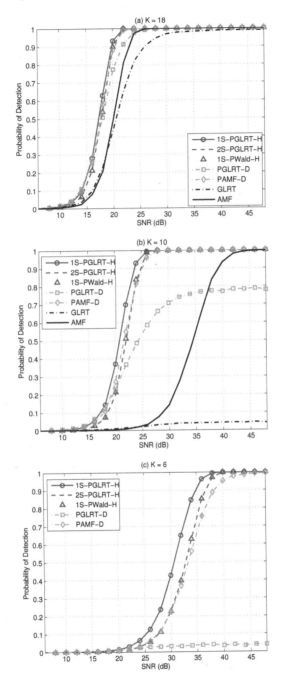

FIGURE 7.2
Detection probability versus SNR for $\cos^2 \phi = 0.8126$ in the mismatched case.

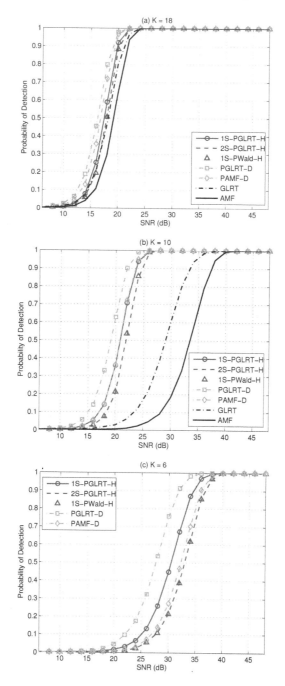

FIGURE 7.3
Detection probability versus SNR for $\cos^2 \phi = 1$ in the matched case.

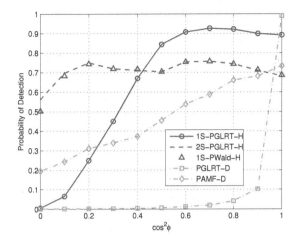

FIGURE 7.4
Detection probability versus $\cos^2 \phi$ for $K = 6$ and SNR $= 35$ dB.

performance degradations, compared to the PGLRT-D. This is expected, because the use of subspace model results in the accumulation of more noise energy. The performance losses of the designed detectors in the matched case can be seen as the costs to achieve robustness in the mismatched case.

In Fig. 7.4, we plot the detection probability with respect to $\cos^2 \phi$. It can be seen that the 2S-PGLRT-H and 1S-PWald-H detectors ensure an increased robustness in most cases (i.e., $\cos^2 \phi < 0.9$ in this example), and the 2S-PGLRT-H and 1S-PWald-H have better performance than the 1S-PGLRT-H in some cases, for instance, when $\cos^2 \phi < 0.4$. Gathering results in Figs. 7.3 and 7.4 reveals that the robustness of the designed detectors in the mismatched case is obtained by sacrificing their detection performance in the matched case.

7.A   Derivations of (7.36)

Using (7.14), we have

$$\frac{\partial \ln f(\mathbf{Y}, \mathbf{Y}_s | H_1)}{\partial \mathbf{b}} = 2\mathbf{S}_r^T \mathbf{R}_r^{-1} \breve{\mathbf{Y}}_r \mathbf{a}_r, \qquad (7.A.1)$$

and

$$\frac{\partial \ln f(\mathbf{Y}, \mathbf{Y}_s | H_1)}{\partial \mathbf{a}_r} = 2\breve{\mathbf{Y}}_r \mathbf{R}_r^{-1} \mathbf{S}_r \mathbf{b}, \qquad (7.A.2)$$

where $\breve{\mathbf{Y}}_r = \mathbf{Y}_r - \mathbf{S}_r \mathbf{ba}_r$. Then, we can obtain

$$
\mathbf{F}_{\Theta_r,\Theta_r} = 4\mathrm{E}\left[\begin{array}{c} \mathbf{S}_r^T \mathbf{R}_r^{-1} \breve{\mathbf{Y}}_r \mathbf{a}_r \mathbf{a}_r^T \breve{\mathbf{Y}}_r^T \mathbf{R}_r^{-1} \mathbf{S}_r \\ \breve{\mathbf{Y}}_r^T \mathbf{R}_r^{-1} \mathbf{S}_r \mathbf{ba}_r^T \breve{\mathbf{Y}}_r^T \mathbf{R}_r^{-1} \mathbf{S}_r \end{array}\right.
$$
$$
\left.\begin{array}{c} \mathbf{S}_r^T \mathbf{R}_r^{-1} \breve{\mathbf{Y}}_r \mathbf{a}_r \mathbf{b}^T \mathbf{S}_r^T \mathbf{R}_r^{-1} \breve{\mathbf{Y}}_r \\ \breve{\mathbf{Y}}_r^T \mathbf{R}_r^{-1} \mathbf{S}_r \mathbf{bb}^T \mathbf{S}_r^T \mathbf{R}_r^{-1} \breve{\mathbf{Y}}_r \end{array}\right] \tag{7.A.3}
$$
$$
= 4\left[\begin{array}{cc} \mathbf{F}_{11} & \mathbf{F}_{12} \\ \mathbf{F}_{21} & \mathbf{F}_{22} \end{array}\right],
$$

where

$$
\mathbf{F}_{11} = \mathrm{E}[\mathbf{S}_r^T \mathbf{R}_r^{-1} \breve{\mathbf{Y}}_r \mathbf{a}_r \mathbf{a}_r^T \breve{\mathbf{Y}}_r^T \mathbf{R}_r^{-1} \mathbf{S}_r] = \mathbf{a}_r^T \mathbf{a}_r \mathbf{S}_r^T \mathbf{R}_r^{-1} \mathbf{S}_r. \tag{7.A.4}
$$

To obtain $\mathbf{F}_{12}$, $\mathbf{F}_{21}$, and $\mathbf{F}_{22}$, we define

$$
\begin{cases} \mathbf{a}_r = [\alpha_1, \alpha_2, \ldots, \alpha_{2J}], \\ \boldsymbol{\theta} = \mathbf{R}_r^{-1} \mathbf{S}_r \mathbf{b}. \end{cases} \tag{7.A.5}
$$

Then, we have

$$
\begin{aligned}
\mathbf{F}_{12} = \mathbf{F}_{21}^T &= \mathrm{E}[\mathbf{S}_r^T \mathbf{R}_r^{-1} \breve{\mathbf{Y}}_r \mathbf{a}_r \boldsymbol{\theta}^T \breve{\mathbf{Y}}_r] \\
&= \mathbf{S}_r^T \mathbf{R}_r^{-1} \mathrm{E}\left\{ \sum_{i=1}^{2J} \alpha_i \breve{\mathbf{y}}_i \left[\breve{\mathbf{y}}_1^T \boldsymbol{\theta}, \breve{\mathbf{y}}_2^T \boldsymbol{\theta}, \ldots, \breve{\mathbf{y}}_{2J}^T \boldsymbol{\theta} \right] \right\} \\
&= \mathbf{S}_r^T \mathbf{R}_r^{-1} \left[\alpha_1 \mathbf{R}_r \boldsymbol{\theta}, \alpha_2 \mathbf{R}_r \boldsymbol{\theta}, \ldots, \alpha_{2J} \mathbf{R}_r \boldsymbol{\theta} \right] \\
&= \mathbf{S}_r^T \mathbf{R}_r^{-1} \mathbf{R}_r \boldsymbol{\theta} \boldsymbol{\alpha}^T = \mathbf{S}_r^T \mathbf{R}_r^{-1} \mathbf{S}_r \mathbf{ba}_r^T,
\end{aligned} \tag{7.A.6}
$$

where $\breve{\mathbf{y}}_i$ is the ith column vector of $\breve{\mathbf{Y}}_r$. In addition,

$$
\begin{aligned}
\mathbf{F}_{22} &= \mathrm{E}\left\{ \left[\begin{array}{c} \boldsymbol{\theta}^T \breve{\mathbf{y}}_1 \\ \boldsymbol{\theta}^T \breve{\mathbf{y}}_2 \\ \cdots \\ \boldsymbol{\theta}^T \breve{\mathbf{y}}_{2J} \end{array}\right] \left[\breve{\mathbf{y}}_1^T \boldsymbol{\theta}, \breve{\mathbf{y}}_2^T \boldsymbol{\theta}, \ldots, \breve{\mathbf{y}}_{2J}^T \boldsymbol{\theta} \right] \right\} \\
&= \boldsymbol{\theta}^T \mathbf{R}_r \boldsymbol{\theta} \mathbf{I}_{2J} \\
&= \mathbf{b}^T \mathbf{S}_r^T \mathbf{R}_r^{-1} \mathbf{S}_r \mathbf{b} \mathbf{I}_{2J}.
\end{aligned} \tag{7.A.7}
$$

Inserting (7.A.4), (7.A.6), and (7.A.7) into (7.A.3), we can obtain (7.36).

Bibliography

[1] O. Besson, L. L. Scharf, and F. Vincent, "Matched direction detectors and estimators for array processing with subspace steering vector uncertainties," *IEEE Transactions on Signal Processing*, vol. 53, no. 12, pp. 4453–4463, December 2005.

[2] O. Besson, L. L. Scharf, and S. Kraut, "Adaptive detection of a signal known only to lie on a line in a known subspace, when primary and secondary data are partially homogeneous," *IEEE Transaction on Signal Processing*, vol. 54, no. 12, pp. 4698–4705, December 2006.

[3] J. Liu, W. Liu, C. Hao, and D. Orlando, "Persymmetric subspace detectors with multiple observations in homogeneous environments," *IEEE Transactions on Aerospace and Electronic Systems*, vol. 56, no. 4, pp. 3276–3284, August 2020.

[4] L. Cai and H. Wang, "A persymmetric multiband GLR algorithm," *IEEE Transactions on Aerospace and Electronic Systems*, vol. 28, no. 3, pp. 806–816, July 1992.

[5] E. Conte, A. De Maio, and G. Ricci, "GLRT-based adaptive detection algorithms for range-spread targets," *IEEE Transactions on Signal Processing*, vol. 49, no. 7, pp. 1336–1348, July 2001.

[6] N. Bon, A. Khenchaf, and R. Garello, "GLRT subspace detection for range and Doppler distributed targets," *IEEE Transactions on Aerospace and Electronic Systems*, vol. 44, no. 2, pp. 678–696, April 2008.

[7] F. Gini and A. Farina, "Vector subspace detection in compound-Gaussian clutter. part I: survey and new results," *IEEE Transactions on Aerospace and Electronic Systems*, vol. 38, no. 4, pp. 1295–1311, October 2002.

[8] ——, "Vector subspace detection in compound-Gaussian clutter. part II: performance analysis," *IEEE Transactions on Aerospace and Electronic Systems*, vol. 38, no. 4, pp. 1312–1323, October 2002.

[9] W. Liu, W. Xie, J. Liu, D. Zou, H. Wang, and Y. Wang, "Detection of a distributed target with direction uncertainty," *IET Radar, Sonar & Navigation*, vol. 8, no. 9, pp. 1177–1183, 2014.

[10] T. L. Marzetta, "A simple derivation of the constrained multiple parameter Cramer-Rao bound," *IEEE Transaction on Signal Processing*, vol. 6, no. 41, pp. 2247–2249, June 1993.

[11] S. M. Kay, *Fundamentals of Statistical Signal Processing: Detection Theory*. Upper Saddle River, NJ: Prentice Hall, 1998.

[12] D. A. Harville, *Matrix Algebra For a Statistician's Perspective*. New York: Springer-Verlag, 1997.

[13] F. Bandiera, D. Orlando, and G. Ricci, "On the CFAR property of GLRT-based direction detectors," *IEEE Transactions on Signal Processing*, vol. 55, no. 8, pp. 4312–4315, August 2007.

[14] J. Carretero-Moya, A. De Maio, J. Gismero-Menoyo, and A. Asensio-Lopez, "Experimental performance analysis of distributed target coherent radar detectors," *IEEE Transactions on Aerospace and Electronic Systems*, vol. 48, no. 3, pp. 2216–2238, July 2012.

[15] F. Bandiera, D. Orlando, and G. Ricci, *Advanced Radar Detection Schemes under Mismatched Signal Models in Synthesis Lectures on Signal Processing*. San Rafael, CA, USA, Morgan & Claypool, 2009.

8

Adaptive Detection with Unknown Steering Vector

The previous chapter assumes that the target signature is partially known, i.e., belonging to a known subspace with unknown coordinates. In practice, the target signature may be completely unknown (e.g., due to severe multipath [1], [2]). In [3], a GLRT-based detector was designed for the problem of detecting a point-like target with unknown signature. In [1], the authors designed several detectors for the distributed target detection with unknown signatures, and first introduced a decision rule of SNT based on the spectral norm [4, p. 295]. The authors in [5] resorted to the Rao and Wald tests to design detectors for solving the same problem. It should be mentioned that the adaptive double subspace signal detection problem in Gaussian noise with unknown covariance matrix was considered in [6], where three SNT detectors were provided.

In this chapter, we consider the distributed target detection problem with unknown signal signatures in Gaussian noise with unknown covariance matrix. Two adaptive detectors are designed by using the persymmetry of the noise covariance matrix. We derive analytical expressions for the probabilities of false alarm of both detectors, which indicate their CFAR properties against the noise covariance matrix. All the theoretical expressions are confirmed by MC simulations. Numerical examples demonstrate that both detectors have better detection performance than their counterparts, especially in the case of limited training data.

8.1 Problem Formulation

Suppose that data are collected from N (temporal, spatial, or spatial-temporal) channels. The problem under consideration is to detect a distributed target across J range cells. The echoes received from these range cells, which are called primary or test data, are denoted by

$$\mathbf{z}_k = \mathbf{b}_k + \mathbf{n}_k, \quad k = 1, \ldots, J, \tag{8.1}$$

where $\mathbf{b}_k$ represents the signal of interest in the kth range cell, $\mathbf{n}_k$ represents noise in the kth range cell. We assume that $\mathbf{n}_k$ has a circularly symmetric,

DOI: 10.1201/9781003340232-8

complex Gaussian distribution with zero mean and unknown covariance matrix $\mathbf{R}$, i.e., $\mathbf{n}_k \sim \mathcal{CN}_N(\mathbf{0}, \mathbf{R})$. In addition, it is assumed that $\mathbf{b}_k$s are unknown (e.g., due to multipath [7]).

As customary, a set of secondary (or training) data $\mathbf{Z}_L = [\mathbf{z}_{L1}, \ldots, \mathbf{z}_{LL}] \in \mathbb{C}^{N \times L}$, which only contains noise $\mathbf{N}_L \in \mathbb{C}^{N \times L}$, is available. It should be pointed out that each column in $\mathbf{N}_L$ shares the same statistical property as $\mathbf{n}_k$ in the primary data and is free of the target echoes. Suppose further that the noise $\mathbf{n}_k$ is IID for $k = 1, \ldots, J + L$.

In practice, the covariance matrix $\mathbf{R}$ has a persymmetric structure, when the receiver uses a symmetrically spaced linear array and/or symmetrically spaced pulse trains [8,9]. When we say $\mathbf{R}$ is persymmetric, it satisfies

$$\mathbf{R} = \mathbf{J}\mathbf{R}^*\mathbf{J}, \tag{8.2}$$

where $\mathbf{J}$ is defined in (1.2).

To this end, we formulate the distributed target detection problem as the following binary hypothesis test:

$$\begin{cases} H_0 : \mathbf{Z} = \mathbf{N}, \mathbf{Z}_L = \mathbf{N}_L, \\ H_1 : \mathbf{Z} = \mathbf{B} + \mathbf{N}, \mathbf{Z}_L = \mathbf{N}_L, \end{cases} \tag{8.3}$$

where

$$\mathbf{Z} = [\mathbf{z}_1, \ldots, \mathbf{z}_J] \in \mathbb{C}^{N \times J}, \tag{8.4}$$

$$\mathbf{N} = [\mathbf{n}_1, \ldots, \mathbf{n}_J] \in \mathbb{C}^{N \times J}, \tag{8.5}$$

$$\mathbf{B} = [\mathbf{b}_1, \ldots, \mathbf{b}_J] \in \mathbb{C}^{N \times J}. \tag{8.6}$$

It should be emphasized here that $\mathbf{B}$ is unknown. In addition, the condition $2L \geqslant N$ has to be satisfied, which is much less restrictive than that in [1,5].

8.2   Per-SNT Detector

8.2.1   Detector Design

It is easy to obtain that the PDF of $\mathbf{Z}$ under the hypothesis H_q ($q = 0, 1$) is

$$f(\mathbf{Z}|H_q) = \pi^{-NJ} \det(\mathbf{R})^{-J} \exp\left[-\mathrm{tr}(\mathbf{R}^{-1}\mathbf{G}_q)\right], \quad q = 0, 1, \tag{8.7}$$

where

$$\mathbf{G}_q = (\mathbf{Z} - q\mathbf{B})(\mathbf{Z} - q\mathbf{B})^\dagger. \tag{8.8}$$

We temporarily assume that the covariance matrix $\mathbf{R}$ is known. Then the detector is given by

$$\Xi_0 = \frac{\max_{\mathbf{B}} f(\mathbf{Z}|H_1)}{f(\mathbf{Z}|H_0)} \underset{H_0}{\overset{H_1}{\gtrless}} \xi_0, \tag{8.9}$$

where ξ_0 is the detection threshold.

Exploiting the persymmetry of $\mathbf{R}$, we have

$$\begin{aligned}
\operatorname{tr}(\mathbf{R}^{-1}\mathbf{G}_q) &= \operatorname{tr}[\mathbf{J}(\mathbf{R}^*)^{-1}\mathbf{J}\mathbf{G}_q] \\
&= \operatorname{tr}[(\mathbf{R}^*)^{-1}\mathbf{J}\mathbf{G}_q\mathbf{J}] \\
&= \operatorname{tr}[\mathbf{R}^{-1}\mathbf{J}\mathbf{G}_q^*\mathbf{J}], \quad q = 0, 1.
\end{aligned} \tag{8.10}$$

Hence, we can rewrite (8.7) as

$$f(\mathbf{Z}|H_q) = \pi^{-NJ}\det(\mathbf{R})^{-J}\exp\left[-\operatorname{tr}(\mathbf{R}^{-1}\mathbf{H}_q)\right], \quad q = 0, 1, \tag{8.11}$$

where

$$\mathbf{H}_q = \frac{1}{2}\left(\mathbf{G}_q + \mathbf{J}\mathbf{G}_q^*\mathbf{J}\right). \tag{8.12}$$

Substituting (8.8) into (8.12), we have (see Appendix 8.A for the detailed derivations)

$$\mathbf{H}_q = (\mathbf{Z}_p - q\mathbf{B}_p)(\mathbf{Z}_p - q\mathbf{B}_p)^\dagger, \tag{8.13}$$

where

$$\mathbf{Z}_p = [\mathbf{z}_{e1}, \ldots \mathbf{z}_{eJ}, \mathbf{z}_{o1}, \ldots, \mathbf{z}_{oJ}] \in \mathbb{C}^{N \times 2J}, \tag{8.14}$$

and

$$\mathbf{B}_p = [\mathbf{b}_{e1}, \ldots \mathbf{b}_{eJ}, \mathbf{b}_{o1}, \ldots, \mathbf{b}_{oJ}] \in \mathbb{C}^{N \times 2J}, \tag{8.15}$$

with

$$\begin{cases}
\mathbf{z}_{ej} = \frac{1}{2}\left(\mathbf{z}_j + \mathbf{J}\mathbf{z}_j^*\right), \\
\mathbf{z}_{oj} = \frac{1}{2}\left(\mathbf{z}_j - \mathbf{J}\mathbf{z}_j^*\right),
\end{cases} \tag{8.16}$$

and

$$\begin{cases}
\mathbf{b}_{ej} = \frac{1}{2}\left(\mathbf{b}_j + \mathbf{J}\mathbf{b}_j^*\right), \\
\mathbf{b}_{oj} = \frac{1}{2}\left(\mathbf{b}_j - \mathbf{J}\mathbf{b}_j^*\right).
\end{cases} \tag{8.17}$$

Now, the maximization of $f(\mathbf{Z}|H_1)$ with respect to $\mathbf{B}_p$ can be achieved at

$$\hat{\mathbf{B}}_p = \mathbf{Z}_p. \tag{8.18}$$

Substituting (8.18) into (8.9), we obtain

$$\max_{\mathbf{B}} f(\mathbf{Z}|H_1) = \pi^{-NJ}\det(\mathbf{R})^{-J}, \tag{8.19}$$

and it is obvious that

$$f(\mathbf{Z}|H_0) = \pi^{-NJ}\det(\mathbf{R})^{-J}\exp\left[-\operatorname{tr}\left(\mathbf{R}^{-1}\mathbf{Z}_p\mathbf{Z}_p^\dagger\right)\right]. \tag{8.20}$$

Applying (8.19) and (8.20) to (8.9), we can derive the detector with known $\mathbf{R}$ as

$$\Xi_1 = \operatorname{tr}(\mathbf{Z}_p^\dagger\mathbf{R}^{-1}\mathbf{Z}_p) \underset{H_0}{\overset{H_1}{\gtrless}} \xi_1, \tag{8.21}$$

where ξ_1 is the detection threshold. This detector cannot be used in practice, since $\mathbf{R}$ is unknown.

Based on the secondary data, we can obtain the estimate (up to a scalar) of the covariance matrix $\mathbf{R}$ as [8]

$$\hat{\mathbf{R}} = \frac{1}{2} \left(\mathbf{R}_1 + \mathbf{J} \mathbf{R}_1^* \mathbf{J} \right) \in \mathbb{C}^{N \times N}, \tag{8.22}$$

where

$$\mathbf{R}_1 = \mathbf{Z}_L \mathbf{Z}_L^\dagger. \tag{8.23}$$

Further, we use $\hat{\mathbf{R}}$ to take the place of $\mathbf{R}$ in (8.21) and then obtain the two-step detector as

$$\Xi_2 = \text{tr} \left(\mathbf{Z}_p^\dagger \hat{\mathbf{R}}^{-1} \mathbf{Z}_p \right) = \sum_{t=1}^{2J} \lambda_t \overset{H_1}{\underset{H_0}{\gtrless}} \xi_2, \tag{8.24}$$

where ξ_2 is the detection threshold, $\lambda_1 \geqslant \ldots \geqslant \lambda_{2J}$ are eigenvalues of $\mathbf{Z}_p^\dagger \hat{\mathbf{R}}^{-1} \mathbf{Z}_p$ with $\mathbf{Z}_p$ and $\hat{\mathbf{R}}$ given by (8.14) and (8.22), respectively.

According to [1], it is reasonable, from a detection point of view, to choose the largest eigenvalue as the detection statistic, because it may have a high probability of not being perturbed by noise. As a result, a spectral-norm-based (spectral norm represents the largest eigenvalue) detector was proposed by the authors in [1]. Other SNT detectors can be found in [5,6]. These SNT detectors behave more like direction detectors which search for a direction of maximum energy. Note that these SNT detectors do not exploit persymmetry.

Now, we choose the largest eigenvalue of the matrix $\mathbf{Z}_p^\dagger \hat{\mathbf{R}}^{-1} \mathbf{Z}_p$ as a detector, i.e.,

$$\Xi_3 = \lambda_1 = \lambda_{\max} \left(\mathbf{Z}_p^\dagger \hat{\mathbf{R}}^{-1} \mathbf{Z}_p \right) \overset{H_1}{\underset{H_0}{\gtrless}} \xi_3, \tag{8.25}$$

where ξ_3 is the detection threshold. This detector is referred to as the per-SNT detector, since it uses the persymmetric structure.

8.2.2 Threshold Setting for Per-SNT

In this section, we derive analytical expressions for the probabilities of false alarm of the per-SNT detector, which can facilitate the threshold setting. To this end, we first introduce a theorem derived by using the results from [10].

Theorem 8.2.1. *Assume that* $\mathbf{X} \sim \mathcal{N}_{p \times n}(\mathbf{0}_{p \times n}, \mathbf{\Omega} \otimes \mathbf{I}_n)$ *and* $\mathbf{Y} \sim \mathcal{N}_{p \times m}(\mathbf{0}_{p \times m}, \mathbf{\Omega} \otimes \mathbf{I}_m)$ *are IID, where* $p \leqslant m$ *and* $\mathbf{\Omega}$ *is a positive definite covariance matrix. Then, the exact and approximate distributions of the largest eigenvalue of* $\mathbf{F} = \mathbf{X}^T (\mathbf{Y} \mathbf{Y}^T)^{-1} \mathbf{X}$ *in different cases are given as follows:*

The Exact Case:

For arbitrary n, the exact distribution is

$$\mathcal{P}(\lambda_{\max}(\mathbf{F}) \leqslant \lambda) = C(z, s, r) \sqrt{\left| \mathbf{A} \left(\frac{\lambda}{1 + \lambda} \right) \right|}, \tag{8.26}$$

where

$$z = \min(p, n), \tag{8.27}$$

$$s = (|n - p| - 1)/2, \tag{8.28}$$

$$r = (m - p - 1)/2, \tag{8.29}$$

$$C(z, s, r) = \pi^{z/2} \prod_{i=1}^{z} \frac{\Gamma(\frac{i+2s+2r+z+2}{2})}{\Gamma(\frac{i}{2})\Gamma(\frac{i+2s+1}{2})\Gamma(\frac{i+2r+1}{2})}. \tag{8.30}$$

When z is even, $\mathbf{A}(\theta)$ is a $z \times z$ skew-symmetric matrix, the (i, j)-element in $\mathbf{A}(\theta)$ is

$$a_{i,j}(\theta) = \mathcal{E}(\theta; s + j, s + i) - \mathcal{E}(\theta; s + i, s + j), \tag{8.31}$$

where $i, j = 1, 2 \ldots, z$, and

$$\mathcal{E}(x; a, b) = \int_0^x t^{a-1}(1 - t)^r \mathcal{B}(t; b, r + 1)dt. \tag{8.32}$$

When z is odd, $\mathbf{A}(\theta)$ is a $(z+1) \times (z+1)$ skew-symmetric matrix, the elements are as (8.31) with the additional elements:

$$a_{i,z+1}(\theta) = \mathcal{B}(\theta; s + i, r + 1), \quad i = 1, \ldots, z, \tag{8.33}$$

$$a_{z+1,j}(\theta) = -a_{j,z+1}(\theta), \quad j = 1, \ldots, z, \tag{8.34}$$

and

$$a_{z+1,z+1}(\theta) = 0. \tag{8.35}$$

Note that $a_{i,j}(\theta) = -a_{j,i}(\theta)$ and $a_{i,i}(\theta) = 0$.

The Approximate Case:

For $m > p$, the approximate distribution with large p and n is

$$\mathcal{P}(\lambda_{\max}(\mathbf{F}) \leqslant \lambda) \simeq \gamma \left(\tau, \frac{\ln \lambda - \mu + \sigma \alpha}{\delta \sigma} \right), \tag{8.36}$$

where $\tau = 46.446, \delta = 0.186054, \alpha = 9.84801$ and

$$\mu = 2 \ln \tan \left(\frac{\varphi + \phi}{2} \right), \tag{8.37}$$

$$\sigma^3 = \frac{16}{(m + n - 1)^2 \sin^2(\varphi + \phi) \sin \varphi \sin \phi}, \tag{8.38}$$

$$\varphi = \arccos \left(\frac{m + n - 2p}{m + n - 1} \right), \tag{8.39}$$

$$\phi = \arccos \left(\frac{m - n}{m + n - 1} \right). \tag{8.40}$$

Proof. See Appendix 8.B.   □

8.2.2.1   Transformation from Complex Domain to Real Domain

For statistical analysis, the collected data are transformed from the complex-valued domain to the real-valued domain. After some equivalent transformations, we can rewrite (8.25) and (8.63) as (see Appendix 8.C for the detailed derivations)

$$\Xi = \lambda_{\max}(\mathbf{Z}_r^T \hat{\mathbf{R}}_r^{-1} \mathbf{Z}_r) \overset{H_1}{\underset{H_0}{\gtrless}} \xi, \tag{8.41}$$

and

$$\Lambda = \frac{\det(\mathbf{Z}_r \mathbf{Z}_r^T + \mathbf{Z}_{Lr} \mathbf{Z}_{Lr}^T)}{\det(\mathbf{Z}_{Lr} \mathbf{Z}_{Lr}^T)} \overset{H_1}{\underset{H_0}{\gtrless}} \zeta, \tag{8.42}$$

respectively, where $\hat{\mathbf{R}}_r$, $\mathbf{Z}_r$, and $\mathbf{Z}_{Lr}$ are given by (8.C.10), (8.C.11), and (8.C.12), respectively, ξ and ζ are the detection thresholds. It has to be emphasized that the condition $2L \geqslant N$ needs to be satisfied to guarantee the nonsingularity of $\hat{\mathbf{R}}_r$ and $\mathbf{Z}_{Lr}\mathbf{Z}_{Lr}^T$. This requirement is less restrictive than that in [1, 5].

Based on [9, eq. (B11)], we have

$$\hat{\mathbf{R}}_r \sim \mathcal{W}_N(2L, \mathbf{R}_r), \tag{8.43}$$

$$\begin{cases} \mathbf{z}_{erj} \sim \mathcal{N}_N(\mathbf{0}, \mathbf{R}_r), \\ \mathbf{z}_{orj} \sim \mathcal{N}_N(\mathbf{0}, \mathbf{R}_r), \end{cases} \tag{8.44}$$

for $j = 1, 2, \ldots, J$ under hypothesis H_0, and

$$\begin{cases} \mathbf{z}_{Lerj} \sim \mathcal{N}_N(\mathbf{0}, \mathbf{R}_r), \\ \mathbf{z}_{Lorj} \sim \mathcal{N}_N(\mathbf{0}, \mathbf{R}_r), \end{cases} \tag{8.45}$$

for $j = 1, 2, \ldots, L$, where $\mathbf{z}_{erj}$, $\mathbf{z}_{orj}$, $\mathbf{z}_{Lerj}$, and $\mathbf{z}_{Lorj}$ are IID, and

$$\begin{aligned} \mathbf{R}_r &= \mathbf{U}\mathbf{R}\mathbf{U}^\dagger \\ &= \mathfrak{Re}(\mathbf{R}) + \mathbf{J}\mathfrak{Im}(\mathbf{R}) \in \mathbb{R}^{N \times N}, \end{aligned} \tag{8.46}$$

with $\mathbf{U}$ being defined in (8.C.1).

8.2.2.2   Probability of False Alarm for Per-SNT

Based on (8.43) and (8.44), we apply the result in Theorem (8.2.1) above to derive analytical expressions for the PFA of the per-SNT in (8.41). Setting $p = N, n = 2J$, and $m = 2L$, one can easily obtain the PFA of the per-SNT as follows.

Exact Expression with $N \leqslant 2L$ and Arbitrary J:

$$\mathcal{P}_{fa}(\Xi \geqslant \xi) = 1 - C(z, s, r) \sqrt{\left| \mathbf{A}\left(\frac{\xi}{1+\xi}\right) \right|}. \tag{8.47}$$

Approximate Expression with $N < 2L$:

$$\mathcal{P}_{fa}(\Xi \geqslant \xi) \simeq 1 - \gamma\left(\tau, \frac{\ln\xi - \mu + \sigma\alpha}{\delta\sigma}\right), \tag{8.48}$$

where z, s, r and $\mu, \sigma, \tau, \delta, \alpha$ are given in the Theorem (8.2.1). When N and J are large, this approximate expression is exact. Further, we can derive the detection threshold of the per-SNT as a function of the PFA, i.e.,

$$\xi \simeq \exp\left\{\sigma\left[\delta\gamma^{-1}(\tau, 1 - y) - \alpha\right] + \mu\right\}, \tag{8.49}$$

where $y = \mathcal{P}_{fa}(\Xi \geqslant \xi)$ is the given PFA. Obviously, the expression in (8.48) is more computationally efficient than that in (8.47).

Note that it can be observed from (8.47) that the per-SNT detector bears the CFAR property against the noise covariance matrix.

8.3 Per-GLRT Detector

8.3.1 Detector Design

Here, we design a GLRT-based detector for the problem (8.3) by exploiting persymmetry. The joint PDF of $\mathbf{Z}$ and $\mathbf{Z}_L$ under hypothesis H_1 or H_0 can be written as

$$f(\mathbf{Z}, \mathbf{Z}_L|H_q) = \frac{\exp\left[-\operatorname{tr}\left(\mathbf{R}^{-1}\mathbf{F_q}\right)\right]}{\pi^{N(J+L)}\det(\mathbf{R})^{J+L}}, \quad q = 0, 1, \tag{8.50}$$

where

$$\mathbf{F}_q = \mathbf{Z}_L\mathbf{Z}_L^\dagger + (\mathbf{Z} - q\mathbf{B})(\mathbf{Z} - q\mathbf{B})^\dagger. \tag{8.51}$$

The one-step GLRT detector is given by

$$\Lambda_0 = \frac{\max\limits_{\mathbf{B}} \max\limits_{\mathbf{R} \in \triangle_{\text{per}}} f(\mathbf{Z}, \mathbf{Z}_L|H_1)}{\max\limits_{\mathbf{R} \in \triangle_{\text{per}}} f(\mathbf{Z}, \mathbf{Z}_L|H_0)} \mathop{\gtrless}\limits_{H_0}^{H_1} \zeta_0, \tag{8.52}$$

where $\triangle_{\text{per}} \triangleq \{\mathbf{R} > \mathbf{0} : \mathbf{R} = \mathbf{JR}^*\mathbf{J}\}$ and ζ_0 is the detection threshold.

Exploiting the persymmetry of $\mathbf{R}$, we have

$$\operatorname{tr}(\mathbf{R}^{-1}\mathbf{F}_q) = \operatorname{tr}[\mathbf{R}^{-1}\mathbf{JF}_q^*\mathbf{J}], \quad q = 0, 1. \tag{8.53}$$

As a result, (8.50) can be rewritten as

$$f(\mathbf{Z}, \mathbf{Z}_L|H_q) = \frac{\exp\left[-\operatorname{tr}\left(\mathbf{R}^{-1}\mathbf{S}_q\right)\right]}{\pi^{N(J+L)}\det(\mathbf{R})^{J+L}}, \quad q = 0, 1, \tag{8.54}$$

where

$$\mathbf{S}_q = \frac{1}{2}\left(\mathbf{F}_q + \mathbf{JF}_q^*\mathbf{J}\right). \tag{8.55}$$

Inserting (8.51) into (8.55), after some algebraic operations, yields

$$\mathbf{S}_q = \mathbf{Z}_{Lp}\mathbf{Z}_{Lp}^{\dagger} + (\mathbf{Z}_p - q\mathbf{B}_p)(\mathbf{Z}_p - q\mathbf{B}_p)^{\dagger}, \tag{8.56}$$

where $\mathbf{Z}_p$ and $\mathbf{B}_p$ are defined in (8.14) and (8.15), respectively, and

$$\mathbf{Z}_{Lp} = [\mathbf{z}_{Le1}, \ldots, \mathbf{z}_{LeL}, \mathbf{z}_{Lo1}, \ldots, \mathbf{z}_{LoL}] \in \mathbb{C}^{N \times 2L}, \tag{8.57}$$

with

$$\begin{cases} \mathbf{z}_{Lej} = \frac{1}{2}\left(\mathbf{z}_{Lj} + \mathbf{J}\mathbf{z}_{Lj}^{*}\right), \\ \mathbf{z}_{Loj} = \frac{1}{2}\left(\mathbf{z}_{Lj} - \mathbf{J}\mathbf{z}_{Lj}^{*}\right). \end{cases} \tag{8.58}$$

Taking the derivative of (8.54) with respect to $\mathbf{B}_p$, and making the result equal to zero, we can obtain that the MLE of $\mathbf{B}_p$ under hypothesis H_1 is the same as (8.18). Plugging (8.18) into (8.54), then taking the derivative with respect to $\mathbf{R}$ and equating the result to zero, yields the MLE of $\mathbf{R}$ under hypothesis H_1 as

$$\check{\mathbf{R}}_1 = \frac{\mathbf{Z}_{Lp}\mathbf{Z}_{Lp}^{\dagger}}{J + L}. \tag{8.59}$$

Substituting (8.18) and (8.59) into (8.54) produces the maximum value of $f(\mathbf{Z}, \mathbf{Z}_L|H_1)$ as

$$\max_{\mathbf{B}}\max_{\mathbf{R}} f(\mathbf{Z}, \mathbf{Z}_L|H_1) = \left[\frac{(J+L)^N}{(e\pi)^N \det(\mathbf{Z}_{Lp}\mathbf{Z}_{Lp}^{\dagger})}\right]^{J+L}. \tag{8.60}$$

Next, taking the derivative of (8.54) with respect to $\mathbf{R}$ then making the result equal to zero, yields the MLE of $\mathbf{R}$ under hypothesis H_0 as

$$\check{\mathbf{R}}_0 = \frac{\mathbf{Z}_p\mathbf{Z}_p^{\dagger} + \mathbf{Z}_{Lp}\mathbf{Z}_{Lp}^{\dagger}}{J + L}. \tag{8.61}$$

Substituting (8.61) into (8.54) produces the maximum value of $f(\mathbf{Z}, \mathbf{Z}_L|H_0)$ as

$$\max_{\mathbf{R}} f(\mathbf{Z}, \mathbf{Z}_L|H_0) = \left[\frac{(J+L)^N}{(e\pi)^N \det(\mathbf{Z}_{Lp}\mathbf{Z}_{Lp}^{\dagger} + \mathbf{Z}_p\mathbf{Z}_p^{\dagger})}\right]^{J+L}. \tag{8.62}$$

Further, inserting (8.60) and (8.62) into (8.52), after some algebra, yields

$$\Lambda_1 = \frac{\det\left(\mathbf{Z}_{Lp}\mathbf{Z}_{Lp}^{\dagger} + \mathbf{Z}_p\mathbf{Z}_p^{\dagger}\right)}{\det(\mathbf{Z}_{Lp}\mathbf{Z}_{Lp}^{\dagger})} \mathop{\gtrless}\limits_{H_0}^{H_1} \zeta_1, \tag{8.63}$$

where ζ_1 is the detection threshold. This detector is referred to as the per-symmetric GLRT (per-GLRT) detector.

8.3.2 Threshold Setting for Per-GLRT

As derived in Appendix 8.D, the PFA of the per-GLRT detector is obtained as follows.

Exact Expression with $J = 1$:

$$\mathcal{P}_{fa}(\Lambda \geqslant \zeta) = \sum_{j=1}^{N} C_{2L-j}^{N-j} \left(\sqrt{\zeta} - 1\right)^{N-j} \left(\sqrt{\zeta}\right)^{-(2L+1-j)}. \qquad (8.64)$$

Approximate Expression with $J > 1$:

For the case of distributed target detection (i.e., $J > 1$), an exact expression for the PFA is unavailable. Based on the result in (8.D.3) of Appendix 8.D, an approximate expression can be obtained as [11, p. 322, eq. (28)]

$$\mathcal{P}_{fa}(\Lambda \geqslant \zeta) = 1 - \mathcal{P}\left(\chi_{2NJ}^2 \leqslant \rho\right) - \frac{\gamma_2}{g^2}\left[\mathcal{P}\left(\chi_{2NJ+4}^2 \leqslant \rho\right) - \mathcal{P}\left(\chi_{2NJ}^2 \leqslant \rho\right)\right]$$

$$- \frac{1}{g^4}\left\{\gamma_4\left[\mathcal{P}\left(\chi_{2NJ+8}^2 \leqslant \rho\right) - \mathcal{P}\left(\chi_{2NJ}^2 \leqslant \rho\right)\right]\right.$$

$$\left. - \gamma_2^2\left[\mathcal{P}\left(\chi_{2NJ+4}^2 \leqslant \rho\right) - \mathcal{P}\left(\chi_{2NJ}^2 \leqslant \rho\right)\right]\right\}, \qquad (8.65)$$

where

$$\rho = g\log\zeta, \qquad (8.66)$$

$$g = 2L - \frac{1}{2}\left(N - 2J + 1\right), \qquad (8.67)$$

$$\gamma_2 = \frac{2NJ\left(N^2 + 4J^2 - 5\right)}{48}, \qquad (8.68)$$

and

$$\gamma_4 = \frac{\gamma_2^2}{2} + \frac{2NJ}{1920}\left[3N^4 + 48J^4 + 40N^2J^2 - 50\left(N^2 + 4J^2\right) + 159\right]. \qquad (8.69)$$

In Appendix 8.E, we prove that the per-GLRT detector also bears the CFAR against the covariance matrix.

It should be emphasized here that the statistical distributions of the detectors under H_1 are difficult to derive. This is because the distribution of the largest eigenvalue of noncentral $\mathbf{F}$ and the noncentral Wilks' Lambda distribution are difficult to analyze in closed-form [12, 13]. Therefore, we cannot provide analytical expressions for the detection probabilities of both detectors.

8.4   Numerical Examples

In this section, numerical examples are given to verify the theoretical results via MC simulations. Assume that the (i, j) element of the noise covariance matrix is chosen as $\mathbf{R}(i, j) = \sigma_c^2 0.9^{|i-j|}$, where σ_c^2 denotes the noise power.

8.4.1   Probability of False Alarm

In Fig. 8.1, we plot the PFA as a function of the detection threshold for the detector per-SNT. The solid line denotes the results calculated by using the exact expression in (8.47). The solid line with "$\star$" represents the approximate results computed by using the approximate expression in (8.48). The number of independent trials for simulation is 10^8. We can observe that the exact analytical results fit the MC ones pretty well, and the approximation accuracy is very high.

The PFA as a function of the detection threshold for the per-GLRT detector is shown in Fig. 8.2. In Fig. 8.2(a), the solid lines denote the results calculated by using the exact expression in (8.64). In Fig. 8.2(b), the solid line represents the approximate results computed by using the approximate expression in (8.65). It can be seen that the exact and approximate results all fit the MC ones pretty well.

Detection threshold as a function of the PFA for the per-SNT detector is presented in Fig. 8.3. The solid line with "$\star$" represents the results computed by using the expression in (8.49), while the solid line with symbol "$\circ$" stands for the MC results. We can observe that the approximation accuracy is high for the detection threshold setting. That is to say, the approximate expression in (8.49) can be effectively used to set the detection threshold for a given PFA in the case of $N < 2L$ and $N \leqslant 2J$.

8.4.2   Detection Performance

Now, we examine the detection probability of the designed detectors. For comparison purposes, the conventional GLRT-based detectors including one-step GLRT (1S-GLRT), two-step GLRT (2S-GLRT), modified-two-step GLRT (M2S-GLRT), and SNT for distributed target detection are considered. These detectors are given in [1, eqs. (18), (25), (27), (30)], respectively. We also consider an asymptotic GLRT (AS-GLRT) detector proposed in [1, eq. (22)] with the known covariance matrix $\mathbf{R}$. This detector is clairvoyant, and just used as a benchmark for any suboptimum detectors.

As to the target signal, we assume that it is a superposition of multiple signals (due to multipath) in each cell. In this situation, signal signature is unknown.

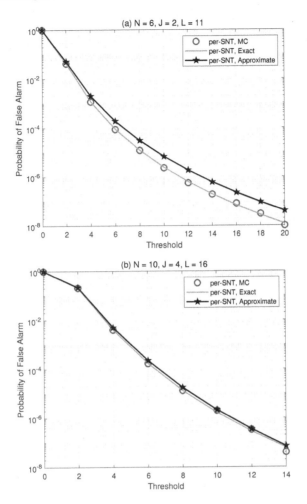

FIGURE 8.1
Probability of false alarm of the per-SNT detector versus threshold for different N, J, and L.

Define

$$\mathbf{t}\,(f_d) = \frac{\left[e^{-\jmath 2\pi f_d \frac{N-1}{2}}, \ldots, e^{-\jmath 2\pi f_d \frac{1}{2}}, e^{\jmath 2\pi f_d \frac{1}{2}}, \ldots, e^{\jmath 2\pi f_d \frac{N-1}{2}} \right]^T}{\sqrt{N}}, \qquad (8.70)$$

for even N, and

$$\mathbf{t}\,(f_d) = \frac{\left[e^{-\jmath 2\pi f_d \frac{N-1}{2}}, \ldots, e^{-\jmath 2\pi f_d}, 1, e^{\jmath 2\pi f_d}, \ldots, e^{\jmath 2\pi f_d \frac{N-1}{2}} \right]^T}{\sqrt{N}}, \qquad (8.71)$$

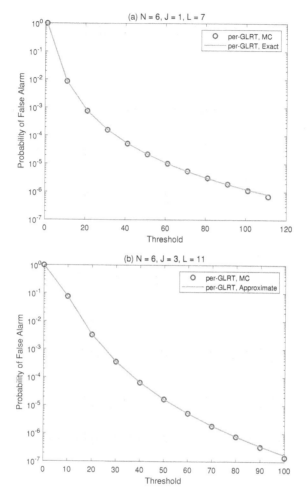

FIGURE 8.2
Probability of false alarm of the per-GLRT detector versus threshold for different N, J, and L.

for odd N, f_d is a normalized Doppler frequency of the signal. The form of the target signal in each cell can be expressed as

$$\mathbf{s} = \sum_{i=1}^{M} s_i \mathbf{t}(f_{di}), \tag{8.72}$$

where M is the number of the multipath signals, s_i and f_{di} are a complex scalar and a Doppler frequency of the ith signal path, respectively. In the following simulations, we assume that the target signal in each cell is formed by superposing three paths (i.e., $M = 3$) and the number of range cell is 3 (i.e., $J = 3$). More precisely, the steering vector of the received target signal

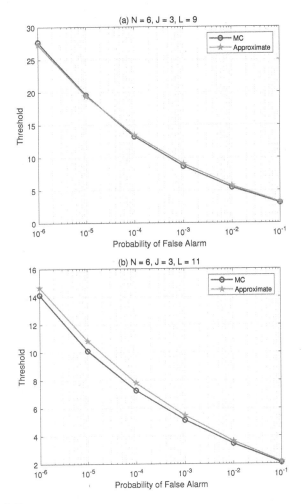

FIGURE 8.3
Threshold versus probability of false alarm. The solid line with "★" represents the results computed by using the expression in (8.49), while the solid line with symbol "o" stands for the MC results.

in each range cell is selected as follows:

$$\begin{cases} \mathbf{s}_1 = 0.9e^{j0.1\pi}\mathbf{t}(0.07) + 0.01e^{j0.12\pi}\mathbf{t}(0.15) + 0.09e^{j0.15\pi}\mathbf{t}(0.23), \\ \mathbf{s}_2 = 0.1e^{j0.2\pi}\mathbf{t}(0.07) + 0.75e^{j0.18\pi}\mathbf{t}(0.15) + 0.15e^{j0.1\pi}\mathbf{t}(0.23), \\ \mathbf{s}_3 = 0.1e^{j0.3\pi}\mathbf{t}(0.07) + 0.1e^{j0.25\pi}\mathbf{t}(0.15) + 0.8e^{j0.32\pi}\mathbf{t}(0.23). \end{cases} \quad (8.73)$$

After normalization, we assume the received target signal in the kth range cell is

$$\mathbf{b}_k = b_k \frac{\mathbf{s}_k}{\sqrt{\|\mathbf{s}_k\|^2}}, \quad (8.74)$$

TABLE 8.1

Energy distribution among the signal
components

$Signal Model$	$\frac{\varrho_1}{\sum_{k=1}^{J}\varrho_k}$	$\frac{\varrho_2}{\sum_{k=1}^{J}\varrho_k}$	$\frac{\varrho_3}{\sum_{k=1}^{J}\varrho_k}$
1	1	0	0
2	1/3	1/3	1/3

for $k = 1, \ldots, J$, where b_k is a real scalar accounting for the power of the
target signal in the kth range cell. The SNR is defined by

$$\mathrm{SNR} = 10 \log_{10} \frac{\frac{1}{J}\sum_{k=1}^{J}\varrho_k}{\sigma_c^2}, \tag{8.75}$$

where $\varrho_k = b_k^2$ accounts for the distribution of the signal energy among the
primary data. Similar to [1], the parameter setting of ϱ_ks is divided into two
cases (see Table 8.1). In the first model, the target signal is only presented
in one range cell, while in the second model the signal energy is uniformly
distributed.

Probabilities of detection as a function of SNR are shown for different L
in Fig. 8.4, where the first signal model in Table 8.1 is adopted. Here, we set
$N = 8$ and the PFA ($\mathcal{P}_{fa}$) is 10^{-3}. For the first signal model, one can see
that the per-SNT generally performs better than its counterparts except for
the AS-GLRT detector, especially in the case of insufficient training data. The
performance gap becomes smaller when the number of training data increases.
This is because the noise covariance matrix estimate is highly accurate by
using sufficient training data, even though the persymmetry is not used. Note
that the conventional detectors cannot work when the training data size L is
less than N. This is the reason why we only provide the performance of the
per-SNT and per-GLRT detectors in Fig. 8.4(c).

In Fig. 8.5, we plot the detection probability versus SNR for the second
signal model. As observed, the detection performance of the per-GLRT is
better than those of its counterparts except for the AS-GLRT detector. When
the number of training data increases, the performance gap between these
detectors also becomes smaller.

In Fig. 8.6, we depict the PD versus the training data size L, where SNR =
11 dB, $N = 8$, and $\mathcal{P}_{fa} = 10^{-3}$. It can be seen that the per-SNT detector
outperforms the competitors in the signal model 1, and the performance of the
per-GLRT detector is better than those of its competitors in the signal model
2. Note that the per-SNT and per-GLRT detectors are more suitable to work
in the limited training data environment than their counterparts. This is to
say, the per-SNT and per-GLRT detectors can achieve the same performance
as their competitor detectors by using less training data.

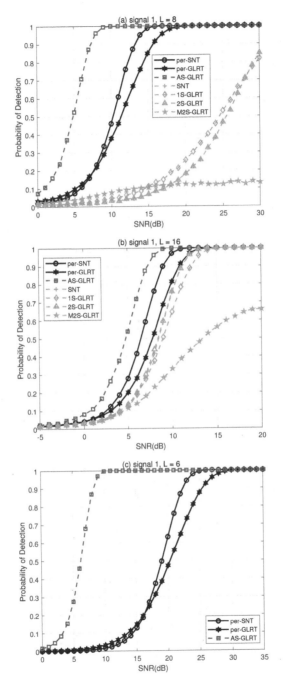

FIGURE 8.4
Probability of detection versus SNR for $N = 8$ and $\mathcal{P}_{fa} = 10^{-3}$ in the signal model 1.

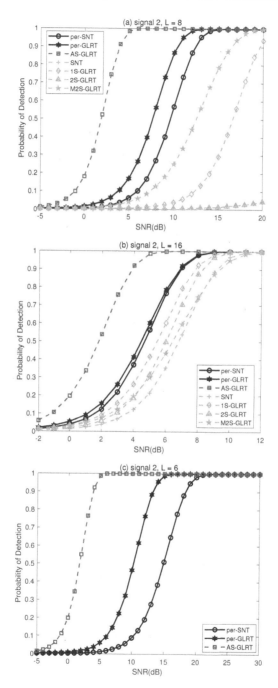

FIGURE 8.5
Probability of detection versus SNR for $N = 8$ and $\mathcal{P}_{fa} = 10^{-3}$ in the signal model 2.

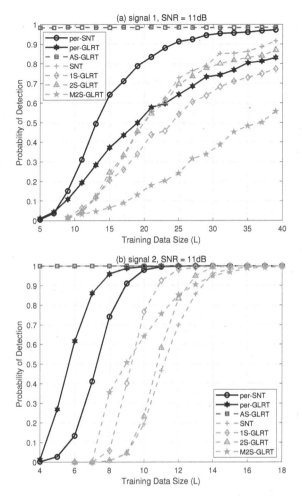

FIGURE 8.6

Probability of detection versus the training data size L for $N = 8$ and $\mathcal{P}_{fa} = 10^{-3}$.

8.4.3 Measured Data

Now numerical examples based on measured data are illustrated to show the performance of the designed detectors. The measured data used here was collected by the MIT Lincoln Laboratory Phase One radar in May 1985 at the Katahdin Hill site, MIT-LL. More detailed introductions about the data are given in Chapter 1. We use the data in the range cells 43–45 in the HH channel as the primary data, and the range cells adjacent to the primary data as the secondary data. The resulting $N(L + J)$ data window, centered on the primary data, is slid one by one along the 30720 time pulses until the end of the dataset. The total number of different data windows is $30720 - 8 + 1 = 30713$,

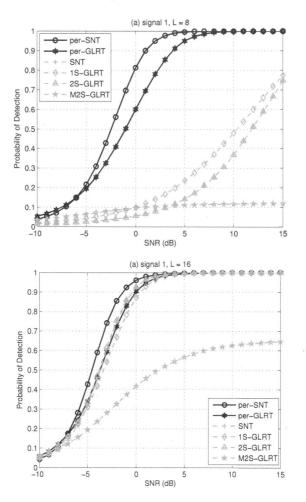

FIGURE 8.7

Probability of detection versus SNR for $N = 8$ and $\mathcal{P}_{fa} = 10^{-2}$ in the signal model 1 on measured data.

which coincides with the total number of trials used to estimate both the PFA and the detection probability of each receiver.

The detection performance based on measured data is shown in Fig. 8.7 for the first signal model. It can be seen from 8.7 that the per-SNT detector performs the best. And the performance gain of the per-SNT detector decreases as L becomes large. Interestingly, the per-GLRT detector has worse performance than some of its counterparts without exploiting persymmetry, when the number of training data is large (i.e., $L = 16$ in this parameter setting).

The detection performance in the second signal model based on measured data is shown in Fig. 8.8. It can be observed that the per-SNT and per-GLRT

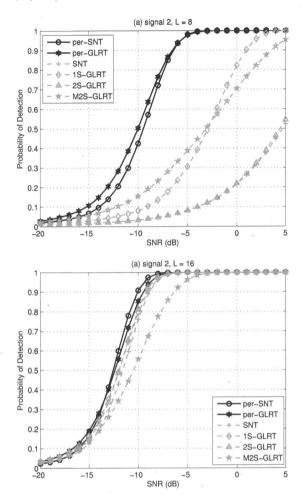

FIGURE 8.8
Probability of detection versus SNR for $N = 8$ and $\mathcal{P}_{fa} = 10^{-2}$ in the signal model 2 on measured data.

detectors have better detection performance than others, especially when the training data size is small.

8.A   Derivations of (8.13)

When $q = 1$, (8.8) becomes

$$\mathbf{G}_1 = (\mathbf{Z} - \mathbf{B})(\mathbf{Z} - \mathbf{B})^{\dagger}. \qquad (8.A.1)$$

Inserting (8.A.1) into (8.12), we have

$$
\begin{aligned}
\mathbf{H}_1 &= \frac{1}{2}(\mathbf{G}_1 + \mathbf{J}\mathbf{G}_1^*\mathbf{J}) \\
&= \frac{1}{2}\left[(\mathbf{Z} - \mathbf{B})(\mathbf{Z} - \mathbf{B})^\dagger + \mathbf{J}(\mathbf{Z} - \mathbf{B})^*(\mathbf{Z} - \mathbf{B})^T\mathbf{J}\right] \\
&= \frac{1}{2}\left[(\mathbf{Z}\mathbf{Z}^\dagger + \mathbf{J}\mathbf{Z}^*\mathbf{Z}^T\mathbf{J}) - (\mathbf{B}\mathbf{Z}^\dagger + \mathbf{J}\mathbf{B}^*\mathbf{Z}^T\mathbf{J}) \right. \\
&\quad \left. -(\mathbf{Z}\mathbf{B}^\dagger + \mathbf{J}\mathbf{Z}^*\mathbf{B}^T\mathbf{J}) + (\mathbf{B}\mathbf{B}^\dagger + \mathbf{J}\mathbf{B}^*\mathbf{B}^T\mathbf{J})\right].
\end{aligned}
\tag{8.A.2}
$$

After some algebra, one can easily obtain

$$
\begin{cases}
\mathbf{Z}_p\mathbf{Z}_p^\dagger = \frac{1}{2}(\mathbf{Z}\mathbf{Z}^\dagger + \mathbf{J}\mathbf{Z}^*\mathbf{Z}^T\mathbf{J}), \\
\mathbf{B}_p\mathbf{B}_p^\dagger = \frac{1}{2}(\mathbf{B}\mathbf{B}^\dagger + \mathbf{J}\mathbf{B}^*\mathbf{B}^T\mathbf{J}), \\
\mathbf{B}_p\mathbf{Z}_p^\dagger = \frac{1}{2}(\mathbf{B}\mathbf{Z}^\dagger + \mathbf{J}\mathbf{B}^*\mathbf{Z}^T\mathbf{J}), \\
\mathbf{Z}_p\mathbf{B}_p^\dagger = \frac{1}{2}(\mathbf{Z}\mathbf{B}^\dagger + \mathbf{J}\mathbf{Z}^*\mathbf{B}^T\mathbf{J}),
\end{cases}
\tag{8.A.3}
$$

where $\mathbf{Z}_p$ and $\mathbf{B}_p$ are defined in (8.14) and (8.15), respectively. Substituting (8.A.3) into (8.A.2), we have

$$
\mathbf{H}_1 = (\mathbf{Z}_p - \mathbf{B}_p)(\mathbf{Z}_p - \mathbf{B}_p)^\dagger.
\tag{8.A.4}
$$

Similarly, when $q = 0$, we have

$$
\mathbf{H}_0 = \mathbf{Z}_p\mathbf{Z}_p^\dagger.
\tag{8.A.5}
$$

8.B   Proof of Theorem 8.2.1

First we consider the exact case. For the random matrix

$$
\mathbf{G} = (\mathbf{X}\mathbf{X}^T + \mathbf{Y}\mathbf{Y}^T)^{-1}\mathbf{X}\mathbf{X}^T,
\tag{8.B.1}
$$

the author in [10] derived the distribution of the largest eigenvalue of $\mathbf{G}$ as

$$
\mathcal{P}(\lambda_{\max}(\mathbf{G}) \leqslant \vartheta) = C(z, s, r)\sqrt{|\mathbf{A}(\vartheta)|},
\tag{8.B.2}
$$

where all the parameters are given in (8.27)–(8.35). Define

$$
\mathbf{L}_1 = \mathbf{X}\mathbf{X}^T(\mathbf{X}\mathbf{X}^T + \mathbf{Y}\mathbf{Y}^T)^{-1},
\tag{8.B.3}
$$

and

$$
\mathbf{L}_2 = \mathbf{X}\mathbf{X}^T(\mathbf{Y}\mathbf{Y}^T)^{-1}.
\tag{8.B.4}
$$

It can be easily obtained that

$$
\lambda_{\max}(\mathbf{L}_1) = \lambda_{\max}(\mathbf{G}),
\tag{8.B.5}
$$

and
$$\lambda_{\max}(\mathbf{L}_2) = \lambda_{\max}(\mathbf{F}). \tag{8.B.6}$$

According to [11, p. 529–530], we have

$$\lambda_{\max}(\mathbf{L}_2) = \frac{\lambda_{\max}(\mathbf{L}_1)}{1 - \lambda_{\max}(\mathbf{L}_1)}. \tag{8.B.7}$$

Substituting (8.B.5) and (8.B.6) into (8.B.7) produces

$$\lambda_{\max}(\mathbf{F}) = \frac{\lambda_{\max}(\mathbf{G})}{1 - \lambda_{\max}(\mathbf{G})}. \tag{8.B.8}$$

Then we obtain

$$\mathcal{P}(\lambda_{\max}(\mathbf{F}) \leqslant \lambda) = \mathcal{P}\left(\lambda_{\max}(\mathbf{G}) \leqslant \frac{\lambda}{1 + \lambda}\right), \tag{8.B.9}$$

and derive the expression in (8.26).

Next, we consider the approximate case where the condition $m > p$ has to be satisfied. In [10, 3. Approximations], it is shown that the logit of $\lambda_{\max}(\mathbf{G})$ approaches the Tracy-Widom law for large p when $n \geqslant p$. Further, the author gave an approximate distribution of $\lambda_{\max}(\mathbf{G})$. Based on these results, we can derive the approximate distribution of $\lambda_{\max}(\mathbf{F})$ as the expression in (8.36).

Denote the PDF of $\lambda_{\max}(\mathbf{G})$ as $\Theta(p, m, n)$. According to [14, p. 84], $\Theta(p, m, n)$ has the same distribution as $\Theta(n, m + n - p, p)$. Thus, for the case of $n < p$, we can obtain the approximate distribution of $\lambda_{\max}(\mathbf{F})$ by using n, $m + n - p$, and p to replace p, m, and n in (8.36), respectively, which leads to the fact that the approximate expression for $n < p$ is the same as (8.36) for $n \geqslant q$.

To summary up, the approximate distribution of $\lambda_{\max}(\mathbf{F})$ for $m > p$ is the expression in (8.36). The proof is completed.

8.C   Derivations of (8.41) and (8.42)

Let us start by defining a unitary matrix as

$$\mathbf{U} = \frac{1}{2}[(\mathbf{I}_N + \mathbf{J}) + \jmath\,(\mathbf{I}_N - \mathbf{J})] \in \mathbb{C}^{N \times N}. \tag{8.C.1}$$

Using (8.14), (8.16), (8.57), (8.58), and (8.C.1), we can obtain

$$\mathbf{U}\mathbf{Z}_p = [\mathbf{z}_{er1}, \ldots, \mathbf{z}_{erJ}, \jmath\,\mathbf{z}_{or1}, \ldots, \jmath\,\mathbf{z}_{orJ}], \tag{8.C.2}$$

$$\mathbf{U}\mathbf{Z}_{Lp} = [\mathbf{z}_{Ler1}, \ldots, \mathbf{z}_{LerL}, \jmath\,\mathbf{z}_{Lor1}, \ldots, \jmath\,\mathbf{z}_{LorL}], \tag{8.C.3}$$

where

$$\mathbf{z}_{erj} = \frac{1}{2}\left[(\mathbf{I}_N + \mathbf{J})\mathfrak{Re}(\mathbf{z}_j) - (\mathbf{I}_N - \mathbf{J})\mathfrak{Im}(\mathbf{z}_j)\right] \in \mathbb{R}^{N\times 1}, \tag{8.C.4}$$

$$\mathbf{z}_{orj} = \frac{1}{2}\left[(\mathbf{I}_N - \mathbf{J})\mathfrak{Re}(\mathbf{z}_j) + (\mathbf{I}_N + \mathbf{J})\mathfrak{Im}(\mathbf{z}_j)\right] \in \mathbb{R}^{N\times 1}, \tag{8.C.5}$$

for $j = 1, 2, \ldots, J$, and

$$\mathbf{z}_{Lerj} = \frac{1}{2}\left[(\mathbf{I}_N + \mathbf{J})\mathfrak{Re}(\mathbf{z}_{Lj}) - (\mathbf{I}_N - \mathbf{J})\mathfrak{Im}(\mathbf{z}_{Lj})\right] \in \mathbb{R}^{N\times 1}, \tag{8.C.6}$$

$$\mathbf{z}_{Lorj} = \frac{1}{2}\left[(\mathbf{I}_N - \mathbf{J})\mathfrak{Re}(\mathbf{z}_{Lj}) + (\mathbf{I}_N + \mathbf{J})\mathfrak{Im}(\mathbf{z}_{Lj})\right] \in \mathbb{R}^{N\times 1}, \tag{8.C.7}$$

for $j = 1, 2, \ldots, L$.
Define

$$\mathbf{V}_1 = \begin{bmatrix} \mathbf{I}_J & \mathbf{0} \\ \mathbf{0} & -\jmath\mathbf{I}_J \end{bmatrix} \in \mathbb{C}^{2J\times 2J}, \tag{8.C.8}$$

and

$$\mathbf{V}_2 = \begin{bmatrix} \mathbf{I}_L & \mathbf{0} \\ \mathbf{0} & -\jmath\mathbf{I}_L \end{bmatrix} \in \mathbb{C}^{2L\times 2L}. \tag{8.C.9}$$

Now we can transform all complex-valued quantities in (8.25) and (8.63) to real-valued ones, i.e.,

$$\begin{aligned} \hat{\mathbf{R}}_r &= \mathbf{U}\hat{\mathbf{R}}\mathbf{U}^\dagger \\ &= \mathfrak{Re}(\hat{\mathbf{R}}) + \mathbf{J}\mathfrak{Im}(\hat{\mathbf{R}}) \in \mathbb{R}^{N\times N}, \end{aligned} \tag{8.C.10}$$

$$\begin{aligned} \mathbf{Z}_r &= \mathbf{U}\mathbf{Z}_p\mathbf{V}_1 \\ &= [\mathbf{z}_{er1}, \ldots, \mathbf{z}_{erJ}, \mathbf{z}_{or1}, \ldots, \mathbf{z}_{orJ}] \in \mathbb{R}^{N\times 2J}, \end{aligned} \tag{8.C.11}$$

and

$$\begin{aligned} \mathbf{Z}_{Lr} &= \mathbf{U}\mathbf{Z}_{Lp}\mathbf{V}_2 \\ &= [\mathbf{z}_{Ler1}, \ldots, \mathbf{z}_{LerL}, \mathbf{z}_{Lor1}, \ldots, \mathbf{z}_{LorL}] \in \mathbb{R}^{N\times 2L}. \end{aligned} \tag{8.C.12}$$

Then (8.25) and (8.63) can be rewritten as (8.41) and (8.42), respectively.

8.D   Derivation of (8.64)

According to (8.44) and (8.45), we have

$$\mathbf{Z}_r\mathbf{Z}_r^T \sim \mathcal{W}_N(2J, \mathbf{R}_r), \tag{8.D.1}$$

and

$$\mathbf{Z}_{Lr}\mathbf{Z}_{Lr}^T \sim \mathcal{W}_N(2L, \mathbf{R}_r). \tag{8.D.2}$$

Based on [11, Lemma 8.4.2] and (8.42), we can obtain

$$\Lambda^{-1} \sim U_{N,2J,2L}. \tag{8.D.3}$$

When $J = 1$, we have

$$\Lambda^{-1} \sim U_{N,2,2L}. \tag{8.D.4}$$

According to [11, Theorem 8.4.6], we obtain

$$\sqrt{U_{N,2,2L}} = \frac{1}{1 + \frac{N}{2L+1-N} F_{2N,2(2L+1-N)}}. \tag{8.D.5}$$

Combining (8.D.4) and (8.D.5) yields

$$\sqrt{\Lambda} - 1 = \frac{\chi_{2N}^2}{\chi_{2(2L-N+1)}^2}. \tag{8.D.6}$$

Similar to the derivations of [15, eq. (10)], we can derive the PFA of the per-GLRT as (8.64).

8.E CFARness of the Per-GLRT

Define

$$\mathbf{C}_r = \mathbf{Z}_{Lr} \mathbf{Z}_{Lr}^T, \tag{8.E.1}$$

$$\bar{\mathbf{Z}}_r = \mathbf{R}_r^{-1/2} \mathbf{Z}_r, \tag{8.E.2}$$

and

$$\bar{\mathbf{C}}_r = \mathbf{R}_r^{-1/2} \mathbf{C}_r \mathbf{R}_r^{-1/2}. \tag{8.E.3}$$

Then, we can rewrite the per-GLRT in (8.42) as

$$\begin{aligned}
\Lambda &= \frac{\det(\mathbf{C}_r + \mathbf{Z}_r \mathbf{Z}_r^T)}{\det(\mathbf{C}_r)} \\
&= \det(\mathbf{C}_r^{-1/2}) \det(\mathbf{C}_r + \mathbf{Z}_r \mathbf{Z}_r^T) \det(\mathbf{C}_r^{-1/2}) \\
&= \det(\mathbf{I}_N + \bar{\mathbf{Z}}_r^T \bar{\mathbf{C}}_r^{-1} \bar{\mathbf{Z}}_r).
\end{aligned} \tag{8.E.4}$$

Under the hypothesis H_0, each column of $\bar{\mathbf{Z}}_r$ is a zero-mean, real Gaussian random vector with the covariance matrix $\mathbf{I}_N$. Considering (8.D.2) and (8.E.3), we have

$$\bar{\mathbf{C}}_r \sim \mathcal{W}_N(2L, \mathbf{I}_N). \tag{8.E.5}$$

So we can conclude that $\bar{\mathbf{Z}}_r$ and $\bar{\mathbf{C}}_r$ are irrelevant to the noise covariance matrix under H_0. That is to say, the per-GLRT detector has the CFAR property against the covariance matrix.

Bibliography

[1] E. Conte, A. De Maio, and C. Galdi, "CFAR detection of multidimensional signals: An invariant approach," *IEEE Transactions on Signal Processing*, vol. 51, no. 1, pp. 142–151, January 2003.

[2] J. Liu, J. Chen, J. Li, and W. Liu, "Persymmetric adaptive detection of distributed targets with unknown steering vectors," *IEEE Transactions on Signal Processing*, vol. 68, pp. 4123–4134, 2020.

[3] R. S. Raghavan, H. F. Qiu, and D. J. McLaughlin, "CFAR detection in clutter with unknown correlation properties," *IEEE Transactions on Aerospace and Electronic Systems*, vol. 31, no. 2, pp. 647–657, April 1995.

[4] R. A. Horn and C. R. Johnson, *Matrix Analysis*. New York, Cambridge University Press, 1985.

[5] W. Liu, W. Xie, and Y. Wang, "Rao and Wald tests for distributed targets detection with unknown signal steering," *IEEE Signal Processing Letters*, vol. 20, no. 11, pp. 1086–1089, November 2013.

[6] W. Liu, W. Xie, J. Liu, and Y. Wang, "Adaptive double subspace signal detection in Gaussian background–Part I: Homogeneous environments," *IEEE Transactions on Signal Processing*, vol. 62, no. 9, pp. 2345–2357, May 2014.

[7] E. Conte and A. De Maio, "Distributed target detection in compound-Gaussian noise with Rao and Wald tests," *IEEE Transactions on Aerospace and Electronic Systems*, vol. 39, no. 2, pp. 568–582, April 2003.

[8] R. Nitzberg, "Application of maximum likelihood estimation of persymmetric covariance matrices to adaptive processing," *IEEE Transactions on Aerospace and Electronic Systems*, vol. AES-16, no. 1, pp. 124–127, January 1980.

[9] L. Cai and H. Wang, "A persymmetric multiband GLR algorithm," *IEEE Transactions on Aerospace and Electronic Systems*, vol. 28, no. 3, pp. 806–816, July 1992.

[10] M. Chiani, "Distribution of the largest root of a matrix for Roy's test in multivariate analysis of variance," *Journal of Multivariate Analysis*, vol. 143, pp. 467–471, 2016.

[11] T. Anderson, *An Introduction to Multivariate Statistical Analysis*, 3rd ed. New York, USA: Wiley, 2003.

[12] A. T. James, "Distribution of matrix variates and latent roots derived from normal samples," *Annales of Mathematical Statistics*, vol. 35, pp. 475–501, 1964.

[13] A. M. Mathai and P. N. Rathie, "The exact distribution of Wilks' criterion," *The Annals of Mathematical Statistics*, vol. 42, no. 3, pp. 1010–1019, 1971.

[14] K. V. Mardia, J. T. Kent, and J. M. Bibby, *Multivariate Analysis*. Academic Press: New York, 1979.

[15] J. Liu, Z.-J. Zhang, Y. Yang, and H. Liu, "A CFAR adaptive subspace detector for first-order or second-order Gaussian signals based on a single observation," *IEEE Transactions on Signal Processing*, vol. 59, no. 11, pp. 5126–5140, November 2011.

9

Adaptive Detection in Interference

This chapter discusses the problem of detecting subspace signals in the presence of subspace interference and Gaussian noise by using multiple observations collected from multiple range cells, bands, and/or coherent processing intervals [1]. We exploit persymmetry to design one-step and two-step detectors, according to the criterion of GLRT. Both the detectors exhibit CFAR properties against the noise covariance matrix. Moreover, the statistical characterizations of the one-step detector are obtained. We derive exact expressions for the PFA of the one-step detector in six cases where the signal subspace dimension is no more than 4 or the number of observations is no more than 2. In other cases, we derive an approximate expression for the PFA of the one-step detector. Numerical examples illustrate that the designed detectors outperform their counterparts, especially when the number of training data is small. In addition, the one-step GLRT detector generally has better detection performance than the two-step GLRT detector.

9.1 Problem Formulation

Assume that multiple test data denoted by a set of $N \times 1$ vectors $\{\dot{\mathbf{x}}_h\}_{h=1}^{H}$ are available, where H is the number of observations, and N is the number of (temporal, spatial, or spatial-temporal) channels. In radar applications, these observations can be collected from multiple range cells, bands, and/or CPIs. When a target is present, $\dot{\mathbf{x}}_h$ can be expressed by

$$\dot{\mathbf{x}}_h = \dot{\mathbf{H}}\boldsymbol{\theta}_h + \dot{\mathbf{J}}\boldsymbol{\phi}_h + \dot{\mathbf{n}}_h, \quad h = 1, 2, \ldots, H, \tag{9.1}$$

where

- $\dot{\mathbf{H}} \in \mathbb{C}^{N \times p}$ is a known full-column-rank target subspace matrix; $\boldsymbol{\theta}_h \in \mathbb{C}^{p \times 1}$ is a deterministic but unknown coordinate vector accounting for both the target reflectivity and channel propagation effects. The subspace model implies that we know the subspace where the target signal lies but we do not know its exact location, since $\boldsymbol{\theta}_h$ is unknown;

- $\dot{\mathbf{J}} \in \mathbb{C}^{N \times q}$ is a known full-column-rank interference subspace matrix, and $\boldsymbol{\phi}_h \in \mathbb{C}^{q \times 1}$ is a deterministic but unknown interference coordinate vector;

DOI: 10.1201/9781003340232-9

- $\dot{\mathbf{n}}_h$ denotes the noise, and $\dot{\mathbf{n}}_h \sim \mathcal{CN}_N(\mathbf{0}, \dot{\mathbf{R}})$ with $\dot{\mathbf{R}}$ being an unknown covariance matrix.

As in [2–9], a standard assumption is that the columns of $\dot{\mathbf{H}}$ and $\dot{\mathbf{J}}$ are linearly independent, and the augmented matrix $\dot{\mathbf{M}} \triangleq [\dot{\mathbf{H}}, \dot{\mathbf{J}}] \in \mathbb{C}^{N \times (p+q)}$ is of full rank. This implies $p+q \leqslant N$. In order to estimate the noise covariance matrix, a set of homogeneous training (secondary) data $\{\dot{\mathbf{x}}_k\}_{k=H+1}^{H+K}$ only containing noise is assumed available, i.e., $\dot{\mathbf{x}}_k \sim \mathcal{CN}_N(\mathbf{0}, \dot{\mathbf{R}}), k = H+1, H+2, \ldots, H+K$. Suppose further that the noise in the test and training data is IID.

The target detection problem can be formulated as the following binary hypotheses test:

$$
\begin{cases}
H_0 : \begin{cases} \dot{\mathbf{x}}_h \sim \mathcal{CN}_N(\dot{\mathbf{J}}\boldsymbol{\phi}_h, \dot{\mathbf{R}}), & h = 1, 2, \ldots, H, \\ \dot{\mathbf{x}}_k \sim \mathcal{CN}_N(\mathbf{0}, \dot{\mathbf{R}}), & k = H+1, \ldots, H+K, \end{cases} \\
H_1 : \begin{cases} \dot{\mathbf{x}}_h \sim \mathcal{CN}_N(\dot{\mathbf{H}}\boldsymbol{\theta}_h + \dot{\mathbf{J}}\boldsymbol{\phi}_h, \dot{\mathbf{R}}), & h = 1, 2, \ldots, H, \\ \dot{\mathbf{x}}_k \sim \mathcal{CN}_N(\mathbf{0}, \dot{\mathbf{R}}), & k = H+1, \ldots, H+K. \end{cases}
\end{cases}
\tag{9.2}
$$

In practice, $\dot{\mathbf{H}}$, $\dot{\mathbf{J}}$, and $\dot{\mathbf{R}}$ have persymmetric structures when a symmetrically spaced linear array with its center at the origin and/or symmetrically spaced pulse trains are used. The justifications for the persymmetric subspace model can be found in Chapter 4. In such cases,

$$
\begin{cases}
\dot{\mathbf{R}} = \mathbf{J}\dot{\mathbf{R}}^*\mathbf{J} \in \mathbb{C}^{N \times N}, \\
\dot{\mathbf{H}} = \mathbf{J}\dot{\mathbf{H}}^* \in \mathbb{C}^{N \times p}, \\
\dot{\mathbf{J}} = \mathbf{J}\dot{\mathbf{J}}^* \in \mathbb{C}^{N \times q},
\end{cases}
\tag{9.3}
$$

where $\mathbf{J}$ is a permutation matrix defined in (1.2).

9.2 GLRT Detection

In this section, one-step and two-step GLRT detectors are designed.

9.2.1 One-Step GLRT

The one-step GLRT for the detection problem in (9.2) is given by

$$
\frac{\max_{\{\boldsymbol{\theta}_h, \boldsymbol{\phi}_h, \dot{\mathbf{R}}\}_{h=1}^{H}} f\left(\dot{\mathbf{X}}_t, \dot{\mathbf{X}}_s | H_1\right)}{\max_{\{\boldsymbol{\phi}_h, \dot{\mathbf{R}}\}_{h=1}^{H}} f\left(\dot{\mathbf{X}}_t, \dot{\mathbf{X}}_s | H_0\right)} \underset{H_0}{\overset{H_1}{\gtrless}} \lambda_0,
\tag{9.4}
$$

where λ_0 is a detection threshold, $f\left(\dot{\mathbf{X}}_t, \dot{\mathbf{X}}_s | H_0\right)$ and $f\left(\dot{\mathbf{X}}_t, \dot{\mathbf{X}}_s | H_1\right)$ are the PDFs of the test and training data under H_0 and H_1, respectively,

$$
\dot{\mathbf{X}}_t = [\dot{\mathbf{x}}_1, \dot{\mathbf{x}}_2, \ldots, \dot{\mathbf{x}}_H] \in \mathbb{C}^{N \times H},
\tag{9.5}
$$

and

$$\dot{\mathbf{X}}_s = [\dot{\mathbf{x}}_{H+1}, \dot{\mathbf{x}}_{H+2}, \ldots, \dot{\mathbf{x}}_{H+K}] \in \mathbb{C}^{N \times K}. \tag{9.6}$$

Due to the independence among the test and training data, $f\left(\dot{\mathbf{X}}_t, \dot{\mathbf{X}}_s | H_0\right)$ can be written as

$$f\left(\dot{\mathbf{X}}_t, \dot{\mathbf{X}}_s | H_0\right) = \left\{ \frac{1}{\pi^N \det(\dot{\mathbf{R}})} \exp\left[-\mathrm{tr}(\dot{\mathbf{R}}^{-1}\dot{\mathbf{T}}_0)\right] \right\}^{K+H}, \tag{9.7}$$

where

$$\dot{\mathbf{T}}_0 = \frac{1}{K+H}\left[\dot{\mathbf{X}}_s \dot{\mathbf{X}}_s^\dagger + \sum_{h=1}^{H}\left(\dot{\mathbf{x}}_h - \dot{\mathbf{J}}\phi_h\right)\left(\dot{\mathbf{x}}_h - \dot{\mathbf{J}}\phi_h\right)^\dagger\right] \in \mathbb{C}^{N \times N}. \tag{9.8}$$

In addition, $f\left(\dot{\mathbf{X}}_t, \dot{\mathbf{X}}_s | H_1\right)$ can be expressed by

$$f\left(\dot{\mathbf{X}}_t, \dot{\mathbf{X}}_s | H_1\right) = \left\{ \frac{1}{\pi^N \det(\dot{\mathbf{R}})} \exp\left[-\mathrm{tr}(\dot{\mathbf{R}}^{-1}\dot{\mathbf{T}}_1)\right] \right\}^{K+H}, \tag{9.9}$$

where

$$\dot{\mathbf{T}}_1 = \frac{1}{K+H}\left[\dot{\mathbf{X}}_s \dot{\mathbf{X}}_s^\dagger + \sum_{h=1}^{H}\left(\dot{\mathbf{x}}_h - \dot{\mathbf{M}}\psi_h\right)\left(\dot{\mathbf{x}}_h - \dot{\mathbf{M}}\psi_h\right)^\dagger\right] \in \mathbb{C}^{N \times N}, \tag{9.10}$$

with

$$\psi_h = [\boldsymbol{\theta}_h^T, \boldsymbol{\phi}_h^T]^T \in \mathbb{C}^{(p+q) \times 1}, \quad h = 1, 2, \ldots, H. \tag{9.11}$$

Using the persymmetric structure in the covariance matrix $\dot{\mathbf{R}}$, we have

$$\begin{aligned} \mathrm{tr}(\dot{\mathbf{R}}^{-1}\dot{\mathbf{T}}_j) &= \mathrm{tr}[\mathbf{J}(\dot{\mathbf{R}}^*)^{-1}\mathbf{J}\dot{\mathbf{T}}_j] \\ &= \mathrm{tr}[\dot{\mathbf{R}}^{-1}\mathbf{J}\dot{\mathbf{T}}_j^*\mathbf{J}], \quad j = 0, 1. \end{aligned} \tag{9.12}$$

As a result, (9.7) can be rewritten as

$$f\left(\dot{\mathbf{X}}_t, \dot{\mathbf{X}}_s | H_0\right) = \left\{ \frac{1}{\pi^N \det(\dot{\mathbf{R}})} \exp\left[-\mathrm{tr}(\dot{\mathbf{R}}^{-1}\mathbf{T}_0)\right] \right\}^{K+H}, \tag{9.13}$$

where

$$\begin{aligned} \mathbf{T}_0 &= \frac{1}{2}\left(\dot{\mathbf{T}}_0 + \mathbf{J}\dot{\mathbf{T}}_0^*\mathbf{J}\right) \\ &= \frac{1}{K+H}\left[\hat{\mathbf{R}} + \left(\dot{\mathbf{X}} - \mathbf{J}\mathbf{\Phi}\right)\left(\dot{\mathbf{X}} - \mathbf{J}\mathbf{\Phi}\right)^\dagger\right] \in \mathbb{C}^{N \times N}, \end{aligned} \tag{9.14}$$

with

$$\hat{\mathbf{R}} = \frac{1}{2}\left[\dot{\mathbf{X}}_s \dot{\mathbf{X}}_s^\dagger + \mathbf{J}\left(\dot{\mathbf{X}}_s \dot{\mathbf{X}}_s^\dagger\right)^*\mathbf{J}\right] \in \mathbb{C}^{N \times N}, \tag{9.15}$$

$$\Phi = [\Re(\phi_1), \dots, \Re(\phi_H), \jmath\Im(\phi_1), \dots, \jmath\Im(\phi_H)] \in \mathbb{C}^{q \times 2H}, \quad (9.16)$$

$$\dot{\mathbf{X}} = [\dot{\mathbf{x}}_{e1}, \dots, \dot{\mathbf{x}}_{eH}, \dot{\mathbf{x}}_{o1}, \dots, \dot{\mathbf{x}}_{oH}] \in \mathbb{C}^{N \times 2H}, \quad (9.17)$$

and

$$\begin{cases} \dot{\mathbf{x}}_{eh} = \frac{1}{2}\left(\dot{\mathbf{x}}_h + \mathbf{J}\dot{\mathbf{x}}_h^*\right) \in \mathbb{C}^{N \times 1}, \\ \dot{\mathbf{x}}_{oh} = \frac{1}{2}\left(\dot{\mathbf{x}}_h - \mathbf{J}\dot{\mathbf{x}}_h^*\right) \in \mathbb{C}^{N \times 1}, \end{cases} \quad (9.18)$$

for $h = 1, 2, \dots, H$. Similarly, we obtain

$$f\left(\dot{\mathbf{X}}_t, \dot{\mathbf{X}}_s | H_1\right) = \left\{ \frac{1}{\pi^N \det(\dot{\mathbf{R}})} \exp\left[-\mathrm{tr}(\dot{\mathbf{R}}^{-1}\mathbf{T}_1)\right] \right\}^{K+H}, \quad (9.19)$$

where

$$\begin{aligned} \mathbf{T}_1 &= \frac{1}{2}\left(\dot{\mathbf{T}}_1 + \mathbf{J}\dot{\mathbf{T}}_1^*\mathbf{J}\right) \\ &= \frac{1}{K+H}\left[\hat{\mathbf{R}} + \left(\dot{\mathbf{X}} - \dot{\mathbf{M}}\Psi\right)\left(\dot{\mathbf{X}} - \dot{\mathbf{M}}\Psi\right)^\dagger\right] \in \mathbb{C}^{N \times N}, \end{aligned} \quad (9.20)$$

with

$$\Psi = [\Re(\psi_1), \dots, \Re(\psi_H), \jmath\Im(\psi_1), \dots, \jmath\Im(\psi_H)] \in \mathbb{C}^{(q+p) \times 2H}. \quad (9.21)$$

Similar to [3], we can derive a detector for the detection problem (9.2) as

$$\Lambda = \frac{\det\left(\mathbf{I}_{2H} + \dot{\mathbf{X}}^\dagger \dot{\mathbf{P}}_1 \dot{\mathbf{X}}\right)}{\det\left(\mathbf{I}_{2H} + \dot{\mathbf{X}}^\dagger \dot{\mathbf{P}}_2 \dot{\mathbf{X}}\right)} \underset{H_1}{\overset{H_0}{\gtrless}} \lambda, \quad (9.22)$$

where λ is a detection threshold,

$$\dot{\mathbf{P}}_1 = \hat{\mathbf{R}}^{-1} - \hat{\mathbf{R}}^{-1}\dot{\mathbf{J}}\left(\dot{\mathbf{J}}^\dagger\hat{\mathbf{R}}^{-1}\dot{\mathbf{J}}\right)^{-1}\dot{\mathbf{J}}^\dagger\hat{\mathbf{R}}^{-1} \in \mathbb{C}^{N \times N}, \quad (9.23)$$

and

$$\dot{\mathbf{P}}_2 = \hat{\mathbf{R}}^{-1} - \hat{\mathbf{R}}^{-1}\dot{\mathbf{M}}\left(\dot{\mathbf{M}}^\dagger\hat{\mathbf{R}}^{-1}\dot{\mathbf{M}}\right)^{-1}\dot{\mathbf{M}}^\dagger\hat{\mathbf{R}}^{-1} \in \mathbb{C}^{N \times N}. \quad (9.24)$$

This detector is referred to as one-step persymmetric GLRT in interference (1S-PGLRT-I). Note that the 1S-PGLRT-I in (9.22) is different from that in [3], since the former exploits the persymmetry, whereas the latter does not.

9.2.2   Two-Step GLRT

Here, we use the two-step method to derive the GLRT detector. In the first step, the noise covariance matrix $\dot{\mathbf{R}}$ is assumed known. The GLRT with known $\dot{\mathbf{R}}$ for the detection problem in (9.2) is given by

$$\frac{\max_{\{\theta_h, \phi_h\}_{h=1}^H} f\left(\dot{\mathbf{X}}_t | H_1\right)}{\max_{\{\phi_h\}_{h=1}^H} f\left(\dot{\mathbf{X}}_t | H_0\right)} \underset{H_0}{\overset{H_1}{\gtrless}} v_0, \quad (9.25)$$

where v_0 is a detection threshold, $f\left(\dot{\mathbf{X}}_t|H_0\right)$ and $f\left(\dot{\mathbf{X}}_t|H_1\right)$ are the PDFs of the test data under H_0 and H_1, respectively,

Similar to (9.19), $f\left(\dot{\mathbf{X}}_t|H_1\right)$ can be recast as

$$f\left(\dot{\mathbf{X}}_t|H_1\right) = \left\{\frac{1}{\pi^N \det(\dot{\mathbf{R}})} \exp\left[-\mathrm{tr}(\dot{\mathbf{R}}^{-1}\mathbf{W}_1)\right]\right\}^H, \qquad (9.26)$$

where

$$\mathbf{W}_1 = \frac{1}{H}\left(\dot{\mathbf{X}} - \dot{\mathbf{M}}\mathbf{\Psi}\right)\left(\dot{\mathbf{X}} - \dot{\mathbf{M}}\mathbf{\Psi}\right)^\dagger \in \mathbb{C}^{N\times N}. \qquad (9.27)$$

Similar to (9.13), $f\left(\dot{\mathbf{X}}_t|H_0\right)$ can be rewritten as

$$f\left(\dot{\mathbf{X}}_t|H_0\right) = \left\{\frac{1}{\pi^N \det(\dot{\mathbf{R}})} \exp\left[-\mathrm{tr}(\dot{\mathbf{R}}^{-1}\mathbf{W}_0)\right]\right\}^H, \qquad (9.28)$$

where

$$\mathbf{W}_0 = \frac{1}{H}\left(\dot{\mathbf{X}} - \dot{\mathbf{J}}\mathbf{\Phi}\right)\left(\dot{\mathbf{X}} - \dot{\mathbf{J}}\mathbf{\Phi}\right)^\dagger \in \mathbb{C}^{N\times N}. \qquad (9.29)$$

Following the derivations in [3], we can obtain the GLRT with known $\dot{\mathbf{R}}$ as

$$\Upsilon_1 = \mathrm{tr}\left[\left(\mathbf{P}_{\dot{\mathbf{R}}^{-1/2}\dot{\mathbf{M}}} - \mathbf{P}_{\dot{\mathbf{R}}^{-1/2}\dot{\mathbf{J}}}\right)\dot{\mathbf{R}}^{-1/2}\dot{\mathbf{X}}\dot{\mathbf{X}}^\dagger\dot{\mathbf{R}}^{-1/2}\right]\underset{H_0}{\overset{H_1}{\gtrless}} v_1, \qquad (9.30)$$

where v_1 is the detection threshold.

In the second step, we use the covariance matrix estimate $\hat{\mathbf{R}}$ to replace $\dot{\mathbf{R}}$ and then obtain the detector as

$$\begin{aligned}\Upsilon &= \mathrm{tr}\left[\left(\mathbf{P}_{\hat{\mathbf{R}}^{-1/2}\dot{\mathbf{M}}} - \mathbf{P}_{\hat{\mathbf{R}}^{-1/2}\dot{\mathbf{J}}}\right)\hat{\mathbf{R}}^{-1/2}\dot{\mathbf{X}}\dot{\mathbf{X}}^\dagger\hat{\mathbf{R}}^{-1/2}\right]\\ &= \mathrm{tr}\left[\left(\dot{\mathbf{P}}_1 - \dot{\mathbf{P}}_2\right)\dot{\mathbf{X}}\dot{\mathbf{X}}^\dagger\right]\underset{H_0}{\overset{H_1}{\gtrless}} v,\end{aligned} \qquad (9.31)$$

where v is the detection threshold. This detector is called two-step persymmetric GLRT in interference (2S-PGLRT-I).

9.3 Probability of False Alarm for 1S-PGLRT-I

To complete the 1S-PGLRT-I detector in (9.22), we provide an approach to calculate the detection threshold for any given PFA. To address this issue, we derive analytical expressions for the probabilities of false alarm in various cases, as shown below. In addition, we will discuss about the PFA of the 2S-PGLRT-I detector in (9.31) at the end of this section.

It is easy to check that under H_0, the column vectors of $\dot{\mathbf{X}}$ are IID and have the following distributions:

$$\begin{cases} \dot{\mathbf{x}}_{eh} \sim \mathbb{CN}_N \left(\mathbf{0}, \frac{\dot{\mathbf{R}}}{2}\right), \\ \dot{\mathbf{x}}_{oh} \sim \mathbb{CN}_N \left(\mathbf{0}, \frac{\dot{\mathbf{R}}}{2}\right), \end{cases} \tag{9.32}$$

for $h = 1, 2, \ldots, H$.

Define two unitary matrices as follows

$$\mathbf{D}_1 = \frac{1}{2}\left[(\mathbf{I}_N + \mathbf{J}) + \jmath(\mathbf{I}_N - \mathbf{J})\right] \in \mathbb{C}^{N \times N}, \tag{9.33}$$

and

$$\mathbf{V}_{2H} = \begin{bmatrix} \mathbf{I}_H & \mathbf{0} \\ \mathbf{0} & -\jmath\mathbf{I}_H \end{bmatrix} \in \mathbb{C}^{2H \times 2H}. \tag{9.34}$$

Using $\mathbf{D}_1$ and $\mathbf{V}_{2H}$, we define

$$\begin{aligned} \bar{\mathbf{X}}_t &= \mathbf{D}_1 \dot{\mathbf{X}} \mathbf{V}_{2H} \\ &= [\dot{\mathbf{x}}_{er1}, \ldots, \dot{\mathbf{x}}_{erH}, \dot{\mathbf{x}}_{or1}, \ldots, \dot{\mathbf{x}}_{orH}] \in \mathbb{R}^{N \times 2H}, \end{aligned} \tag{9.35}$$

where

$$\begin{aligned} \dot{\mathbf{x}}_{erh} &= \mathbf{D}_1 \dot{\mathbf{x}}_{eh} \\ &= \frac{1}{2}\left[(\mathbf{I}_N + \mathbf{J})\Re(\dot{\mathbf{x}}_h) - (\mathbf{I}_N - \mathbf{J})\Im(\dot{\mathbf{x}}_h)\right] \in \mathbb{R}^{N \times 1}, \end{aligned} \tag{9.36}$$

and

$$\begin{aligned} \dot{\mathbf{x}}_{orh} &= -\jmath\mathbf{D}_1 \dot{\mathbf{x}}_{oh} \\ &= \frac{1}{2}\left[(\mathbf{I}_N - \mathbf{J})\Re(\dot{\mathbf{x}}_h) + (\mathbf{I}_N + \mathbf{J})\Im(\dot{\mathbf{x}}_h)\right] \in \mathbb{R}^{N \times 1}, \end{aligned} \tag{9.37}$$

for $h = 1, 2, \ldots, H$. It follows from (9.32) that

$$\bar{\mathbf{X}}_t \sim \mathcal{N}_{N \times 2H}\left(\mathbf{0}, \mathbf{R} \otimes \mathbf{I}_{2H}\right), \tag{9.38}$$

where

$$\mathbf{R} = \frac{1}{2}\mathbf{D}_1 \dot{\mathbf{R}} \mathbf{D}_1^\dagger = \frac{1}{2}\left[\Re(\dot{\mathbf{R}}) + \mathbf{J}\Im(\dot{\mathbf{R}})\right] \in \mathbb{R}^{N \times N}. \tag{9.39}$$

Let

$$\begin{cases} \bar{\mathbf{R}} = \mathbf{D}_1 \hat{\mathbf{R}} \mathbf{D}_1^\dagger = \Re(\hat{\mathbf{R}}) + \mathbf{J}\Im(\hat{\mathbf{R}}) \in \mathbb{R}^{N \times N}, \\ \bar{\mathbf{J}} = \mathbf{D}_1 \dot{\mathbf{J}} = \Re(\dot{\mathbf{J}}) - \Im(\dot{\mathbf{J}}) \in \mathbb{R}^{N \times q}, \\ \bar{\mathbf{M}} = \mathbf{D}_1 \dot{\mathbf{M}} = \Re(\dot{\mathbf{M}}) - \Im(\dot{\mathbf{M}}) \in \mathbb{R}^{N \times (p+q)}, \end{cases} \tag{9.40}$$

and then, the 1S-PGLRT-I in (9.22) can be recast to

$$\Lambda = \frac{\det\left(\mathbf{I}_{2H} + \bar{\mathbf{X}}_t^T \mathbf{P}_1 \bar{\mathbf{X}}_t\right)}{\det\left(\mathbf{I}_{2H} + \bar{\mathbf{X}}_t^T \mathbf{P}_2 \bar{\mathbf{X}}_t\right)} \underset{H_0}{\overset{H_1}{\gtrless}} \lambda, \tag{9.41}$$

where

$$\mathbf{P}_1 = \bar{\mathbf{R}}^{-1} - \bar{\mathbf{R}}^{-1}\bar{\mathbf{J}}\left(\bar{\mathbf{J}}^T\bar{\mathbf{R}}^{-1}\bar{\mathbf{J}}\right)^{-1}\bar{\mathbf{J}}^T\bar{\mathbf{R}}^{-1} \in \mathbb{R}^{N\times N}, \tag{9.42}$$

and

$$\mathbf{P}_2 = \bar{\mathbf{R}}^{-1} - \bar{\mathbf{R}}^{-1}\bar{\mathbf{M}}\left(\bar{\mathbf{M}}^T\bar{\mathbf{R}}^{-1}\bar{\mathbf{M}}\right)^{-1}\bar{\mathbf{M}}^T\bar{\mathbf{R}}^{-1} \in \mathbb{R}^{N\times N}. \tag{9.43}$$

Note that all the quantities in (9.41) are real-valued.

Before deriving the statistical characterizations of Λ, we first introduce the following theorem:

Theorem 9.3.1. *Assume that*

$$\begin{cases} \mathbf{Y} \sim \mathcal{N}_{Q\times J}(\mathbf{0}, \mathbf{I}_Q \otimes \mathbf{I}_J), \\ \mathbf{S} \sim \mathcal{W}_Q(L, \mathbf{I}_Q), \end{cases} \tag{9.44}$$

where J is a positive integer, $L \geqslant Q$, and $\mathbf{Q}$ is defined as

$$\mathbf{Q} \triangleq \mathbf{S}^{-1} - \mathbf{S}^{-1}\mathbf{B}\left(\mathbf{B}^T\mathbf{S}^{-1}\mathbf{B}\right)^{-1}\mathbf{B}^T\mathbf{S}^{-1} \in \mathbb{R}^{Q\times Q} \tag{9.45}$$

with $\mathbf{B} \in \mathbb{R}^{Q\times p}$ being a full-column-rank matrix and $Q > p$. Then,

$$\frac{\det\left(\mathbf{I}_J + \mathbf{Y}^T\mathbf{Q}\mathbf{Y}\right)}{\det\left(\mathbf{I}_J + \mathbf{Y}^T\mathbf{S}^{-1}\mathbf{Y}\right)} \sim U_{p,J,L+p-Q}$$

$$\overset{\text{d}}{=} \prod_{j=1}^{p} \beta\left(\frac{L+p-Q-j+1}{2}, \frac{J}{2}\right). \tag{9.46}$$

Proof. The proof is similar to that in [10] and hence is omitted here for brevity. $\square$

As derived in Appendix 9.A, the test statistic Λ in (9.41) can be recast to

$$\Lambda = \frac{\det\left(\mathbf{I}_{2H} + \mathbf{X}_2^T\mathbf{R}_{22}^{-1}\mathbf{X}_2\right)}{\det\left(\mathbf{I}_{2H} + \mathbf{X}_2^T\mathbf{P}\mathbf{X}_2\right)} \overset{H_1}{\underset{H_0}{\gtrless}} \lambda, \tag{9.47}$$

where

$$\mathbf{P} = \mathbf{R}_{22}^{-1} - \mathbf{R}_{22}^{-1}\mathbf{H}_2\left(\mathbf{H}_2^T\mathbf{R}_{22}^{-1}\mathbf{H}_2\right)^{-1}\mathbf{H}_2^T\mathbf{R}_{22}^{-1} \in \mathbb{R}^{M\times M} \tag{9.48}$$

with $M = N - q$ given in (9.A.6). It is easy to derive that

$$\mathbf{X}_2 \sim \mathcal{N}_{M\times 2H}(\mathbf{0}, \mathbf{I}_M \otimes \mathbf{I}_{2H}), \quad \text{under } H_0, \tag{9.49}$$

and

$$\mathbf{R}_{22} \sim \mathcal{W}_M(2K, \mathbf{I}_M). \tag{9.50}$$

According to Theorem 9.3.1, we can obtain that the test statistic Λ in (9.47) under H_0 is distributed as

$$\Lambda^{-1} \sim U_{p,2H,2K+p-M}$$

$$\overset{\text{d}}{=} \prod_{j=1}^{p} \beta\left(\frac{2K+p-M-j+1}{2}, H\right). \tag{9.51}$$

It is obvious that the distribution of Λ is irrelevant to $\dot{\mathbf{R}}$. Therefore, the 1S-PGLRT-I detector exhibits the CFAR property against the noise covariance matrix. Based on the statistical property in (9.51), we derive analytical expressions for the PFA of the 1S-PGLRT-I detector in the following seven cases.

9.3.1 p is 1

When $p = 1$, the target steering vector is exactly known. In such a case, we have

$$\Lambda^{-1} \sim \beta\left(\frac{2K - M + 1}{2}, H\right). \tag{9.52}$$

As a result,

$$\Lambda - 1 \stackrel{\mathrm{d}}{=} \frac{\chi^2_{2H}}{\chi^2_{2K-M+1}}. \tag{9.53}$$

Further, the 1S-PGLRT-I can be equivalently cast as

$$\frac{\chi^2_{2H}}{\chi^2_{2K-M+1}} \underset{H_0}{\overset{H_1}{\gtrless}} \lambda - 1. \tag{9.54}$$

Then, the PFA for $p = 1$ can be derived as

$$P_{\mathrm{FA}} = \sum_{j=1}^{H} \frac{\Gamma\left(\frac{2K+2H-2j-M+1}{2}\right)}{\Gamma\left(H-j+1\right)\Gamma\left(\frac{2K-M+1}{2}\right)}(\lambda-1)^{H-j}\lambda^{-\frac{2K+2H-2j-M+1}{2}}. \tag{9.55}$$

9.3.2 p is 2

When $p = 2$, the distribution of Λ^{-1} is $\Lambda^{-1} \sim U_{2,2H,2K+2-M}$. According to [11, p. 311, Theorem 8.4.6], we have

$$\sqrt{\Lambda} - 1 \stackrel{\mathrm{d}}{=} \frac{\chi^2_{4H}}{\chi^2_{2(2K-M+1)}}. \tag{9.56}$$

Hence, the 1S-PGLRT-I can be equivalently cast as

$$\frac{\chi^2_{4H}}{\chi^2_{2(2K-M+1)}} \underset{H_0}{\overset{H_1}{\gtrless}} \sqrt{\Lambda} - 1. \tag{9.57}$$

The PFA for $p = 2$ is given by

$$P_{\mathrm{FA}} = \sum_{j=1}^{2H} C_{2K+2H-M-j}^{2H-j}(\sqrt{\Lambda}-1)^{2H-j}\lambda^{-\frac{2K+2H-j-M+1}{2}}. \tag{9.58}$$

9.3.3 p is 3

When $p = 3$, the distribution of Λ^{-1} is $\Lambda^{-1} \sim U_{3,2H,2K+3-M}$. The PDF of $U_{3,2H,2K+3-M}$ can be obtained according to the result in [12, eq. (4.2)]. Further, we can derive the PFA of the 1S-PGLRT-I with $p = 3$ as

$$
\begin{aligned}
P_{\text{FA}} = 1 - & \frac{1}{2B(g_1 - 1, 2H)B(\frac{1}{2}(g_1 - 2), H)} \left\{ \sum_{m=1}^{H-1} (-1)^{m-1} C_{2H-1}^{2m-1} C_{H-1}^{m} \right. \\
& \times \left[\frac{1}{(\frac{1}{2}g_1 + m - 1)^2} + \frac{-\ln\lambda}{\frac{1}{2}g_1 + m - 1} \lambda^{-(\frac{1}{2}g_1 + m - 1)} \right. \\
& \left. - \frac{\lambda^{-(\frac{1}{2}g_1 + m - 1)}}{(\frac{1}{2}g_1 + m - 1)^2} \right] + 2 \sum_{\substack{m=0 \\ l \neq 2m-1}}^{H-1} \sum_{l=0}^{2H-1} \frac{(-1)^{l+m}}{2m - l - 1} C_{2H-1}^{l} C_{H-1}^{m} \\
& \times \left[\frac{2}{g_1 + l - 1} - \frac{2}{g_1 + 2m - 2} - \frac{2\lambda^{-\frac{1}{2}(g_1 + l - 1)}}{g_1 + l - 1} \right. \\
& \left. \left. + \frac{2\lambda^{-\frac{1}{2}(g_1 + 2m - 2)}}{g_1 + 2m - 2} \right] \right\}.
\end{aligned}
\tag{9.59}
$$

where $g_1 = 2K + 3 - M$.

9.3.4 p is 4

When $p = 4$, the distribution of Λ^{-1} is $\Lambda^{-1} \sim U_{4,2H,2K+4-M}$. According to the result in [12, eq. (4.3)], we can derive the PFA of the 1S-PGLRT-I with $p = 4$ as (9.60) shown on the next page,

$$
\begin{aligned}
P_{\text{FA}} = 1 - & \frac{1}{2B(f_1 - 1, 2H)B(f_1 - 3, 2H)} \left\{ \sum_{l=0}^{2H-3} C_{2H-1}^{l} C_{2H-1}^{l+2} \right. \\
& \times \left[\frac{4}{(f_1 + l - 1)^2} + \frac{-2\ln\lambda}{f_1 + l - 1} \lambda^{-\frac{1}{2}(f_1 + l - 1)} - \frac{4\lambda^{-\frac{1}{2}(f_1 + l - 1)}}{(f_1 + l - 1)^2} \right] \\
& + 2 \sum_{\substack{m=0 \\ l \neq m-2}}^{2H-1} \sum_{l=0}^{2H-1} \frac{(-1)^{l+m}}{m - l - 2} C_{2H-1}^{l} C_{2H-1}^{m} \\
& \times \left[\frac{2}{f_1 + l - 1} - \frac{2}{f_1 + m - 3} - \frac{2\lambda^{-\frac{1}{2}(f_1 + l - 1)}}{f_1 + l - 1} \right. \\
& \left. \left. + \frac{2\lambda^{-\frac{1}{2}(f_1 + m - 3)}}{f_1 + m - 3} \right] \right\}.
\end{aligned}
\tag{9.60}
$$

where

$$f_1 = 2K + 4 - M. \tag{9.61}$$

9.3.5 $H = 1$

When $H = 1$, the data model reduces to the point-like target case. In this case, we have

$$\Lambda^{-1} \sim U_{p,2,2K+p-M}. \tag{9.62}$$

Using [11, p. 311, Theorem 8.4.6], we have

$$\sqrt{\Lambda} - 1 \overset{\mathrm{d}}{=} \frac{\chi_{2p}^2}{\chi_{2(2K-M+1)}^2}. \tag{9.63}$$

Therefore, the PFA of the 1S-PGLRT-I for $H = 1$ is given by

$$P_{\mathrm{FA}} = \sum_{j=1}^{p} C_{2K+p-M-j}^{p-j}(\sqrt{\lambda}-1)^{p-j}\lambda^{-\frac{2K+p-j-M+1}{2}}, \tag{9.64}$$

Note that the result in (9.64) with $q = 0$ is consistent with that in [10]. In other words, (9.64) includes the result in [10] as a special case.

9.3.6 $H = 2$

When $H = 2$, we have

$$\Lambda^{-1} \sim U_{p,4,2K+p-M} = U_{4,p,2K+4-M}, \tag{9.65}$$

where the second line is obtained using Theorem 2.1 in [12]. Similar to the derivations of (9.60), we can derive the PFA of the 1S-PGLRT-I with $H = 2$ as (9.66) shown on the next page,

$$P_{\mathrm{FA}} = 1 - \frac{1}{2B(f_1-1,p)B(f_1-3,p)} \left\{ \sum_{l=0}^{p-3} C_{p-1}^l C_{p-1}^{l+2} \right.$$

$$\times \left[\frac{4}{(f_1+l-1)^2} + \frac{-2\ln\lambda}{f_1+l-1}\lambda^{-\frac{1}{2}(f_1+l-1)} - \frac{4\lambda^{-\frac{1}{2}(f_1+l-1)}}{(f_1+l-1)^2} \right]$$

$$+ 2\sum_{m=0}^{p-1}\sum_{\substack{l=0 \\ l\neq m-2}}^{p-1} \frac{(-1)^{l+m}}{m-l-2} C_{p-1}^l C_{p-1}^m$$

$$\times \left[\frac{2}{f_1+l-1} - \frac{2}{f_1+m-3} - \frac{2\lambda^{-\frac{1}{2}(f_1+l-1)}}{f_1+l-1} \right.$$

$$\left.\left. + \frac{2\lambda^{-\frac{1}{2}(f_1+m-3)}}{f_1+m-3} \right] \right\}. \tag{9.66}$$

where f_1 is defined in (9.61).

9.3.7 Arbitrary H and p

In the general cases where H and p are arbitrary positive integers, the distribution of Λ^{-1} is a product of multiple beta distributions. The exact and explicit expression for the PFA is very complicated in the general cases of arbitrary H and p. As an alternative, we seek to obtain an approximate but simple expression for the PFA in the general cases. Based on the result in [13, eq. (4.1)], the PFA of the 1S-PGLRT-I detector in the general cases can be approximated by

$$
P_{\text{FA}} \approx 1 - \left[\left(1 - \frac{r_2}{k_1^2} - \frac{r_4}{k_1^4} + \frac{r_2^2}{k_1^4} \right) F_z(2Hp) \right.
$$
$$
\left. + \left(\frac{r_2}{k_1^2} - \frac{r_2^2}{k_1^4} \right) F_z(2Hp+4) + \frac{r_4}{k_1^4} F_z(2Hp+8) \right], \tag{9.67}
$$

where

$$
\begin{cases}
k_1 = \frac{1}{2}(4K + 2H + p - 2M - 1), \\
z = k_1 \ln \lambda, \\
r_2 = \frac{Hp}{24}(p^2 + 4H^2 - 5),
\end{cases} \tag{9.68}
$$

$$
r_4 = \frac{r_2^2}{2} + \frac{Hp}{960} \left[3p^4 + 48H^4 + 40p^2 H^2 - 50(p^2 + 4H^2) + 159 \right], \tag{9.69}
$$

and

$$
F_z(2\nu) = \int_0^z \frac{y^{\nu-1} \exp(-\frac{y}{2})}{2^\nu \Gamma(\nu)} dy
$$
$$
= \exp\left(-\frac{z}{2}\right) \sum_{k=0}^{\nu-1} \frac{1}{k!} \left(\frac{z}{2}\right)^k. \tag{9.70}
$$

We summarize the theoretical expressions for the probabilities of false alarm of the 1S-PGLRT-I detector in Table 9.1.

As to the detection probability of the 1S-PGLRT-I detector in (9.22), we cannot derive an exact expression for it. This is because the noncentral Wilks' distribution is difficult to handle analytically [11].

As to the PFA and detection probability of the 2S-PGLRT-I detector in (9.31), we cannot derive them. However, we can prove that the 2S-PGLRT-I detector exhibits the CFAR property against the noise covariance matrix, by using the approaches in [14]. The detailed derivations are omitted for brevity.

TABLE 9.1
Analytical expressions for P_{FA} of 1S-PGLRT-I

	$p=1$	$p=2$	$p=3$	$p=4$	$H=1$	$H=2$	other
P_{FA}	(9.55)	(9.58)	(9.59)	(9.60)	(9.64)	(9.66)	(9.67)

9.4   Numerical Examples

In this section, numerical examples are provided to confirm the above theoretical results. Assume that a pulsed Doppler radar is used, where the number of pulses in a CPI is N. The target subspace matrix $\dot{\mathbf{H}}$ and interference subspace matrix $\dot{\mathbf{J}}$ are selected as

$$\dot{\mathbf{H}} = [\mathbf{t}(0.15), \mathbf{t}(0.18), \dots, \mathbf{t}(0.15 + (p-1)0.03)], \tag{9.71}$$

and

$$\dot{\mathbf{J}} = [\mathbf{t}(-0.03), \mathbf{t}(0), \dots, \mathbf{t}(-0.03 + (q-1)0.03)], \tag{9.72}$$

respectively, where

$$\mathbf{t}(f_d) = \frac{1}{\sqrt{N}} \left[e^{-j2\pi f_d \frac{(N-1)}{2}}, \dots, e^{-j2\pi f_d \frac{1}{2}}, e^{j2\pi f_d \frac{1}{2}}, \dots, e^{j2\pi f_d \frac{(N-1)}{2}} \right]^T, \tag{9.73}$$

for even N, and

$$\mathbf{t}(f_d) = \frac{1}{\sqrt{N}} \left[e^{-j2\pi f_d \frac{(N-1)}{2}}, \dots, e^{-j2\pi f_d}, \ 1, \ e^{j2\pi f_d}, \dots, e^{j2\pi f_d \frac{(N-1)}{2}} \right]^T, \tag{9.74}$$

for odd N. The (i,j)th element of the noise covariance matrix is chosen as $\dot{\mathbf{R}}(i,j) = \sigma^2 0.9^{|i-j|}$, where σ^2 is the noise power. Without loss of generality, we set $\boldsymbol{\theta}_h = \sigma_1^2[\theta_1, \dots, \theta_p]^T$ and $\boldsymbol{\phi}_h = \sigma_2^2[\phi_1, \dots, \phi_q]^T$, where $h = 1, 2, \dots, H$, σ_1^2 and σ_2^2 are the signal and interference powers, respectively, θ_i and ϕ_j are randomly sampled from $\mathcal{CN}_1(0,1)$ for $i = 1, \dots, p$ and $j = 1, \dots, q$. The SNR and INR are defined as SNR $= 10\log_{10}\frac{\sigma_1^2}{\sigma^2}$, and INR $= 10\log_{10}\frac{\sigma_2^2}{\sigma^2}$, respectively. The INR is set to be 30 dB.

In Fig. 9.1, the detection probability as a function of SNR is presented. Here, we select $N = 12$, $p = 2$, $q = 3$, $H = 3$, and $P_{\mathrm{FA}} = 10^{-3}$. For comparison purposes, we consider four conventional detectors, i.e., the GLRT and AMF developed in Eq. (10) and Section IV.A of [3], respectively, the persymmetric GLRT (PGLRT) in [10, eq. (28)], and the persymmetric AMF (PAMF) in [15, eq. (14)]. All the detection probability curves are obtained using MC counting techniques. Note that the GLRT and AMF in [3] do not exploit persymmetry, and the PGLRT and PAMF are designed for $H = 1$ without taking the interference into account. It can be seen that the designed detectors outperform their counterparts, and the 1S-PGLRT-I outperforms the 2S-PGLRT-I, especially in the cases where the number of training data is small (for example, $K = 13$ in this parameter setting). When the training data size increases, the performance gains of the designed detectors over their counterparts become small. This is because the estimation accuracy of the noise covariance matrix is high, even when the persymmetry is not exploited.

The detection probability versus K is plotted in Fig. 9.2, where $N = 12$, $p = 2$, $q = 3$, $H = 3$, SNR $= 3$ dB, and $P_{\mathrm{FA}} = 10^{-3}$. These results highlight

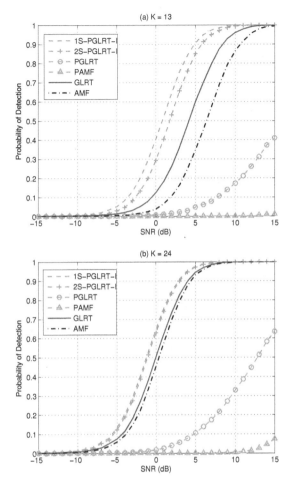

FIGURE 9.1
Detection probability versus SNR for $N = 12$, $p = 2$, $q = 3$, $H = 3$, and $P_{\mathrm{FA}} = 10^{-3}$.

that the designed detectors clearly outperform their counterparts in the cases of small training data size. The designed detectors can work in the sample-starved case (for example, $K = 8$), whereas the GLRT and AMF in [3] cannot work for $K < N$.

9.A   Derivations of (9.47)

Using the result in [16, p. 54], we have

$$\bar{\mathbf{X}}_t^T \mathbf{P}_2 \bar{\mathbf{X}}_t = \bar{\mathbf{X}}_t^T \mathbf{P}_1 \bar{\mathbf{X}}_t - \bar{\mathbf{X}}_t^T \mathbf{P}_1 \bar{\mathbf{H}} \left(\bar{\mathbf{H}}^T \mathbf{P}_1 \bar{\mathbf{H}} \right)^{-1} \bar{\mathbf{H}}^T \mathbf{P}_1 \bar{\mathbf{X}}_t, \qquad (9.A.1)$$

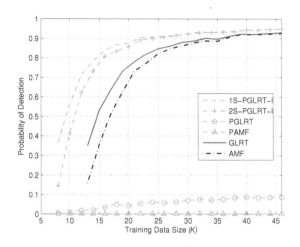

FIGURE 9.2
Detection probability versus K for $N = 12$, $p = 2$, $q = 3$, $H = 3$, SNR = 3 dB, and $P_{\text{FA}} = 10^{-3}$.

where

$$\bar{\mathbf{H}} = \mathbf{D}_1 \dot{\mathbf{H}} = \Re(\dot{\mathbf{H}}) - \Im(\dot{\mathbf{H}}) \in \mathbb{R}^{N \times p}. \tag{9.A.2}$$

Define

$$\mathbf{J}_{\parallel} \triangleq \bar{\mathbf{R}}^{-1/2} \bar{\mathbf{J}} \left(\bar{\mathbf{J}}^T \bar{\mathbf{R}}^{-1} \bar{\mathbf{J}} \right)^{-1/2} \in \mathbb{R}^{N \times q}, \tag{9.A.3}$$

and then,

$$\mathbf{J}_{\parallel}^T \mathbf{J}_{\parallel} = \mathbf{I}_{q \times q}. \tag{9.A.4}$$

It means that the column vectors of $\mathbf{J}_{\parallel}$ are q unit vectors, and they are orthogonal to each other. Hence, we can construct an orthogonal matrix $\mathbf{U}$ such that

$$\mathbf{U} = \left[\mathbf{J}_{\parallel}, \mathbf{J}_{\perp} \right] \in \mathbb{R}^{N \times N}, \tag{9.A.5}$$

where $\mathbf{J}_{\perp} \in \mathbb{R}^{N \times M}$, and

$$\begin{cases} M = N - q, \\ \mathbf{J}_{\perp}^T \mathbf{J}_{\perp} = \mathbf{I}_M, \\ \mathbf{J}_{\perp}^T \mathbf{J}_{\parallel} = \mathbf{0}_{M \times q}. \end{cases} \tag{9.A.6}$$

We now employ the orthogonal matrix $\mathbf{U}$ to define

$$\begin{cases} \mathbf{X} = \mathbf{U}^T \mathbf{R}^{-1/2} \bar{\mathbf{X}}_t \in \mathbb{R}^{N \times 2H}, \\ \mathbf{H} = \mathbf{U}^T \mathbf{R}^{-1/2} \bar{\mathbf{H}} \in \mathbb{R}^{N \times p}, \\ \tilde{\mathbf{P}}_1 = \mathbf{U}^T \mathbf{R}^{1/2} \mathbf{P}_1 \mathbf{R}^{1/2} \mathbf{U} \in \mathbb{R}^{N \times N}. \end{cases} \tag{9.A.7}$$

Then, we obtain

$$\bar{\mathbf{X}}_t^T \mathbf{P}_1 \bar{\mathbf{H}} \left(\bar{\mathbf{H}}^T \mathbf{P}_1 \bar{\mathbf{H}} \right)^{-1} \bar{\mathbf{H}}^T \mathbf{P}_1 \bar{\mathbf{X}}_t = \mathbf{X}^T \tilde{\mathbf{P}}_1 \mathbf{H} \left(\mathbf{H}^T \tilde{\mathbf{P}}_1 \mathbf{H} \right)^{-1} \mathbf{H}^T \tilde{\mathbf{P}}_1 \mathbf{X}, \tag{9.A.8}$$

and

$$\bar{\mathbf{X}}_t^\dagger \mathbf{P}_1 \bar{\mathbf{X}}_t = \mathbf{X}^\dagger \tilde{\mathbf{P}}_1 \mathbf{X}. \tag{9.A.9}$$

Let

$$\mathbf{J}_0 = \mathbf{U}^T \mathbf{R}^{-1/2} \bar{\mathbf{J}} \in \mathbb{R}^{N \times q}. \tag{9.A.10}$$

Using (9.A.3) and (9.A.5), we have

$$\mathbf{J}_0 = \mathbf{E}_q \left(\bar{\mathbf{J}}^T \bar{\mathbf{R}}^{-1} \bar{\mathbf{J}} \right)^{1/2}, \tag{9.A.11}$$

where

$$\mathbf{E}_q = [\mathbf{I}_q, \mathbf{0}_{q \times M}]^T \in \mathbb{R}^{N \times q}. \tag{9.A.12}$$

Define

$$\mathbf{R} = \mathbf{U}^T \mathbf{R}^{-1/2} \bar{\mathbf{R}} \mathbf{R}^{-1/2} \mathbf{U} \in \mathbb{R}^{N \times N}, \tag{9.A.13}$$

and then, $\tilde{\mathbf{P}}_1$ in (9.A.7) can be written as

$$\begin{aligned}
\tilde{\mathbf{P}}_1 &= \mathbf{R}^{-1} - \mathbf{R}^{-1} \mathbf{J}_0 \left(\mathbf{J}_0{}^T \mathbf{R}^{-1} \mathbf{J}_0 \right)^{-1} \mathbf{J}_0{}^T \mathbf{R}^{-1} \\
&= \mathbf{R}^{-1} - \mathbf{R}^{-1} \mathbf{E}_q \left(\mathbf{E}_q^T \mathbf{R}^{-1} \mathbf{E}_q \right)^{-1} \mathbf{E}_q^T \mathbf{R}^{-1}, \\
&= \begin{bmatrix} \mathbf{0}_{q \times q} & \mathbf{0}_{q \times M} \\ \mathbf{0}_{M \times q} & \mathbf{R}_{22}^{-1} \end{bmatrix},
\end{aligned} \tag{9.A.14}$$

where $\mathbf{R}_{22}$ is a submatrix of $\mathbf{R}$, i.e.,

$$\mathbf{R} = \begin{bmatrix} \mathbf{R}_{11} & \mathbf{R}_{21}^T \\ \mathbf{R}_{21} & \mathbf{R}_{22} \end{bmatrix}, \tag{9.A.15}$$

with $\mathbf{R}_{11} \in \mathbb{R}^{q \times q}$, $\mathbf{R}_{21} \in \mathbb{R}^{M \times q}$, and $\mathbf{R}_{22} \in \mathbb{R}^{M \times M}$.

Write

$$\mathbf{X} = \left[\mathbf{X}_1^T, \mathbf{X}_2^T \right]^T \tag{9.A.16}$$

with $\mathbf{X}_1 \in \mathbb{R}^{q \times 2H}$ and $\mathbf{X}_2 \in \mathbb{R}^{M \times 2H}$, and

$$\mathbf{H} = \left[\mathbf{H}_1^T, \mathbf{H}_2^T \right]^T \tag{9.A.17}$$

with $\mathbf{H}_1 \in \mathbb{R}^{q \times p}$ and $\mathbf{H}_2 \in \mathbb{R}^{M \times p}$. Applying (9.A.14)–(9.A.17) to (9.A.8) and (9.A.9) produces

$$\begin{aligned}
\mathbf{X}^T \tilde{\mathbf{P}}_1 \mathbf{H} \left(\mathbf{H}^T \tilde{\mathbf{P}}_1 \mathbf{H} \right)^{-1} \mathbf{H}^T \tilde{\mathbf{P}}_1 \mathbf{X} \\
= \mathbf{X}_2^T \mathbf{R}_{22}^{-1} \mathbf{H}_2 \left(\mathbf{H}_2^T \mathbf{R}_{22}^{-1} \mathbf{H}_2 \right)^{-1} \mathbf{H}_2{}^T \mathbf{R}_{22}^{-1} \mathbf{X}_2,
\end{aligned} \tag{9.A.18}$$

and

$$\mathbf{X}_t^\dagger \mathbf{P}_1 \mathbf{X}_t = \mathbf{X}_2^T \mathbf{R}_{22}^{-1} \mathbf{X}_2, \tag{9.A.19}$$

respectively. Using (9.A.1), (9.A.8), (9.A.18), and (9.A.19), we can rewrite (9.41) as (9.47).

Bibliography

[1] J. Liu, T. Jian, and W. Liu, "Persymmetric detection of subspace signals based on multiple observations in the presence of subspace interference," *Signal Processing*, vol. 183, 107964, 2021.

[2] L. L. Scharf and B. Friedlander, "Matched subspace detectors," *IEEE Transactions on Signal Processing*, vol. 42, no. 8, pp. 2146–2157, August 1994.

[3] F. Bandiera, A. De Maio, A. S. Greco, and G. Ricci, "Adaptive radar detection of distributed targets in homogeneous and partially homogeneous noise plus subspace interference," *IEEE Transactions on Signal Processing*, vol. 55, no. 4, pp. 1223–1237, April 2007.

[4] W. Liu, Y. Wang, J. Liu, L. Huang, and C. Hao, "Performance analysis of adaptive detectors for point targets in subspace interference and Gaussian noise," *IEEE Transactions on Aerospace and Electronic Systems*, vol. 54, no. 1, pp. 429–441, Feburary 2018.

[5] W. Liu, J. Liu, L. Huang, D. Zou, and Y. Wang, "Rao tests for distributed target detection in interference and noise," *Signal Processing*, vol. 117, pp. 333–342, 2015.

[6] A. De Maio and D. Orlando, "Adaptive radar detection of a subspace signal embedded in subspace structured plus Gaussian interference via invariance," *IEEE Transactions on Signal Processing*, vol. 64, no. 8, pp. 2156–2167, April 2016.

[7] D. Ciuonzo, A. De Maio, and D. Orlando, "On the statistical invariance for adaptive radar detection in partially homogeneous disturbance plus structured interference," *IEEE Transaction on Signal Processing*, vol. 5, no. 65, pp. 1222–1234, March 1 2017.

[8] D. Ciuonzo, A. D. Maio, and D. Orlando, "A unifying framework for adaptive radar detection in homogeneous plus structured interference–Part I: On the maximal invariant statistic," *IEEE Transaction on Signal Processing*, vol. 11, no. 64, pp. 2894–2906, June 1 2016.

[9] ——, "A unifying framework for adaptive radar detection in homogeneous plus structured interference–Part II: Detectors design," *IEEE Transaction on Signal Processing*, vol. 11, no. 64, pp. 2907–2919, June 1 2016.

[10] J. Liu, S. Sun, and W. Liu, "One-step persymmetric GLRT for subspace signals," *IEEE Transaction on Signal Processing*, vol. 14, no. 67, pp. 3639–3648, July 15 2019.

[11] T. Anderson, *An Introduction to Multivariate Statistical Analysis*, 3rd ed. New York, USA: Wiley, 2003.

[12] K. C. Sreedharan Pillai and A. K. Gupta, "On the exact distribution of Wilks's criterion," *Biometrika*, vol. 56, pp. 109–118, 1969.

[13] M. Schatzoff, "Exact distribution of Wilks's likelihood ratio criterion," *Biometrika*, vol. 53, pp. 347–358, 1966.

[14] W. Liu, W. Xie, J. Liu, and Y. Wang, "Adaptive double subspace signal detection in Gaussian background–Part I: Homogeneous environments," *IEEE Transactions on Signal Processing*, vol. 62, no. 9, pp. 2345–2357, May 2014.

[15] J. Liu, W. Liu, Y. Gao, S. Zhou, and X.-G. Xia, "Persymmetric adaptive detection of subspace signals: Algorithms and performance analysis," *IEEE Transactions on Signal Processing*, vol. 66, no. 23, pp. 6124–6136, December 1 2018.

[16] H. Yanai, K. Takeuchi, and Y. Takane, *Projection Matrices, Generalized Inverse Matrices, and Singular Value Decomposition (Statistics for Social and Behavioral Sciences)*. New York, USA: Springer, 2011.

10

Adaptive Detection in Partially Homogeneous Environments

In order to account for power variations, a slightly generalized noise model has been introduced in [1] and widely investigated by the radar community [2–6] which is referred to as PHE. The PHE assumes that the covariance matrix of the primary data and that of secondary data coincide only up to a scaling factor [7]. Partially homogeneous scenarios fit in those situations where the maximum spacing between any two secondary range cells is small compared with the scale over which power levels change (adjacent range cells that are not in the immediate vicinity of the CUT) [3].

In this chapter, we will have a full review of the PHE by introducing the one-step and two-step GLRTs, as well as the Rao and the Wald test for the radar targets detection problem. To this end, we borrow the hypothesis testing in (6.1) and extend it to PHE by defining $\mathbf{n}_j \sim \mathcal{CN}_N(\mathbf{0}, \mathbf{R})$ for $j = 1, \ldots, J$, $J > 1$, $\mathbf{n}_k \sim \mathcal{CN}_N(\mathbf{0}, \gamma\mathbf{R})$ for $k = J + 1, J + 2, \ldots, J + K$, namely

$$\begin{cases} \mathrm{E}[\mathbf{n}_j \mathbf{n}_j^\dagger] = \mathbf{R} \in \mathbb{C}^{N \times N}, \; j = 1, \ldots, J, \\ \mathrm{E}[\mathbf{n}_k \mathbf{n}_k^\dagger] = \gamma\mathbf{R} \in \mathbb{C}^{N \times N}, \; k = J + 1, J + 2, \ldots, J + K, \end{cases} \quad (10.1)$$

where $\gamma > 0$ is the unknown power scaling factor. At the same time, we assume that the covariance $\mathbf{R}$ and steering vector $\mathbf{p}$ exhibit persymmetric structures so that $\mathbf{R} = \mathbf{J}\mathbf{R}^*\mathbf{J}$ with $\mathbf{J}$ the permutation matrix, and $\mathbf{p} = \mathbf{J}\mathbf{p}^*$. Numerical examples show that the detection schemes accounting for the power varying environments significantly outperform their conventional counterparts.

10.1 Detector Design

10.1.1 One-Step GLRT

The one-step GLRT for the distributed targets under PHE is given by

$$\Xi_0 = \frac{\max_{\{\gamma, \mathbf{a}, \mathbf{R}\}} f_1(\mathbf{Y}, \mathbf{Y}_1 | \gamma, \mathbf{a}, \mathbf{R})}{\max_{\{\gamma, \mathbf{R}\}} f_0(\mathbf{Y}, \mathbf{Y}_1 | \gamma, \mathbf{R})} \underset{H_0}{\overset{H_1}{\gtrless}} \xi_0, \quad (10.2)$$

DOI: 10.1201/9781003340232-10

where ξ_0 is the detection threshold, and

$$\begin{cases} f_0(\mathbf{Y}, \mathbf{Y}_1 | \gamma, \mathbf{R}) = \frac{\gamma^{-NK}}{[\pi^N \det(\mathbf{R})]^{K+J}} \exp\left[-\operatorname{tr}\left(\mathbf{R}^{-1}\mathbf{T}_0\right)\right], \\ f_1(\mathbf{Y}, \mathbf{Y}_1 | \gamma, \mathbf{a}, \mathbf{R}) = \frac{\gamma^{-NK}}{[\pi^N \det(\mathbf{R})]^{K+J}} \exp\left[-\operatorname{tr}\left(\mathbf{R}^{-1}\mathbf{T}_1\right)\right], \end{cases} \qquad (10.3)$$

with

$$\begin{cases} \mathbf{T}_0 = \frac{1}{\gamma}\mathbf{Y}_1\mathbf{Y}_1^{\dagger} + \mathbf{Y}\mathbf{Y}^{\dagger}, \\ \mathbf{T}_1 = \frac{1}{\gamma}\mathbf{Y}_1\mathbf{Y}_1^{\dagger} + (\mathbf{Y} - \mathbf{p}\mathbf{a}^T)(\mathbf{Y} - \mathbf{p}\mathbf{a}^T)^{\dagger}. \end{cases} \qquad (10.4)$$

Now, notice that the PDF of $\mathbf{Y}$ and $\mathbf{Y}_1$ under both hypotheses can be rewritten exploiting the persymmetric property of $\mathbf{R}$ and $\mathbf{p}$ as follows

$$\begin{cases} f_0(\mathbf{Y}, \mathbf{Y}_1 | \gamma, \mathbf{R}) = \frac{\gamma^{-NK}}{[\pi^N \det(\mathbf{R})]^{K+J}} \exp\left[-\operatorname{tr}\left(\mathbf{R}^{-1}\mathbf{T}_{p0}\right)\right], \\ f_1(\mathbf{Y}, \mathbf{Y}_1 | \gamma, \mathbf{a}, \mathbf{R}) = \frac{\gamma^{-NK}}{[\pi^N \det(\mathbf{R})]^{K+J}} \exp\left[-\operatorname{tr}\left(\mathbf{R}^{-1}\mathbf{T}_{p1}\right)\right], \end{cases} \qquad (10.5)$$

where

$$\begin{cases} \mathbf{T}_{p0} = \frac{1}{\gamma}\hat{\mathbf{R}} + \mathbf{Y}_p\mathbf{Y}_p^{\dagger}, \\ \mathbf{T}_{p1} = \frac{1}{\gamma}\hat{\mathbf{R}} + (\mathbf{Y}_p - \mathbf{p}\mathbf{a}_p^T)(\mathbf{Y}_p - \mathbf{p}\mathbf{a}_p^T)^{\dagger}, \end{cases} \qquad (10.6)$$

with $\hat{\mathbf{R}}$, $\mathbf{Y}_p$, and $\mathbf{a}_p$ given by (6.8), (6.10), and (6.11), respectively.

As a consequence, substituting (10.5) into (10.2) and maximizing the numerator and the denominator over $\mathbf{R}$ (under the constraint that $\mathbf{R}$ is persymmetric), the GLRT can be recast as

$$\Xi_1 = \frac{\min_{\{\gamma\}} \gamma^{\frac{NK}{J+K}} \det\left(\mathbf{T}_{p0}\right)}{\min_{\{\gamma, \mathbf{a}_p\}} \left[\gamma^{\frac{NK}{J+K}} \det\left(\mathbf{T}_{p1}\right)\right]} \underset{H_0}{\overset{H_1}{\gtrless}} \xi_1. \qquad (10.7)$$

The minimization of the denominator over $\mathbf{a}_p$ can be performed following the lead of [1]. More precisely, the minimum is achieved when

$$\hat{\mathbf{a}}_p^T = \frac{\mathbf{p}^{\dagger}\hat{\mathbf{R}}^{-1}\mathbf{Y}_p}{\mathbf{p}^{\dagger}\hat{\mathbf{R}}^{-1}\mathbf{p}}. \qquad (10.8)$$

It follows that the GLRT can be recast as

$$\Xi_1 = \frac{\min_{\{\gamma\}} \gamma^{\frac{NK}{J+K}} \det\left(\mathbf{A}_0 + \frac{1}{\gamma}\hat{\mathbf{R}}\right)}{\min_{\{\gamma\}} \gamma^{\frac{NK}{J+K}} \det\left(\mathbf{A}_1 + \frac{1}{\gamma}\hat{\mathbf{R}}\right)} \underset{H_0}{\overset{H_1}{\gtrless}} \xi_1, \qquad (10.9)$$

where

$$\begin{cases} \mathbf{A}_0 = \mathbf{Y}_p\mathbf{Y}_p^{\dagger}, \\ \mathbf{A}_1 = (\mathbf{Y}_p - \mathbf{p}\hat{\mathbf{a}}_p^T)(\mathbf{Y}_p - \mathbf{p}\hat{\mathbf{a}}_p^T)^{\dagger}. \end{cases} \qquad (10.10)$$

In order to come up with the GLRT-based detector for the PHE, it still remains to minimize both numerator and denominator of (10.9) with respect to γ. To this end, we denote by r_0 and r_1 the ranks of $\hat{\mathbf{R}}^{-\frac{1}{2}}\mathbf{A}_0\hat{\mathbf{R}}^{-\frac{1}{2}}$ and $\hat{\mathbf{R}}^{-\frac{1}{2}}\mathbf{A}_1\hat{\mathbf{R}}^{-\frac{1}{2}}$, respectively. More precisely, $r_0 = \min(N, 2J)$, and $r_1 = \min(N-1, 2J)$. By *Proposition 2* of [8], it is not difficult to show that under the constraint $\min(r_0, r_1) > \frac{NJ}{J+K}$, the GLRT for PHE can be recast as

$$\Xi = \frac{\hat{\gamma}_0^{\frac{NK}{J+K}}\det\left(\mathbf{A}_0 + \frac{1}{\hat{\gamma}_0}\hat{\mathbf{R}}\right)}{\hat{\gamma}_1^{\frac{NK}{J+K}}\det\left(\mathbf{A}_1 + \frac{1}{\hat{\gamma}_1}\hat{\mathbf{R}}\right)} \overset{H_1}{\underset{H_0}{\gtrless}} \xi, \tag{10.11}$$

where $\hat{\gamma}_i, i = 0, 1$, is the unique positive solution of equation

$$\sum_{t=1}^{r_i} \frac{\mu_{t,i}\gamma}{\mu_{t,i}\gamma + 1} = \frac{NJ}{J+K}, \quad i = 0, 1, \tag{10.12}$$

with $\mu_{t,i}$ the non-zero eigenvalues of $\hat{\mathbf{R}}^{-\frac{1}{2}}\mathbf{A}_i\hat{\mathbf{R}}^{-\frac{1}{2}}$. Note that (10.12) can be solved by resorting to the Matlab function "roots" which evaluates the eigenvalues of a companion matrix of order $(r_i + 1) \times (r_i + 1)$, at most $i = 0, 1$.

10.1.2 Two-Step GLRT

We derive an *ad hoc* detector based upon the two-step GLRT-based design procedure. In order to facilitate the derivations of the test, we move the power scaling factor from the secondary data to primary data. Precisely, we define $\mathbf{\Sigma} = \gamma\mathbf{R}$, which implies that $\mathbf{R} = \lambda\mathbf{\Sigma}$ with $\lambda = \frac{1}{\gamma}$. We first derive the GLRT based on primary data, assuming that $\mathbf{\Sigma}$ is known, then, a fully adaptive detector is obtained by replacing the unknown matrix $\mathbf{\Sigma}$ with the structured covariance matrix estimate $\hat{\mathbf{R}}$. Under the assumption that $\mathbf{\Sigma}$ is known, the GLRT is given by

$$\Xi_0 = \frac{\max_{\{\mathbf{a},\lambda\}} f_1(\mathbf{Y}|\mathbf{a},\lambda,\mathbf{\Sigma})}{\max_{\{\lambda\}} f_0(\mathbf{Y}|\lambda,\mathbf{\Sigma})} \overset{H_1}{\underset{H_0}{\gtrless}} \eta_0, \tag{10.13}$$

where $f_1(\mathbf{Y}|\mathbf{a},\lambda,\mathbf{\Sigma})$ and $f_0(\mathbf{Y}|\lambda,\mathbf{\Sigma})$ are the PDF of $\mathbf{Y}$ under H_1 and H_0 given by

$$\begin{cases} f_1(\mathbf{Y}|\mathbf{a},\lambda,\mathbf{\Sigma}) = \left[\frac{1}{\pi^N \det(\lambda\mathbf{\Sigma})}\right]^J \exp\left[-\operatorname{tr}\left(\mathbf{\Sigma}^{-1}\frac{1}{\lambda}(\mathbf{Y}-\mathbf{pa}^T)(\mathbf{Y}-\mathbf{pa}^T)^\dagger\right)\right], \\ f_0(\mathbf{Y}|\lambda,\mathbf{\Sigma}) = \left[\frac{1}{\pi^N \det(\lambda\mathbf{\Sigma})}\right]^J \exp\left[-\operatorname{tr}\left(\mathbf{\Sigma}^{-1}\frac{1}{\lambda}\mathbf{Y}\mathbf{Y}^\dagger\right)\right], \end{cases}$$

$$\tag{10.14}$$

respectively. Again, exploiting the persymmetry of $\mathbf{\Sigma}$, the above equation can be expressed in persymmetric form, i.e.,

$$
\begin{cases}
f_1(\mathbf{Y}|\mathbf{a}_p, \lambda, \mathbf{\Sigma}) = \left[\frac{1}{\pi^N \det(\lambda \mathbf{\Sigma})}\right]^J \exp\left[-\operatorname{tr}\left(\mathbf{\Sigma}^{-1}\frac{1}{\lambda}(\mathbf{Y}_p - \mathbf{pa}_p^T)(\mathbf{Y}_p - \mathbf{pa}_p^T)^\dagger)\right)\right], \\
f_0(\mathbf{Y}|\lambda, \mathbf{\Sigma}) = \left[\frac{1}{\pi^N \det(\lambda \mathbf{\Sigma})}\right]^J \exp\left[-\operatorname{tr}\left(\mathbf{\Sigma}^{-1}\frac{1}{\lambda}\mathbf{Y}_p\mathbf{Y}_p^\dagger\right)\right],
\end{cases}
$$
$$(10.15)$$

where $\mathbf{Y}_p$ and $\mathbf{a}_p$ given by (6.10) and (6.11).

The optimization problem under the H_0 hypothesis consists in evaluating the MLE of λ, which is given by

$$
\hat{\lambda} = \frac{1}{NJ}\operatorname{tr}\left[\mathbf{\Sigma}^{-1}\mathbf{Y}_p\mathbf{Y}_p^\dagger\right]. \tag{10.16}
$$

As a consequence, the compressed likelihood function under H_0 can be written as follows

$$
f_0(\mathbf{Y}|\hat{\lambda}, \mathbf{\Sigma}) = \left(\frac{NJ}{e\pi}\right)^{NJ} \left[\det\{\mathbf{\Sigma}\}\right]^{-J} \left[\operatorname{tr}\left(\mathbf{\Sigma}^{-1}\mathbf{Y}_p\mathbf{Y}_p^\dagger\right)\right]^{-NJ}. \tag{10.17}
$$

On the other hand, under the H_1 hypothesis, it is not difficult to show that

$$
f_1(\mathbf{Y}|\hat{\mathbf{a}}_p, \hat{\lambda}, \mathbf{\Sigma}) = \left(\frac{NJ}{e\pi}\right)^{NJ} \left[\det\{\mathbf{\Sigma}\}\right]^{-J}
$$
$$
\times \left\{\operatorname{tr}\left[\mathbf{Y}_p^\dagger \mathbf{\Sigma}^{-1/2}\left(\mathbf{I}_N - \mathbf{P}_{\mathbf{p}_\Sigma}\right)\mathbf{\Sigma}^{-1/2}\mathbf{Y}_p\right]\right\}^{-NJ} \tag{10.18}
$$

, where $\mathbf{P}_{\mathbf{p}_\Sigma} = \mathbf{p}_\Sigma\left(\mathbf{p}_\Sigma^\dagger\mathbf{p}_\Sigma\right)^{-1}\mathbf{p}_\Sigma^\dagger$ is the projection matrix onto the space spanned by $\mathbf{p}_\Sigma = \mathbf{\Sigma}^{-1/2}\mathbf{p}$.

Summarizing, the GLRT given by (10.13) is equivalent to

$$
\Xi_1 = \frac{\operatorname{tr}\left(\mathbf{\Sigma}^{-1}\mathbf{Y}_p\mathbf{Y}_p^\dagger\right)}{\operatorname{tr}\left[\mathbf{Y}_p^\dagger\mathbf{\Sigma}^{-1/2}\left(\mathbf{I}_N - \mathbf{P}_{\mathbf{p}_\Sigma}\right)\mathbf{\Sigma}^{-1/2}\mathbf{Y}_p\right]} \underset{H_0}{\overset{H_1}{\gtrless}} \eta_1. \tag{10.19}
$$

In order to come up with a fully adaptive detector, we can plug $\hat{\mathbf{R}}$ in place of $\mathbf{\Sigma}$, which yields

$$
\Xi = \frac{\mathbf{p}^\dagger\hat{\mathbf{R}}^{-1}\mathbf{Y}_p\mathbf{Y}_p^\dagger\hat{\mathbf{R}}^{-1}\mathbf{p}}{\operatorname{tr}\left(\mathbf{Y}_p^\dagger\hat{\mathbf{R}}^{-1}\mathbf{Y}_p\right)\mathbf{p}^\dagger\hat{\mathbf{R}}^{-1}\mathbf{p}^\dagger} \underset{H_0}{\overset{H_1}{\gtrless}} \eta, \tag{10.20}
$$

where η is the detection threshold.

As a final remark, the detectors (10.11) and (10.20), referred to in the sequel as the persymmetric one-step GLRT for PHE (P1S-GLRT-PHE) and persymmetric two-step GLRT for PHE (P2S-GLRT-PHE), respectively, ensure the CFAR property with respect to $\mathbf{R}$ and γ.

10.1.3 Rao and Wald Tests

In this subsection, we introduce the Rao and Wald tests for PHE and denote by $\boldsymbol{\theta}_r = [a_{1r}, a_{1i}, \ldots, a_{Jr}, a_{Ji}]^T \in \mathbb{R}^{2J \times 1}$ the signal parameter vector, with $a_{jr} = \Re(a_j)$ and $a_{ji} = \Im(a_j)$, $j = 1, \ldots, J$, $\boldsymbol{\theta}_s = [\gamma, \, \mathbf{g}^T(\mathbf{R})]^T$ the $(N^2 + 1)$ dimensional nuisance parameter vector, where $\mathbf{g}(\mathbf{R})$ is an N^2 dimensional vector that contains in univocal way the elements of $\mathbf{R}$, and $\boldsymbol{\theta} = [\boldsymbol{\theta}_r^T, \boldsymbol{\theta}_s^T]^T$ the overall unknown parameters.

The Rao test for the problem at hand can be written as follows:

$$\Lambda_0 = \left[\frac{\partial \ln f_1(\mathbf{Y}, \mathbf{Y}_1 | \gamma, \mathbf{a}, \mathbf{R})}{\partial \boldsymbol{\theta}_r} \right]^T_{\boldsymbol{\theta} = \hat{\boldsymbol{\theta}}_0} \mathbf{J}_{rr}(\boldsymbol{\theta}) \Big|_{\boldsymbol{\theta} = \hat{\boldsymbol{\theta}}_0}$$
$$\times \left[\frac{\partial \ln f_1(\mathbf{Y}, \mathbf{Y}_1 | \gamma, \mathbf{a}, \mathbf{R})}{\partial \boldsymbol{\theta}_r} \right]_{\boldsymbol{\theta} = \hat{\boldsymbol{\theta}}_0} \underset{H_0}{\overset{H_1}{\gtrless}} \eta_0. \tag{10.21}$$

where η_0 is the detection threshold and $\mathbf{J}_{rr}(\boldsymbol{\theta})$ is given by (5.16), $\hat{\boldsymbol{\theta}}_0$ is the estimate of $\boldsymbol{\theta}_0$ under H_0.

Following the lead of [9–11], we have

$$\mathbf{J}_{rr}(\boldsymbol{\theta}) = (2\mathbf{p}^\dagger \mathbf{R}^{-1} \mathbf{p})^{-1} \mathbf{I}_{2J}, \tag{10.22}$$

$$\frac{\partial \ln f(\mathbf{Y}, \mathbf{Y}_1 | \boldsymbol{\theta}, H_1)}{\partial \boldsymbol{\theta}_r} = \begin{bmatrix} 2\Re\{\mathbf{p}^\dagger \mathbf{R}^{-1}(\mathbf{y}_1 - a_1 \mathbf{p})\} \\ 2\Im\{\mathbf{p}^\dagger \mathbf{R}^{-1}(\mathbf{y}_1 - a_1 \mathbf{p})\} \\ \vdots \\ 2\Re\{\mathbf{p}^\dagger \mathbf{R}^{-1}(\mathbf{y}_J - a_J \mathbf{p})\} \\ 2\Im\{\mathbf{p}^\dagger \mathbf{R}^{-1}(\mathbf{y}_J - a_J \mathbf{p})\} \end{bmatrix}. \tag{10.23}$$

Moreover, from [2], we have $\hat{\mathbf{R}}_0 = \frac{1}{K+1} \left[\mathbf{Y}_p \mathbf{Y}_p^\dagger + \frac{1}{\hat{\gamma}_0} \hat{\mathbf{R}} \right]$ where $\hat{\mathbf{R}}_0$ and $\hat{\gamma}_0$ are the MLEs of $\mathbf{R}$ and γ under H_0, respectively, and $\hat{\gamma}_0$ is the same as in (10.11). Gathering the above results, the Rao test can be recast as

$$\Lambda = \frac{\sum_{j=1}^{J} \left| \mathbf{p}^\dagger (\hat{\mathbf{R}}/\hat{\gamma}_0 + \mathbf{Y}_p \mathbf{Y}_p^\dagger) \mathbf{y}_j \right|^2}{\mathbf{p}^\dagger (\hat{\mathbf{R}}/\hat{\gamma}_0 + \mathbf{Y}_p \mathbf{Y}_p^\dagger) \mathbf{p}} \underset{H_0}{\overset{H_1}{\gtrless}} \eta. \tag{10.24}$$

The Wald test has the following expression [12]:

$$\Lambda_0 = \hat{\boldsymbol{\theta}}_{r,1}^T \left[\mathbf{J}_{rr}(\hat{\boldsymbol{\theta}}_1) \right]^{-1} \hat{\boldsymbol{\theta}}_{r,1} \underset{H_0}{\overset{H_1}{\gtrless}} \eta_0, \tag{10.25}$$

where $\hat{\boldsymbol{\theta}}_{r,1}$ and $\hat{\boldsymbol{\theta}}_1$ are the MLEs of $\boldsymbol{\theta}_r$ and $\boldsymbol{\theta}$, respectively, and η_0 is the detection threshold.

It can be shown that under H_1, the estimate of $\mathbf{a}_p$, denoted by $\hat{\mathbf{a}}_p$, is given by (10.8), thus we have $\hat{a}_{ej} = \frac{\mathbf{p}^\dagger \hat{\mathbf{R}}^{-1} \mathbf{y}_{ej}}{\mathbf{p}^\dagger \hat{\mathbf{R}}^{-1} \mathbf{p}}$, $\hat{a}_{oj} = \frac{\mathbf{p}^\dagger \hat{\mathbf{R}}^{-1} \mathbf{y}_{oj}}{\mathbf{p}^\dagger \hat{\mathbf{R}}^{-1} \mathbf{p}}$, $j = 1, \ldots, J$, and

$\hat{\boldsymbol{\theta}}_{r,1} = [\hat{a}_{e1}, -j\hat{a}_{o1}, \ldots, \hat{a}_{eJ}, -j\hat{a}_{oJ}]$. The estimate of $\boldsymbol{\theta}_s$ under H_1 is given by $\hat{\boldsymbol{\theta}}_{s,1} = [\hat{\gamma}_1, \mathbf{g}(\hat{\mathbf{R}}_1)]$, where $\hat{\mathbf{R}}_1 = \frac{1}{K+1}\left[(\mathbf{Y}_p - \mathbf{p}\hat{\mathbf{a}}_p^T)(\mathbf{Y}_p - \mathbf{p}\hat{\mathbf{a}}_p^T)^\dagger + \frac{1}{\hat{\gamma}_1}\hat{\mathbf{R}}\right]$ and $\hat{\gamma}_1$ is the same as in (10.11). Hence, the Wald test has the form

$$\Lambda = \frac{\hat{\gamma}_1 \sum_{j=1}^{J}\left[(\mathbf{p}^\dagger\hat{\mathbf{R}}^{-1}\mathbf{y}_{ej})^2 - (\mathbf{p}^\dagger\hat{\mathbf{R}}^{-1}\mathbf{y}_{oj})^2\right]}{\mathbf{p}^\dagger\hat{\mathbf{R}}^{-1}\mathbf{p}} \underset{H_0}{\overset{H_1}{\gtrless}} \eta, \qquad (10.26)$$

where we have used the following equality $\mathbf{p}^\dagger\hat{\mathbf{R}}_1^{-1}\mathbf{p} = (K+1)\hat{\gamma}_1\mathbf{p}^\dagger\hat{\mathbf{R}}^{-1}\mathbf{p}$. As a final remark, both the Rao test and the Wald test, referred to as the persymmetric Rao detector for PHE (P-RAO-PHE) and the persymmetric Wald detector for PHE (P-WALD-PHE), respectively, guarantee the CFAR property with respect to $\mathbf{R}$ and γ. The proofs of such statements, which are not reported here for the sake of brevity, follow the lead of [1] and references therein.

10.2   Numerical Examples

In this section, we present some numerical examples to show the performances of the P1S-GLRT-PHE, P2S-GLRT-PHE, P-RAO-PHE, and P-WALD-PHE. Moreover, we compare our detectors to those detectors for distributed targets designed without any assumption on the structure of the noise covariance matrix, namely the GLRT (referred to as GLRT-PHE in this subsection), the GASD derived in [8], as well as the Rao and Wald tests derived in [6]. Since closed form expressions for both P_{fa} and P_d are not available, we resort to standard MC counting techniques. More precisely, in order to evaluate the threshold necessary to ensure a preassigned $P_{fa} = 0.001$ and the values of P_d, we resort to $100/P_{fa}$ and 10^4 independent trials, respectively. The SNR is defined as $\text{SNR} = \sum_{j=1}^{J}|a_j|^2\mathbf{v}_m(\theta_{az})^\dagger\mathbf{R}^{-1}\mathbf{v}_m(\theta_{az})$, where θ_{az} is the target azimuthal angle and

$$\begin{aligned}\mathbf{v}_m(\theta_{az}) = &\left[\exp\left(j\pi\frac{d}{\lambda}(N-1)\cos\theta_{az}\right), \cdots, 1,\right.\\ &\left.\cdots, \exp\left(-j\pi\frac{d}{\lambda}(N-1)\cos\theta_{az}\right)\right]/\sqrt{N},\end{aligned} \qquad (10.27)$$

with d the element spacing, λ the working wavelength. We consider a clutter-dominated environment and the (i,j)th element of the clutter covariance matrix of the primary data is chosen as $\mathbf{R}(i,j) = \sigma^2 0.9^{|i-j|}$, where σ^2 is the clutter power. The illustrative examples assume that $J = 4$, $\gamma = 5$, varying N and K.

In Fig. 10.1(a) we plot P_d versus SNR assuming $N = 10$ and $K = N + 2$, in order to reflect the severity of the environment. It is clear that when a small number of secondary data is available, the persymmetric detectors significantly

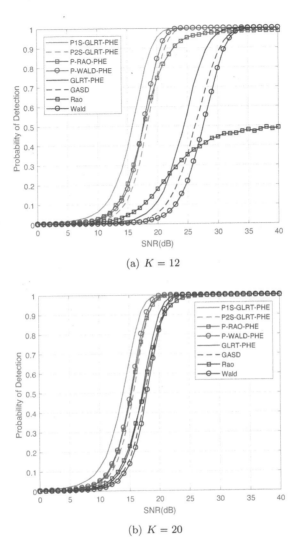

(a) $K = 12$

(b) $K = 20$

FIGURE 10.1
Detection probability versus SNR for $N = 10$, $J = 4$, and $\gamma = 5$.

outperform their unstructured counterparts. Specifically, the best performance is attained by the P1S-GLRT-PHE, while the P-WALD-PHE, P2S-GLRT-PHE, and the P-RAO-PHE experience a loss of about 1.5 dB, 2.2 dB, and 3.6 dB at $P_d = 0.9$, respectively; such a loss increases to about 9.3 dB, 11.5 dB and 12.1 dB for the GLRT-PHE, GASD, and Wald, respectively. Note that the Rao test performs poorly in this scenario. The use of a priori knowledge about the structure of the disturbance covariance matrix is a suitable strategy for restoring system performance when only a small amount of training data is

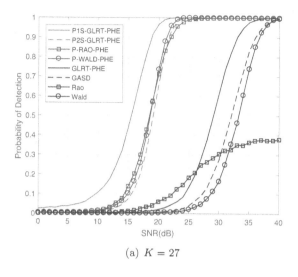

(a) $K = 27$

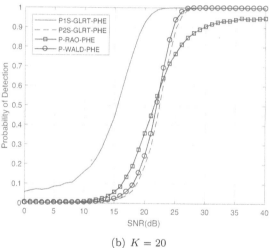

(b) $K = 20$

FIGURE 10.2
Detection probability versus SNR for $N = 25$, $J = 4$, and $\gamma = 5$.

available. Nevertheless, the above mentioned loss can be reduced by increasing K due to the fact that the estimate of $\mathbf{R}$ becomes more reliable, as shown in Fig. 10.1(b), where we plot P_d versus SNR for the same system parameters as in Fig. 10.1(a), but for $K = 2N$. In this case, the best performance is still achieved by the P1S-GLRT-PHE, while the P-WALD-PHE and GLRT-PHE experience a loss of about 1 dB, 3.1 dB at $P_d = 0.9$, respectively.

In Fig. 10.2(a), we plot P_d versus SNR assuming $N = 25$, $K = N + 2$. Inspection highlights that the P1S-GLRT-PHE ensures the best performance

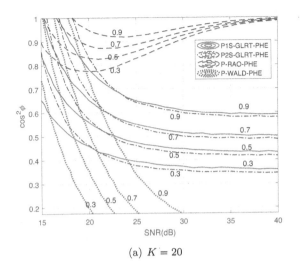

(a) $K = 20$

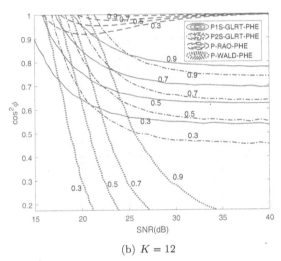

(b) $K = 12$

FIGURE 10.3
Contour of constant detection probability in the mismatched case for $N = 10$, $J = 4$, and $\gamma = 5$.

and the gain with respect to the P-WALD-PHE, and the GLRT-PHE are 2.3 dB, and 11.2 dB, respectively. It turns out that the performance gain of persymmetric detectors with respect to their unstructured counterparts increases. Finally, in Fig. 10.2(b) we show the performances of the proposed detectors when $K < N$, in particular we set $N = 25$ and $K = 20$; the remaining parameters are the same as in Fig. 10.2(a).

In Fig. 10.3, we plot contours of constant detection probability for the P-RAO-PHE, the P-WALD-PHE, the P1S-GLRT-PHE, and the

P2S-GLRT-PHE for $N = 10$ and different values of K. The contours of constant detection probability are represented as a function of the squared cosine of the mismatch angle. It can be seen that the P-RAO-PHE is most selective, while the P-WALD-PHE has the least mismatch rejection capabilities. This is in accordance with the mismatched detection performance of the unstructured Rao and Wald tests. Meanwhile,the P1S-GLRT-PHE possesses better rejection capabilities of mismatched signals than both the P2S-GLRT-PHE and the P-WALD-PHE when SNR > 25 dB.

In order to complete our analysis, notice that the persymmetric detectors are computationally more efficient than their unstructured counterparts due to the fact that the calculations of their statistics can be transferred from the complex domain to the real domain (see Appendix A in [13]).

Bibliography

[1] S. Kraut and L. L. Scharf, "The CFAR adaptive subspace detector is a scale-invariant GLRT," *IEEE Transactions on Signal Processing*, vol. 47, no. 9, pp. 2538–2541, 1999.

[2] M. Casillo, A. De Maio, S. Iommelli, and L. Landi, "A persymmetric GLRT for adaptive detection in partially-homogeneous environment," *IEEE Signal Processing Letters*, vol. 14, no. 12, pp. 1016–1019, December 2007.

[3] C. Hao, D. Orlando, G. Foglia, X. Ma, S. Yan, and C. Hou, "Persymmetric adaptive detection of distributed targets in partially-homogeneous environment," *Digital Signal Processing*, vol. 24, pp. 42–51, January 2014.

[4] A. De Maio and S. Iommelli, "Coincidence of the Rao test, Wald test, and GLRT in partially homogeneous environment," *IEEE Signal Processing Letters*, vol. 15, no. 4, pp. 385–388, April 2008.

[5] S. Lei, Z. Zhao, Z. Nie, and Q.-H. Liu, "Adaptive polarimetric detection method for target in partially homogeneous background," *Signal Processing*, vol. 106, pp. 301–311, 2015.

[6] C. Hao, X. Ma, X. Shang, and L. Cai, "Adaptive detection of distributed targets in partially homogeneous environment with Rao and Wald tests," *Signal Processing*, vol. 92, pp. 926–930, 2012.

[7] S. Kraut, L. L. Scharf, and L. T. McWhorter, "Adaptive subspace detectors," *IEEE Transactions on Signal Processing*, vol. 49, no. 1, pp. 1–16, January 2001.

[8] E. Conte, A. De Maio, and G. Ricci, "GLRT-based adaptive detection algorithms for range-spread targets," *IEEE Transactions on Signal Processing*, vol. 49, no. 7, pp. 1336–1348, July 2001.

[9] A. D. Maio, "Rao test for adaptive detection in Gaussian interference with unknown covariance matrix," *IEEE Transactions on Signal Processing*, vol. 55, no. 7, pp. 3577–3584, July 2007.

[10] D. Orlando and G. Ricci, "A Rao test with enhanced selectivity properties in homogeneous scenarios," *IEEE Transactions on Signal Processing*, vol. 58, no. 10, pp. 5385–5390, October 2010.

[11] Z. Wang, M. Li, H. Chen, Y. Lu, R. Cao, P. Zhang, L. Zuo, and Y. Wu, "Persymmetric detectors of distributed targets in partially homogeneous disturbance," *Signal Processing*, vol. 128, pp. 382–388, 2016.

[12] S. M. Kay, *Fundamentals of Statistical Signal Processing: Detection Theory*. New Jersey, USA, Prentice Hall, 1998, vol. 2.

[13] L. Cai and H. Wang, "A persymmetric multiband GLR algorithm," *IEEE Transactions on Aerospace and Electronic Systems*, vol. 28, no. 3, pp. 806–816, July 1992.

11

Robust Detection in Partially Homogeneous Environments

In this chapter, we discuss the target detection problems in the presence of steering vector uncertainties in PHEs. We exploit persymmetry to study the adaptive detection problem with multiple observations in PHEs where noise shares the same covariance matrix up to different power levels between the test and training data. A persymmetric subspace model is designed for taking into account steering vector mismatches [1]. Based on the persymmetric subspace model, we design adaptive detectors in PHEs, according to the criteria of two-step versions of GLRT, Wald test, and Rao test. It is found that the GLRT and Wald test coincide, while the Rao test does not exist. The designed detector is proved to exhibit a CFAR property against both the covariance matrix structure and the scaling factor. Numerical examples show that the designed detector, compared to its counterparts, is more robust to steering vector mismatches.

11.1 Problem Formulation

Assume that the received data are collected by a pulsed Doppler radar using N pulse trains. We want to detect a target using J observations which may be collected from multiple range cells, radar bands, and/or CPIs. These observations, referred to as primary (test) data, are denoted by

$$\mathbf{y}_j = a_j^* \mathbf{q} + \mathbf{n}_j, \quad j = 1, \ldots, J, \qquad (11.1)$$

where a_j^* is an unknown complex scalar accounting for the target reflectivity and channel propagation effects, $\mathbf{q}$ is the target signal steering vector, and $\mathbf{n}_j$ denotes noise in the jth observations. Assume that the noise $\mathbf{n}_j$ has a circularly symmetric, complex Gaussian distribution with zero mean and covariance matrix $\alpha \mathbf{R}$, i.e., $\mathbf{n}_j \sim \mathcal{CN}_N(\mathbf{0}, \alpha \mathbf{R})$ for $j = 1, 2, \ldots, J$. The unknown scaling factor α accounts for the environmental non-homogeneity [2].

To estimate the covariance matrix $\mathbf{R}$, one often imposes a standard assumption that a set of secondary (training) data $\{\mathbf{y}_k\}_{k=J+1}^{J+K}$ (free of target

DOI: 10.1201/9781003340232-11

echoes) is available, i.e.,

$$\mathbf{y}_k = \mathbf{n}_k \sim \mathcal{CN}_N(\mathbf{0}, \mathbf{R}), \tag{11.2}$$

for $k = J+1, J+2, \ldots, J+K$. Assume that the noise $\mathbf{n}_k$ for $k = 1, 2, \ldots, J+K$ are IID.

When the pulsed Doppler radar uses symmetrically spaced pulse trains, the covariance matrix $\mathbf{R}$ and the steering vector $\mathbf{q}$ are persymmetric [3–5]. The persymmetry of $\mathbf{R}$ means

$$\mathbf{R} = \mathbf{J}\mathbf{R}^*\mathbf{J}, \tag{11.3}$$

where $\mathbf{J}$ is defined in (1.2), and the persymmetry of $\mathbf{q}$ implies $\mathbf{q} = \mathbf{J}\mathbf{q}^*$. As a specific example, the steering vector $\mathbf{q}$ in the pulsed Doppler radar with symmetrically spaced pulse trains can have the form as $\mathbf{t}$ given by

$$\mathbf{t}(\bar{f}) = \left[e^{-j2\pi \bar{f} \frac{(N-1)}{2}}, \ldots, e^{-j 2\pi \bar{f} \frac{1}{2}}, e^{j 2\pi \bar{f} \frac{1}{2}}, \ldots, e^{j2\pi \bar{f} \frac{(N-1)}{2}} \right]^T, \tag{11.4}$$

for even N, and

$$\mathbf{t}(\bar{f}) = \left[e^{-j2\pi \bar{f} \frac{(N-1)}{2}}, \ldots, e^{-j 2\pi \bar{f}}, 1, e^{j 2\pi \bar{f}}, \ldots, e^{j2\pi \bar{f} \frac{(N-1)}{2}} \right]^T, \tag{11.5}$$

for odd N, where $\bar{f}$ is a normalized Doppler frequency.

Denote by $\mathbf{p}$ the nominal target steering vector, and we select it to be persymmetric (i.e., $\mathbf{p} = \mathbf{J}\mathbf{p}^*$). In practice, mismatches often exist between the assumed target steering vector $\mathbf{p}$ and the true steering vector $\mathbf{q}$ (e.g., due to pointing errors in grid search [6,7]). The mismatch angle ϕ between the true steering vector $\mathbf{q}$ and the assumed one $\mathbf{p}$ is defined by [7]

$$\cos^2 \phi = \frac{|\mathbf{p}^\dagger \mathbf{R}^{-1} \mathbf{q}|^2}{(\mathbf{p}^\dagger \mathbf{R}^{-1} \mathbf{p})(\mathbf{q}^\dagger \mathbf{R}^{-1} \mathbf{q})}. \tag{11.6}$$

In the matched case, $\mathbf{q} = \mathbf{p}$ and $\cos^2 \phi = 1$. In the mismatched case, $\mathbf{q} \neq \mathbf{p}$ and $\cos^2 \phi < 1$.

In order to take into account the mismatches in the target steering vector, we design a persymmetric subspace model for the target steering vector[1]. More specifically, we assume that

$$\mathbf{p} = \mathbf{S}\mathbf{b}, \tag{11.7}$$

where $\mathbf{S} \in \mathbb{C}^{N \times q}$ is a known full-column-rank matrix, q is the subspace dimension, and $\mathbf{b} \in \mathbb{R}^{q \times 1}$ is an unknown coordinate vector. We make two assumptions for the persymmetric subspace model (11.7): 1) each column vector of $\mathbf{S}$ is persymmetric; 2) $\mathbf{b}$ is a real-valued q-dimensional vector. They are sufficient but not necessary conditions for guaranteeing the persymmetry of $\mathbf{S}\mathbf{b}$.

---

[1]As in [6,8,9], the subspace matrix is constructed by using the nominal steering vector and the ones close to the nominal steering vector.

In addition, both assumptions play a vital role in the derivations of robust detectors in the next section.

In the persymmetric subspace model, we know the subspace where the target signal belongs to, but we do not know its exact location, since **b** is unknown. Note that the true steering vector **q** is not guaranteed to lie in the chosen subspace, since **q** is unknown in practice.

The detection problem to be solved can be formulated as the following binary hypothesis test

$$\begin{cases} H_0 : \mathbf{y}_j \sim \mathcal{CN}_N(\mathbf{0}, \alpha\mathbf{R}), \ j = 1, \dots, J, \\ H_1 : \mathbf{y}_j \sim \mathcal{CN}_N(a_j^* \mathbf{Sb}, \alpha\mathbf{R}), \ j = 1, \dots, J, \end{cases} \tag{11.8}$$

where a_j, **b**, α, and **R** are all unknown. Note that we assume a set of training data $\mathbf{y}_k \sim \mathcal{CN}_N(\mathbf{0}, \mathbf{R})$, $k = J+1, J+2, \dots, J+K$, is available. When $J = 1$, the problem in (11.8) becomes the point-like target detection problem, and the unknown parameter a_j can be absorbed into **b**. A difference between the data model (7.2) in Chapter 7 and this chapter is the unknown scaling factor α, which accounts for the environmental non-homogeneity [2].

For mathematical tractability, we assume in the detection problem (11.8) that $K \geq \lceil \frac{N}{2} \rceil$, which is much less restrictive than the one (i.e., $K \geq N$) required in [10–12]. This case is more interesting and also practically motivated, since the number of training data is often limited in practice.

11.2 Robust Detection

11.2.1 GLRT

In the first step. we assume that the covariance matrix **R** is known. The GLRT detector with known **R** is given by

$$\Phi_0 = \frac{\max\limits_{\{\mathbf{a},\mathbf{b},\alpha\}} f(\mathbf{Y}|H_1)}{\max\limits_{\{\alpha\}} f(\mathbf{Y}|H_0)} \underset{H_0}{\overset{H_1}{\gtrless}} \phi_0, \tag{11.9}$$

where ϕ_0 is a detection threshold, $f(\mathbf{Y}|H_1)$ and $f(\mathbf{Y}|H_0)$ denote the PDFs of the primary data $\mathbf{Y} = [\mathbf{y}_1, \mathbf{y}_2, \dots, \mathbf{y}_J]$ under H_1 and H_0, respectively, and

$$\mathbf{a} = [a_1, a_2, \dots, a_J]^T \in \mathbb{C}^{J \times 1}. \tag{11.10}$$

Due to the independence among the primary data, the PDF of **Y** under H_1 can be written as

$$f(\mathbf{Y}|H_1) = \frac{1}{\pi^{NJ} \alpha^{NJ} \det(\mathbf{R})^J} \exp\left[-\operatorname{tr}(\alpha^{-1}\mathbf{R}^{-1}\mathbf{F}_1)\right], \tag{11.11}$$

where

$$\mathbf{F}_1 = \left(\mathbf{Y} - \mathbf{Sba}^\dagger\right)\left(\mathbf{Y} - \mathbf{Sba}^\dagger\right)^\dagger \in \mathbb{C}^{N \times N}. \tag{11.12}$$

In order to exploit the persymmetry, we first express the PDF of $\mathbf{Y}$ as another form. Using the persymmetry structure in the covariance matrix $\mathbf{R}$, we have

$$\begin{aligned}
\operatorname{tr}(\mathbf{R}^{-1}\mathbf{F}_1) &= \operatorname{tr}[\mathbf{J}(\mathbf{R}^*)^{-1}\mathbf{J}\mathbf{F}_1] \\
&= \operatorname{tr}[(\mathbf{R}^*)^{-1}\mathbf{J}\mathbf{F}_1\mathbf{J}] \\
&= \operatorname{tr}[\mathbf{R}^{-1}\mathbf{J}\mathbf{F}_1^*\mathbf{J}].
\end{aligned} \tag{11.13}$$

As a result, we can rewrite (11.11) as

$$f(\mathbf{Y}|H_1) = \frac{1}{\pi^{NJ}\alpha^{NJ}\det(\mathbf{R})^J}\exp\left[-\operatorname{tr}(\alpha^{-1}\mathbf{R}^{-1}\mathbf{H}_1)\right], \tag{11.14}$$

where

$$\mathbf{H}_1 = \left(\mathbf{Y}_p - \mathbf{Sba}_p^\dagger\right)\left(\mathbf{Y}_p - \mathbf{Sba}_p^\dagger\right)^\dagger \in \mathbb{C}^{N \times N}, \tag{11.15}$$

$$\mathbf{a}_p = [a_{e1}, \ldots, a_{eJ}, a_{o1}, \ldots, a_{oJ}]^T \in \mathbb{C}^{2J \times 1}, \tag{11.16}$$

with

$$\begin{cases} a_{ej} = \frac{1}{2}(a_j + a_j^*) = \mathfrak{Re}(a_j), \\ a_{oj} = \frac{1}{2}(a_j - a_j^*) = \jmath\,\mathfrak{Im}(a_j), \end{cases} \tag{11.17}$$

and

$$\mathbf{Y}_p = [\mathbf{y}_{e1}, \ldots \mathbf{y}_{eJ}, \mathbf{y}_{o1}, \ldots, \mathbf{y}_{oJ}] \in \mathbb{C}^{N \times 2J}, \tag{11.18}$$

with

$$\begin{cases} \mathbf{y}_{ej} = \frac{1}{2}\left(\mathbf{y}_j + \mathbf{J}\mathbf{y}_j^*\right) \in \mathbb{C}^{N \times 1}, \\ \mathbf{y}_{oj} = \frac{1}{2}\left(\mathbf{y}_j - \mathbf{J}\mathbf{y}_j^*\right) \in \mathbb{C}^{N \times 1}. \end{cases} \tag{11.19}$$

Similar to (11.14), $f(\mathbf{Y}|H_0)$ can be written as

$$f(\mathbf{Y}|H_0) = \frac{1}{\pi^{NJ}\alpha^{NJ}\det(\mathbf{R})^J}\exp\left[-\operatorname{tr}(\alpha^{-1}\mathbf{R}^{-1}\mathbf{Y}_p\mathbf{Y}_p^\dagger)\right]. \tag{11.20}$$

It is worth noting that the above derivations can be easily obtained from Appendix A of [13].

Next, we transform all the variables in (11.14) and (11.20) from the complex-valued domain to the real-valued domain. We proceed by defining two unitary matrices as

$$\mathbf{D} = \frac{1}{2}[(\mathbf{I}_N + \mathbf{J}) + \jmath\,(\mathbf{I}_N - \mathbf{J})] \in \mathbb{C}^{N \times N}, \tag{11.21}$$

and

$$\mathbf{V} = \begin{bmatrix} \mathbf{I}_J & \mathbf{0} \\ \mathbf{0} & -\jmath\mathbf{I}_J \end{bmatrix} \in \mathbb{C}^{2J \times 2J}. \tag{11.22}$$

We employ them to transform $\mathbf{Y}_p$ as [13, eqs. (B7) and (B8)]

$$
\begin{aligned}
\mathbf{Y}_r &= \mathbf{D}\mathbf{Y}_p\mathbf{V} \\
&= [\mathbf{y}_{er1}, \ldots, \mathbf{y}_{erJ}, \mathbf{y}_{or1}, \ldots, \mathbf{y}_{orJ}] \in \mathbb{R}^{N \times 2J},
\end{aligned}
\tag{11.23}
$$

where

$$
\begin{cases}
\mathbf{y}_{erj} = \mathbf{D}\mathbf{y}_{ej} \in \mathbb{R}^{N \times 1}, \\
\mathbf{y}_{orj} = -\jmath\mathbf{D}\mathbf{y}_{oj} \in \mathbb{R}^{N \times 1},
\end{cases}
\tag{11.24}
$$

for $j = 1, 2, \ldots, J$. It is worth noting that $\mathbf{y}_{erj}$ and $\mathbf{y}_{orj}$ are real-valued vectors. Further, we make the following transformations [13, eqs. (A30)–(A32)]:

$$
\mathbf{S}_r = \mathbf{D}\mathbf{S} = \mathfrak{Re}(\mathbf{S}) - \mathfrak{Im}(\mathbf{S}) \in \mathbb{R}^{N \times q},
\tag{11.25}
$$

$$
\mathbf{R}_r = \mathbf{D}\mathbf{R}\mathbf{D}^\dagger = [\mathfrak{Re}(\mathbf{R}) + \mathbf{J}\mathfrak{Im}(\mathbf{R})] \in \mathbb{R}^{N \times N},
\tag{11.26}
$$

and

$$
\begin{aligned}
\mathbf{a}_r &= \mathbf{V}^\dagger\mathbf{a}_p \\
&= [\mathfrak{Re}(a_1), \ldots, \mathfrak{Re}(a_J), -\mathfrak{Im}(a_1), \ldots, -\mathfrak{Im}(a_J)]^T \in \mathbb{C}^{2J \times 1}.
\end{aligned}
\tag{11.27}
$$

Due to (11.26), we have

$$
\det(\mathbf{R}_r) = \det(\mathbf{D}\mathbf{R}\mathbf{D}^\dagger) = \det(\mathbf{R}).
\tag{11.28}
$$

Then, we can rewrite $f(\mathbf{Y}|H_1)$ and $f(\mathbf{Y}|H_0)$ as

$$
f(\mathbf{Y}|H_1) = \frac{\exp\left\{-\operatorname{tr}\left[\alpha^{-1}\mathbf{R}_r^{-1}\left(\mathbf{Y}_r - \mathbf{S}_r\mathbf{b}\mathbf{a}_r^T\right)\left(\mathbf{Y}_r - \mathbf{S}_r\mathbf{b}\mathbf{a}_r^T\right)^T\right]\right\}}{\pi^{NJ}\alpha^{NJ}\det(\mathbf{R}_r)^J},
\tag{11.29}
$$

and

$$
f(\mathbf{Y}|H_0) = \frac{\exp\left[-\operatorname{tr}\left(\alpha^{-1}\mathbf{R}_r^{-1}\mathbf{Y}_r\mathbf{Y}_r^T\right)\right]}{\pi^{NJ}\alpha^{NJ}\det(\mathbf{R}_r)^J},
\tag{11.30}
$$

respectively.

It is obvious that the MLEs of α under H_1 and H_0 are

$$
\hat{\alpha} = \frac{1}{NJ}\operatorname{tr}\left[\mathbf{R}_r^{-1}\left(\mathbf{Y}_r - \mathbf{S}_r\mathbf{b}\mathbf{a}_r^T\right)\left(\mathbf{Y}_r - \mathbf{S}_r\mathbf{b}\mathbf{a}_r^T\right)^T\right],
\tag{11.31}
$$

and

$$
\hat{\omega} = \frac{1}{NJ}\operatorname{tr}\left(\mathbf{R}_r^{-1}\mathbf{Y}_r\mathbf{Y}_r^T\right),
\tag{11.32}
$$

respectively. Hence, the GLRT with known covariance matrix can be equivalently written as

$$
\Phi_1 = \frac{\operatorname{tr}\left(\mathbf{R}_r^{-1}\mathbf{Y}_r\mathbf{Y}_r^T\right)}{\min_{\{\mathbf{a}_r, b\}}\operatorname{tr}\left[\mathbf{R}_r^{-1}\left(\mathbf{Y}_r - \mathbf{S}_r\mathbf{b}\mathbf{a}_r^T\right)\left(\mathbf{Y}_r - \mathbf{S}_r\mathbf{b}\mathbf{a}_r^T\right)^T\right]} \overset{H_1}{\underset{H_0}{\gtrless}} \phi_1,
\tag{11.33}
$$

where ϕ_1 is a detection threshold.

It can be checked that the value of $\mathbf{a}_r$ minimizing the denominator of Φ_1 is

$$\hat{\mathbf{a}}_r = \frac{\mathbf{Y}_r^T \mathbf{R}_r^{-1} \mathbf{S}_r \mathbf{b}}{\mathbf{b}^T \mathbf{S}_r^T \mathbf{R}_r^{-1} \mathbf{S}_r \mathbf{b}}. \tag{11.34}$$

Taking (11.34) into the determinant of Φ_1 leads to

$$\begin{aligned}
\rho &= \min_{\{\mathbf{b}\}} \operatorname{tr}\left[\mathbf{R}_r^{-1} \mathbf{Y}_r \mathbf{Y}_r^T - \frac{\mathbf{Y}_r^T \mathbf{R}_r^{-1} \mathbf{S}_r \mathbf{b} \mathbf{b}^T \mathbf{S}_r^T \mathbf{R}_r^{-1} \mathbf{Y}_r}{\mathbf{b}^T \mathbf{S}_r^T \mathbf{R}_r^{-1} \mathbf{S}_r \mathbf{b}} \right] \\
&= \operatorname{tr}\left(\mathbf{R}_r^{-1} \mathbf{Y}_r \mathbf{Y}_r^T \right) - \rho_1,
\end{aligned} \tag{11.35}$$

where

$$\rho_1 = \max_{\{\mathbf{b}\}} \frac{\mathbf{b}^T \mathbf{S}_r^T \mathbf{R}_r^{-1} \mathbf{Y}_r \mathbf{Y}_r^T \mathbf{R}_r^{-1} \mathbf{S}_r \mathbf{b}}{\mathbf{b}^T \mathbf{S}_r^T \mathbf{R}_r^{-1} \mathbf{S}_r \mathbf{b}}. \tag{11.36}$$

According to the Rayleigh-Ritz theorem [14, p. 176], the optimal value of $\mathbf{b}$ is given be

$$\hat{\mathbf{b}} = \mathcal{P}_e \left\{ \left(\mathbf{S}_r^T \mathbf{R}_r^{-1} \mathbf{S}_r \right)^{-1} \mathbf{S}_r^T \mathbf{R}_r^{-1} \mathbf{Y}_r \mathbf{Y}_r^T \mathbf{R}_r^{-1} \mathbf{S}_r \right\}. \tag{11.37}$$

As a result, we have

$$\rho_1 = \lambda_{\max} \left\{ \left(\mathbf{S}_r^T \mathbf{R}_r^{-1} \mathbf{S}_r \right)^{-1} \mathbf{S}_r^T \mathbf{R}_r^{-1} \mathbf{Y}_r \mathbf{Y}_r^T \mathbf{R}_r^{-1} \mathbf{S}_r \right\}. \tag{11.38}$$

Using (11.35) and (11.38), we obtain the GLRT with known covariance matrix as

$$\Phi_2 = \frac{\lambda_{\max} \left\{ \left(\mathbf{S}_r^T \mathbf{R}_r^{-1} \mathbf{S}_r \right)^{-1} \mathbf{S}_r^T \mathbf{R}_r^{-1} \mathbf{Y}_r \mathbf{Y}_r^T \mathbf{R}_r^{-1} \mathbf{S}_r \right\}}{\operatorname{tr}\left(\mathbf{R}_r^{-1} \mathbf{Y}_r \mathbf{Y}_r^T \right)} \underset{H_0}{\overset{H_1}{\gtrless}} \phi_2, \tag{11.39}$$

where ϕ_2 is a detection threshold.

In the second step, based on the training data we exploit the persymmetry to obtain the MLE of $\mathbf{R}$ as [13]

$$\hat{\mathbf{R}} = \frac{1}{2} \left[\mathbf{Y}_s \mathbf{Y}_s^\dagger + \mathbf{J} \left(\mathbf{Y}_s \mathbf{Y}_s^\dagger \right)^* \mathbf{J} \right] \in \mathbb{C}^{N \times N}, \tag{11.40}$$

with

$$\mathbf{Y}_s = [\mathbf{y}_{J+1}, \mathbf{y}_{J+2}, \ldots, \mathbf{y}_{J+K}] \in \mathbb{C}^{N \times K}. \tag{11.41}$$

Replacing $\mathbf{R}$ with $\hat{\mathbf{R}}$ in (11.39), we derive the final GLRT as

$$\Phi = \frac{\lambda_{\max} \left\{ \left(\mathbf{S}_r^T \hat{\mathbf{R}}_r^{-1} \mathbf{S}_r \right)^{-1} \mathbf{S}_r^T \hat{\mathbf{R}}_r^{-1} \mathbf{Y}_r \mathbf{Y}_r^T \hat{\mathbf{R}}_r^{-1} \mathbf{S}_r \right\}}{\operatorname{tr}\left(\hat{\mathbf{R}}_r^{-1} \mathbf{Y}_r \mathbf{Y}_r^T \right)} \underset{H_0}{\overset{H_1}{\gtrless}} \phi, \tag{11.42}$$

where ϕ is a detection threshold, and

$$\begin{aligned}
\hat{\mathbf{R}}_r &= \mathbf{D} \hat{\mathbf{R}} \mathbf{D}^\dagger \\
&= \mathfrak{Re}(\hat{\mathbf{R}}) + \mathbf{J} \mathfrak{Im}(\hat{\mathbf{R}}) \in \mathbb{R}^{N \times N}.
\end{aligned} \tag{11.43}$$

This detector is referred to as two-step persymmetric GLRT in partially homogeneous environments (2S-PGLRT-PH).

11.2.2 Wald Test

Now we aim to derive the Wald test for the detection problem in (11.8). In the first step, the covariance matrix $\mathbf{R}$ and the scaling factor α are assumed to be known. Let $\boldsymbol{\Theta}$ be the parameter vector partitioned as

$$\boldsymbol{\Theta} = [\mathbf{b}^T, \mathbf{a}_r^T]^T. \tag{11.44}$$

The FIM $\mathbf{F}(\boldsymbol{\Theta})$ associated with $f(\mathbf{Y}|H_1)$ can be expressed as

$$\mathbf{F}(\boldsymbol{\Theta}) = \mathrm{E}\left\{\frac{\partial \ln f(\mathbf{Y}|H_1)}{\partial \boldsymbol{\Theta}} \frac{\partial \ln f(\mathbf{Y}|H_1)}{\partial \boldsymbol{\Theta}^T}\right\}, \tag{11.45}$$

where $f(\mathbf{Y}|H_1)$ is defined in (11.29). According to [15, p. 188], the Wald test in the real-valued domain is given as

$$\Upsilon_0 = \left(\hat{\boldsymbol{\Theta}}_1 - \hat{\boldsymbol{\Theta}}_0\right)^T \mathbf{F}(\hat{\boldsymbol{\Theta}}_1) \left(\hat{\boldsymbol{\Theta}}_1 - \hat{\boldsymbol{\Theta}}_0\right) \underset{H_0}{\overset{H_1}{\gtrless}} \upsilon_0, \tag{11.46}$$

where υ_0 is a detection threshold, $\hat{\boldsymbol{\Theta}}_1$ and $\hat{\boldsymbol{\Theta}}_0$ are the MLEs of $\boldsymbol{\Theta}$ under H_1 and H_0, respectively. As derived in Appendix 11.A, we have

$$\mathbf{F}(\boldsymbol{\Theta}) = \frac{4}{\alpha}\begin{bmatrix} \mathbf{a}_r^T\mathbf{a}_r\mathbf{S}_r^T\mathbf{R}_r^{-1}\mathbf{S}_r & \mathbf{S}_r^T\mathbf{R}_r^{-1}\mathbf{S}_r\mathbf{b}\mathbf{a}_r^T \\ (\mathbf{S}_r^T\mathbf{R}_r^{-1}\mathbf{S}_r\mathbf{b}\mathbf{a}_r^T)^T & \mathbf{b}^T\mathbf{S}_r^T\mathbf{R}_r^{-1}\mathbf{S}_r\mathbf{b}\mathbf{I}_{2J} \end{bmatrix}. \tag{11.47}$$

Applying (11.47) to (11.46) produces the Wald test with known parameters as

$$\Upsilon_1 = \frac{16}{\alpha}\mathbf{a}_r^T\mathbf{a}_r\mathbf{b}^T\mathbf{S}_r^T\mathbf{R}_r^{-1}\mathbf{S}_r\mathbf{b}. \tag{11.48}$$

The MLEs of α, $\mathbf{a}_r$, and $\mathbf{b}$ under H_1 are given in (11.31), (11.34), and (11.37), respectively. Substituting these estimates into (11.48), with the constant scalar dropped, we can rewrite the Wald test as

$$\Upsilon_2 = \frac{\lambda_{\max}\left\{\left(\mathbf{S}_r^T\mathbf{R}_r^{-1}\mathbf{S}_r\right)^{-1}\mathbf{S}_r^T\mathbf{R}_r^{-1}\mathbf{Y}_r\mathbf{Y}_r^T\mathbf{R}_r^{-1}\mathbf{S}_r\right\}}{\mathrm{tr}\left(\mathbf{R}_r^{-1}\mathbf{Y}_r\mathbf{Y}_r^T\right)}, \tag{11.49}$$

where (11.38) is used. Replacing $\mathbf{R}$ with $\hat{\mathbf{R}}$ in (11.49), we can obtain the adaptive Wald test as

$$\Upsilon = \frac{\lambda_{\max}\left\{\left(\mathbf{S}_r^T\hat{\mathbf{R}}_r^{-1}\mathbf{S}_r\right)^{-1}\mathbf{S}_r^T\hat{\mathbf{R}}_r^{-1}\mathbf{Y}_r\mathbf{Y}_r^T\hat{\mathbf{R}}_r^{-1}\mathbf{S}_r\right\}}{\mathrm{tr}\left(\hat{\mathbf{R}}_r^{-1}\mathbf{Y}_r\mathbf{Y}_r^T\right)} \underset{H_0}{\overset{H_1}{\gtrless}} \upsilon, \tag{11.50}$$

where υ is a detection threshold. Interestingly, the Wald test in (11.50) is identical to the 2S-PGLRT-PH in (11.42).

11.2.3 Rao Test

According to [15, p. 189], the Rao test in the real-valued domain is given by

$$\Xi = \frac{\partial \ln f(\mathbf{Y}|H_1)}{\partial \mathbf{\Theta}}\bigg|^T_{\mathbf{\Theta}=\hat{\mathbf{\Theta}}_0} \mathbf{F}^{-1}(\hat{\mathbf{\Theta}}_0) \frac{\partial \ln f(\mathbf{Y}|H_1)}{\partial \mathbf{\Theta}}\bigg|_{\mathbf{\Theta}=\hat{\mathbf{\Theta}}_0}, \tag{11.51}$$

where $\hat{\mathbf{\Theta}}_0$ is the MLE of $\mathbf{\Theta}$ under H_0. According to the matrix inversion formula in [16, p. 188], we can obtain

$$\det(\mathbf{F}(\mathbf{\Theta})) = 0, \tag{11.52}$$

where $\mathbf{F}(\mathbf{\Theta})$ is defined in (11.47). It means that $\mathbf{F}(\mathbf{\Theta})$ is singular. That is to say, the Rao test does not exist.

In summary, we have found that the 2S-PGLRT-PH and Wald test coincide for the detection problem (11.8), and the Rao test does not exist.

11.3 CFARness Analysis

In this part, we aim to prove that the detector in (11.42) bears the CFAR properties against both α and $\mathbf{R}$. Based on [13, eq. (B11)], we have

$$\hat{\mathbf{R}}_r \sim \mathcal{W}_N\left(2K, \frac{\mathbf{R}_r}{2}\right), \tag{11.53}$$

and

$$\begin{cases} \mathbf{y}_{erj} \sim \mathcal{N}_N(\mathbf{0}, \frac{\alpha}{2}\mathbf{R}_r), \\ \mathbf{y}_{orj} \sim \mathcal{N}_N(\mathbf{0}, \frac{\alpha}{2}\mathbf{R}_r), \end{cases} \tag{11.54}$$

where $j = 1, 2, \ldots, J$, $\mathbf{y}_{erj}$ and $\mathbf{y}_{orj}$ are independent random vectors defined in (11.24), and $\mathbf{R}_r$ is defined in (11.26).

First, we examine the distributions of the dominator and numerator of the 2S-PGLRT-PH detector in (11.42). Under H_0, we can use (11.23) and (11.54) to obtain

$$\bar{\mathbf{Y}} \triangleq (\alpha\mathbf{R}_r)^{-1/2}\mathbf{Y}_r \sim \mathcal{N}_{N \times 2J}\left(\mathbf{0}, \frac{\mathbf{I}_N}{2} \otimes \mathbf{I}_{2J}\right). \tag{11.55}$$

Define

$$\mathbf{Q} \triangleq \mathbf{R}_r^{-1/2}\hat{\mathbf{R}}_r\mathbf{R}_r^{-1/2} \in \mathbb{C}^{N \times N}. \tag{11.56}$$

Under H_0, we have

$$\mathbf{Q} \sim \mathcal{W}_N(2K, \mathbf{I}_N/2). \tag{11.57}$$

As a result,

$$\mathbf{Y}_r^T\hat{\mathbf{R}}_r^{-1}\mathbf{Y}_r = \alpha\bar{\mathbf{Y}}^T\mathbf{Q}^{-1}\bar{\mathbf{Y}}. \tag{11.58}$$

According to (11.55) and (11.57), we know that the term $\mathbf{Y}_r^T \hat{\mathbf{R}}_r^{-1} \mathbf{Y}_r$ is independent of the noise covariance matrix $\mathbf{R}$.

Using (11.55) and (11.57), we have

$$\mathbf{Y}_r^T \hat{\mathbf{R}}_r^{-1} \mathbf{S}_r \left(\mathbf{S}_r^T \hat{\mathbf{R}}_r^{-1} \mathbf{S}_r \right)^{-1} \mathbf{S}_r^T \hat{\mathbf{R}}_r^{-1} \mathbf{Y}_r$$
$$= \alpha \bar{\mathbf{Y}}^T \mathbf{Q}^{-1} \bar{\mathbf{S}} \left(\bar{\mathbf{S}}^T \mathbf{Q}^{-1} \bar{\mathbf{S}} \right)^{-1} \bar{\mathbf{S}}^T \mathbf{Q}^{-1} \bar{\mathbf{Y}}, \tag{11.59}$$

where

$$\bar{\mathbf{S}} = \mathbf{R}_r^{-1/2} \mathbf{S}_r \in \mathbb{R}^{N \times q}. \tag{11.60}$$

According to (11.58) and (11.59), the 2S-PGLRT-PH in (11.42) can be rewritten as

$$\Upsilon = \frac{\lambda_{\max} \{\mathbf{M}\}}{\text{tr} \left(\bar{\mathbf{Y}}^T \mathbf{Q}^{-1} \bar{\mathbf{Y}} \right)} \overset{H_1}{\underset{H_0}{\gtrless}} v, \tag{11.61}$$

where

$$\mathbf{M} = \bar{\mathbf{Y}}^T \mathbf{Q}^{-1} \bar{\mathbf{S}} \left(\bar{\mathbf{S}}^T \mathbf{Q}^{-1} \bar{\mathbf{S}} \right)^{-1} \bar{\mathbf{S}}^T \mathbf{Q}^{-1} \bar{\mathbf{Y}} \in \mathbb{C}^{2J \times 2J}. \tag{11.62}$$

It can be seen that the Υ under H_0 is independent of the scaling factor α. This is to say, the 2S-PGLRT-PH exhibits the CFAR property against α.

Next, we show that $\mathbf{M}$ is independent of the noise covariance matrix $\mathbf{R}$. According to the QR decomposition, $\bar{\mathbf{S}}$ can be written as

$$\bar{\mathbf{S}} = \mathbf{W}\mathbf{G}, \tag{11.63}$$

where $\mathbf{W} \in \mathbb{R}^{N \times q}$ is a slice of an orthogonal matrix, and $\mathbf{G} \in \mathbb{R}^{q \times q}$ is a nonsingular upper triangular matrix. Taking (11.63) into (11.59), we have

$$\mathbf{M} = \bar{\mathbf{Y}}^T \mathbf{Q}^{-1} \mathbf{W} \left(\mathbf{W}^T \mathbf{Q}^{-1} \mathbf{W} \right)^{-1} \mathbf{W}^T \mathbf{Q}^{-1} \bar{\mathbf{Y}}. \tag{11.64}$$

It is known that an orthogonal matrix $\mathbf{U} \in \mathbb{R}^{N \times N}$ exists such that

$$\mathbf{U}^T \mathbf{W} = \mathbf{E}, \tag{11.65}$$

where

$$\mathbf{E} \triangleq [\mathbf{I}_q, \mathbf{0}_{q \times (N-q)}]^T \in \mathbb{C}^{N \times q}. \tag{11.66}$$

Define

$$\begin{cases} \mathbf{Q}_1 = \mathbf{U}^T \mathbf{Q} \mathbf{U} \in \mathbb{C}^{N \times N}, \\ \breve{\mathbf{Y}} = \mathbf{U}^T \bar{\mathbf{Y}} \in \mathbb{C}^{N \times 2J}. \end{cases} \tag{11.67}$$

Then, $\mathbf{M}$ can be recast to

$$\mathbf{M} = \breve{\mathbf{Y}}^T \mathbf{Q}_1^{-1} \mathbf{E} \left(\mathbf{E}^T \mathbf{Q}_1^{-1} \mathbf{E} \right)^{-1} \mathbf{E}^T \mathbf{Q}_1^{-1} \breve{\mathbf{Y}}. \tag{11.68}$$

Based on (11.55), (11.57), and (11.67), we can obtain

$$\begin{cases} \mathbf{Q}_1 \sim \mathcal{W}_N \left(2K, \frac{\mathbf{I}_N}{2} \right), \\ \breve{\mathbf{Y}} \sim \mathcal{N}_{N \times 2J} \left(\mathbf{0}, \frac{\mathbf{I}_N}{2} \otimes \mathbf{I}_{2J} \right), \end{cases} \tag{11.69}$$

which are irrelative to $\mathbf{R}$. It is indicated that $\mathbf{M}$ is irrelative to the noise covariance matrix $\mathbf{R}$. In summary, the detector in (11.42) exhibits the CFAR property against both α and $\mathbf{R}$.

Combining (11.55), (11.57), (11.68), and (11.69), we can know that the distribution of the test statistic Υ in (11.61) under H_0 is irrelevant to the subspace matrix $\mathbf{S}$. It means that the detection threshold of the 2S-PGLRT-PH is independent of $\mathbf{S}$.

11.4 Numerical Examples

Assume that a pulsed Doppler radar is used, which transmits symmetrically spaced pulse trains. The number of pulses in a CPI is $N = 9$. The (i,j)th element of the noise covariance matrix is chosen as $\mathbf{R}(i,j) = \sigma^2 \rho^{|i-j|}$, where σ^2 is the noise power, and $\rho = 0.9$ is the noise correlation coefficient. The scaling factor is set to be 2. We choose the number of range cells the target occupies to be $J = 4$. We select that $q = 3$, and the subspace matrix $\mathbf{S}$ is created using q steering vectors at $\tilde{f}_d - 0.03$, $\tilde{f}_d$, and $\tilde{f}_d + 0.03$, where $\tilde{f}_d$ is the nominal Doppler frequency. The SNR in decibel is defined as

$$\text{SNR} = 10 \log_{10} \left(\mathbf{a}^\dagger \mathbf{a} \mathbf{q}^\dagger \mathbf{R}^{-1} \mathbf{q} \right). \tag{11.70}$$

The mismatch angle ϕ between the true steering vector $\mathbf{q}$ and the assumed one $\mathbf{p}$ is defined in (11.6).

For comparison purposes, the conventional P1SGLRT [17, eq. (15)], P2SGLRT [17, eq. (25)], PRao test [18, eq. 11], PWald test [18, eq. 20], and the GLRT without exploiting persymmetry in [10] are considered. Due to the lack of analytical expressions, we use MC counting techniques to simulate the PFA and detection probability in the following.

In Fig. 11.1, we depict the detection probability curves with respect to SNR for the mismatched case where $\cos^2 \phi = 0.5427$. In such a case, the nominal and true Doppler frequencies of target are 0.16 and 0.18, respectively. Here, the PFA is set to be 10^{-3}. It is shown that the 2S-PGLRT-PH detector has stronger robustness than its competitors to the target steering vector mismatches. In addition, the performance of the 2S-PGLRT-PH detector decreases as the number of training data decreases. It is worth noting that the GLRT without using persymmetry in [10] cannot work in the case where $K < N$. Hence, the performance of the GLRT is not provided in Fig. 11.1(c) where $K = 6$. As indicated in [17] and [18], the P1SGLRT, PRao test and PWald cannot work with $K = 6$ when $N = 9$. Hence, their performance is not provided in Fig. 11.1(c). It can be seen from Fig. 11.1(c) that the 2S-PGLRT-PH detector significantly outperforms the P2SGLRT in the case where the training data size is very limited (i.e., $K = 6$ for $N = 9$).

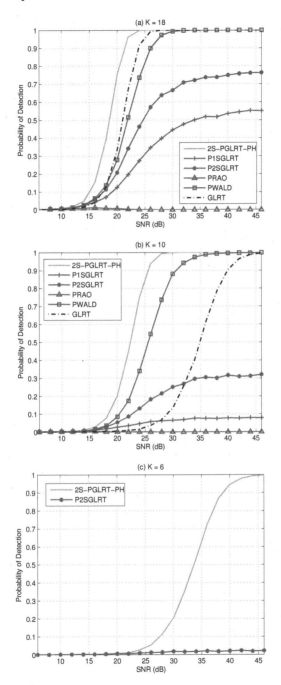

FIGURE 11.1
Detection probability versus SNR for the mismatched case with $\cos^2 \phi = 0.5427$. The probability of false alarm is 10^{-3}.

We consider the mismatched case of $\cos^2 \phi = 0.8126$ in Fig. 11.2, where the other parameters are the same as those in Fig. 11.1. In such a case, the nominal and true Doppler frequencies of target are 0.165 and 0.18, respectively. As the mismatches reduce (i.e., $\cos^2 \phi$ increases from 0.5427 to be 0.8126), the 2S-PGLRT-PH detector still outperforms its counterparts, but the performance gap becomes narrow. In particular, the 2S-PGLRT-PH detector has detection performance worse than some of its competitors in the matched case (i.e., $\cos^2 \phi = 1$), as shown in Fig. 11.3 where both the nominal and true Doppler frequencies of target are 0.18. This is to say, the 2S-PGLRT-PH detector obtains improved robustness in the mismatched case at the price of performance degradation in the matched case.

In Fig. 11.4, the detection probability as a function of $\cos^2 \phi$ is presented. The nominal Doppler frequency of target is 0.15, and we change the value of true Doppler frequency to alter the mismatch degree. We can observe that the 2S-PGLRT-PH detector has strong robustness to the target steering vector mismatches. Specifically, the performance of the 2S-PGLRT-PH detector keeps almost the same, when the mismatched degree increases from $\cos^2 \phi = 1$ to $\cos^2 \phi = 0.2$. However, its counterparts suffer from obvious performance degradations with the increases in the mismatches.

11.A   Derivations of (11.47)

Using (11.29), we have

$$\frac{\partial \ln f(\mathbf{Y}|H_1)}{\partial \mathbf{b}} = \frac{2}{\alpha} \mathbf{S}_r^T \mathbf{R}_r^{-1} \breve{\mathbf{Y}}_r \mathbf{a}_r, \tag{11.A.1}$$

$$\frac{\partial \ln f(\mathbf{Y}|H_1)}{\partial \mathbf{a}_r} = \frac{2}{\alpha} \breve{\mathbf{Y}}_r^T \mathbf{R}_r^{-1} \mathbf{S}_r \mathbf{b}, \tag{11.A.2}$$

where

$$\breve{\mathbf{Y}}_r = \mathbf{Y}_r - \mathbf{S}_r \mathbf{b} \mathbf{a}_r^T. \tag{11.A.3}$$

Then,

$$\begin{aligned}
\mathbf{F}(\boldsymbol{\Theta}) &= \frac{4}{\alpha^2} \mathrm{E} \left[\begin{array}{cc} \mathbf{S}_r^T \mathbf{R}_r^{-1} \breve{\mathbf{Y}}_r \mathbf{a}_r \mathbf{a}_r^T \breve{\mathbf{Y}}_r^T \mathbf{R}_r^{-1} \mathbf{S}_r & \mathbf{S}_r^T \mathbf{R}_r^{-1} \breve{\mathbf{Y}}_r \mathbf{a}_r \mathbf{b}^T \mathbf{S}_r^T \mathbf{R}_r^{-1} \breve{\mathbf{Y}}_r \\ \breve{\mathbf{Y}}_r^T \mathbf{R}_r^{-1} \mathbf{S}_r \mathbf{b} \mathbf{a}_r^T \breve{\mathbf{Y}}_r^T \mathbf{R}_r^{-1} \mathbf{S}_r & \breve{\mathbf{Y}}_r^T \mathbf{R}_r^{-1} \mathbf{S}_r \mathbf{b} \mathbf{b}^T \mathbf{S}_r^T \mathbf{R}_r^{-1} \breve{\mathbf{Y}}_r \end{array} \right] \\
&= \frac{4}{\alpha^2} \left[\begin{array}{cc} \mathbf{F}_{11} & \mathbf{F}_{12} \\ \mathbf{F}_{21} & \mathbf{F}_{22} \end{array} \right],
\end{aligned} \tag{11.A.4}$$

where

$$\begin{aligned}
\mathbf{F}_{11} &= \mathrm{E}[\mathbf{S}_r^T \mathbf{R}_r^{-1} \breve{\mathbf{Y}}_r \mathbf{a}_r \mathbf{a}_r^T \breve{\mathbf{Y}}_r^T \mathbf{R}_r^{-1} \mathbf{S}_r] \\
&= \alpha \mathbf{a}_r^T \mathbf{a}_r \mathbf{S}_r^T \mathbf{R}_r^{-1} \mathbf{S}_r.
\end{aligned} \tag{11.A.5}$$

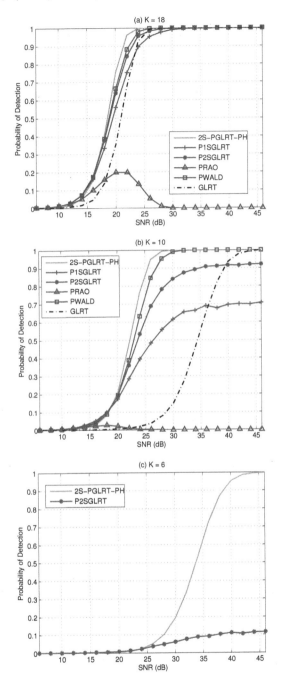

FIGURE 11.2

Detection probability versus SNR for the mismatched case with $\cos^2 \phi = 0.8126$. The probability of false alarm is 10^{-3}.

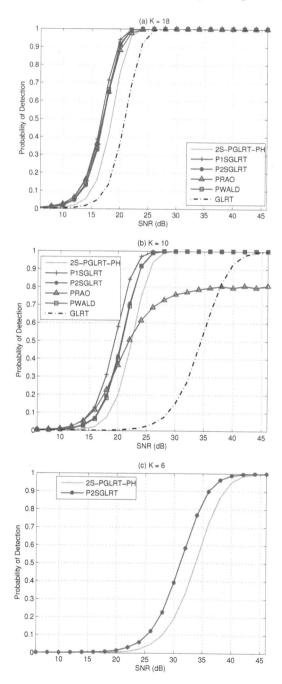

FIGURE 11.3

Detection probability versus SNR for the matched case with $\cos^2 \phi = 1$. The probability of false alarm is 10^{-3}.

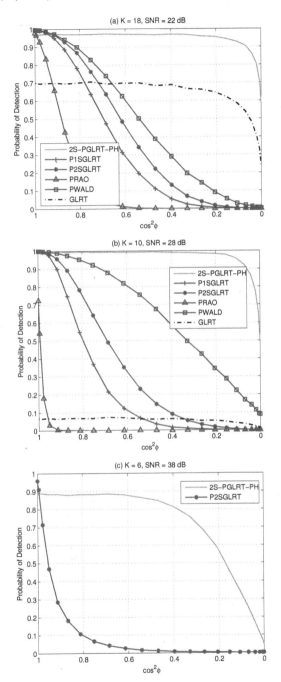

FIGURE 11.4

Detection probability versus $\cos^2 \phi$ for different K. The probability of false alarm is 10^{-3}.

To obtain explicit expressions for $\mathbf{F}_{12}$, $\mathbf{F}_{21}$, and $\mathbf{F}_{22}$, we denote the i-th element of $\mathbf{a}_r$ by γ_i, i.e., $\mathbf{a}_r = [\gamma_1, \gamma_2, \ldots, \gamma_{2J}]^T$, and define

$$\boldsymbol{\theta} = \mathbf{R}_r^{-1}\mathbf{S}_r\mathbf{b}. \tag{11.A.6}$$

Then, we can obtain

$$
\begin{aligned}
\mathbf{F}_{12} = \mathbf{F}_{21}^T &= \mathrm{E}[\mathbf{S}_r^T\mathbf{R}_r^{-1}\mathbf{\breve{Y}}_r\mathbf{a}_r\boldsymbol{\theta}^T\mathbf{\breve{Y}}_r] \\
&= \mathbf{S}_r^T\mathbf{R}_r^{-1}\mathrm{E}\left\{\sum_{i=1}^{2J}\gamma_i\mathbf{y}_i\left[\mathbf{y}_1^T\boldsymbol{\theta}, \mathbf{y}_2^T\boldsymbol{\theta}, \ldots, \mathbf{y}_{2J}^T\boldsymbol{\theta}\right]\right\} \\
&= \alpha\mathbf{S}_r^T\mathbf{R}_r^{-1}\left[\gamma_1\mathbf{R}_r\boldsymbol{\theta}, \gamma_2\mathbf{R}_r\boldsymbol{\theta}, \ldots, \gamma_{2J}\mathbf{R}_r\boldsymbol{\theta}\right] \\
&= \alpha\mathbf{S}_r^T\mathbf{R}_r^{-1}\mathbf{R}_r\boldsymbol{\theta}\mathbf{a}_r^T \\
&= \alpha\mathbf{S}_r^T\mathbf{R}_r^{-1}\mathbf{S}_r\mathbf{b}\mathbf{a}_r^T,
\end{aligned}
\tag{11.A.7}
$$

where $\mathbf{y}_i$ is the i-th column vector of $\mathbf{\breve{Y}}_r$. And,

$$
\begin{aligned}
\mathbf{F}_{22} &= \mathrm{E}\left\{\begin{bmatrix} \boldsymbol{\theta}^T\mathbf{y}_1 \\ \boldsymbol{\theta}^T\mathbf{y}_2 \\ \vdots \\ \boldsymbol{\theta}^T\mathbf{y}_{2J} \end{bmatrix}\left[\mathbf{y}_1^T\boldsymbol{\theta}, \mathbf{y}_2^T\boldsymbol{\theta}, \ldots, \mathbf{y}_{2J}^T\boldsymbol{\theta}\right]\right\} \\
&= \alpha\boldsymbol{\theta}^T\mathbf{R}\boldsymbol{\theta}\mathbf{I}_{2J} \\
&= \alpha\mathbf{b}^T\mathbf{S}_r^T\mathbf{R}_r^{-1}\mathbf{S}_r\mathbf{b}\mathbf{I}_{2J}.
\end{aligned}
\tag{11.A.8}
$$

Inserting (11.A.5), (11.A.7), and (11.A.8) into (11.A.4) yields (11.47).

Bibliography

[1] J. Liu, T. Jian, W. Liu, C. Hao, and D. Orlando, "Persymmetric adaptive detection with improved robustness to steering vector mismatches," *Signal Processing*, vol. 176, p. 107669, 2020.

[2] S. Kraut, L. L. Scharf, and R. W. Butler, "The adaptive coherence estimator: A uniformly-most-powerful-invariant adaptive detection statistic," *IEEE Transactions on Signal Processing*, vol. 53, no. 2, pp. 417–438, Febuary 2005.

[3] H. L. Van Trees, *Optimum Array Processing, Part IV of Detection, Estimation, and Modulation Theory.* New Yok, Wiley-Interscience, 2002.

[4] A. De Maio, "Maximum likelihood estimation of structured persymmetric covariance matrices," *Signal Processing*, vol. 83, no. 3, pp. 633–640, March 2003.

[5] E. Conte and A. De Maio, "Exploiting persymmetry for cfar detection in compound-Gaussian clutter," *IEEE Transactions on Aerospace and Electronic Systems*, vol. 39, no. 2, pp. 719–724, April 2003.

[6] F. Bandiera, D. Orlando, and G. Ricci, *Advanced Radar Detection Schemes under Mismatched Signal Models*. Morgan & Claypool, 2009.

[7] E. J. Kelly, "Performance of an adaptive detection algorithm; rejection of unwanted signals," *IEEE Transactions on Aerospace and Electronic Systems*, vol. AES-25, no. 2, pp. 122–133, March 1989.

[8] J. Carretero-Moya, A. De Maio, J. Gismero-Menoyo, and A. Asensio-Lopez, "Experimental performance analysis of distributed target coherent radar detectors," *IEEE Transactions on Aerospace and Electronic Systems*, vol. 48, no. 3, pp. 2216–2238, July 2012.

[9] F. Gini and A. Farina, "Vector subspace detection in compound-Gaussian clutter. part II: performance analysis," *IEEE Transactions on Aerospace and Electronic Systems*, vol. 38, no. 4, pp. 1312–1323, October 2002.

[10] O. Besson, L. L. Scharf, and S. Kraut, "Adaptive detection of a signal known only to lie on a line in a known subspace, when primary and secondary data are partially homogeneous," *IEEE Transaction on Signal Processing*, vol. 54, no. 12, pp. 4698–4705, December 2006.

[11] E. Conte, A. De Maio, and G. Ricci, "GLRT-based adaptive detection algorithms for range-spread targets," *IEEE Transactions on Signal Processing*, vol. 49, no. 7, pp. 1336–1348, July 2001.

[12] X. Shuai, L. Kong, and J. Yang, "Adaptive detection for distributed targets in Gaussian noise with Rao and Wald tests," *Science China Information Sciences*, vol. 55, no. 6, pp. 1290–1300, 2012.

[13] L. Cai and H. Wang, "A persymmetric multiband GLR algorithm," *IEEE Transactions on Aerospace and Electronic Systems*, vol. 28, no. 3, pp. 806–816, July 1992.

[14] R. A. Horn and C. R. Johnson, *Matrix Analysis*. Cambridge University Press, New Yok, 1985.

[15] S. M. Kay, *Fundamentals of Statistical Signal Processing: Detection Theory*. Upper Saddle River, NJ: Prentice Hall, 1998.

[16] D. A. Harville, *Matrix Algebra for a Statistician's Perspective*. New York: Springer-Verlag, 1997.

[17] C. Hao, D. Orlando, G. Foglia, X. Ma, S. Yan, and C. Hou, "Persymmetric adaptive detection of distributed targets in partially-homogeneous environment," *Digital Signal Processing*, vol. 24, pp. 42–51, January 2014.

[18] Z. Wang, M. Li, H. Chen, Y. Lu, R. Cao, P. Zhang, L. Zuo, and Y. Wu, "Persymmetric detectors of distributed targets in partially homogeneous disturbance," *Signal Processing*, vol. 128, pp. 382–388, 2016.

12

Joint Exploitation of Persymmetry and Symmetric Spectrum

In this chapter, we jointly exploit the persymmetry and the symmetric spectrum so as to come up with adaptive detectors with enhanced detection performance. To this end, we show that the original detection problem is equivalent to another decision problem under the assumption of a persymmetric structured ICM, dealing with independent circularly symmetric complex vectors. Then, we reformulate the latter decision problem exploiting the symmetric spectrum for the clutter, which leads to a real-valued covariance matrix and allows us to transfer the data from the complex domain to the real domain [1]. From a practical point of view, the first step allows us to obtain an intermediate form of the problem where the number of primary and secondary data is doubled, and the second step is a further transformation coming up with an equivalent decision problem where data are real and data volume is double [2].

This chapter devises detection schemes with the above-mentioned joint properties based upon the principles of two-step GLRT, Rao test, and Wald test in HE and PHE. Numerical examples carried out with MC simulations show that the devised detection schemes outperform the detectors that do not exploit the two types of prior information jointly.

12.1 Problem Formulation

Let us begin by considering the detection problem at hand formulated by (6.1) when $J = 1$. Specifically, we can write the decision problem as follows

$$H_0 : \begin{cases} \mathbf{y} = \mathbf{n}, \\ \mathbf{y}_k = \mathbf{n}_k, \ k = 2, \ldots, K+1, \end{cases} \tag{12.1a}$$

$$H_1 : \begin{cases} \mathbf{y} = a\,\mathbf{p} + \mathbf{n}, \\ \mathbf{y}_k = \mathbf{n}_k, \ k = 2, \ldots, K+1, \end{cases} \tag{12.1b}$$

where $\mathbf{y} = \mathbf{y}_1 \in \mathbb{C}^{N \times 1}$ denotes the primary data, while $\mathbf{Y}_1 = [\mathbf{y}_2, \mathbf{y}_3, \ldots, \mathbf{y}_{K+1}] \in \mathbb{C}^{N \times K}$ denotes the secondary data. The complex-valued scalar $a = a_r + ja_i$ with $a_r = \Re(a)$ and $a_i = \Im(a)$ represents the target

DOI: 10.1201/9781003340232-12

reflectivity and channel propagation effects, $\mathbf{p} = \mathbf{p}_r + \mathbf{p}_i$ with $\mathbf{p}_r = \mathfrak{Re}(\mathbf{p})$ and $\mathbf{p}_i = \mathfrak{Im}(\mathbf{p})$ is the target signal steering vector. $\mathbf{n}$ and $\mathbf{n}_k$, $k = 2, \dots, K+1$ are independent identically complex Gaussian distributed vectors with zero-mean and unknown covariance matrices given by

$$E[\mathbf{n}\mathbf{n}^\dagger] = \mathbf{M} \in \mathbb{R}^{N \times N}, \ E[\mathbf{n}_k \mathbf{n}_k^\dagger] = \gamma \mathbf{M}, \tag{12.2}$$

with $\mathbf{M}$ the positive definite matrix and $\gamma > 0$ the unknown power scaling factor.

For an active system utilizing symmetrically spaced linear arrays and/or pulsed trains, both $\mathbf{M}$ and $\mathbf{p}$ have the persymmetric property (See 6.1.1). Now, following the lead of *Appendix B* of [3], we can recast problem (12.1) in terms of

$$\begin{cases} \mathbf{y}_e = \frac{1}{2}(\mathbf{y} + \mathbf{J}\mathbf{y}^*), & \mathbf{y}_o = \frac{1}{2}(\mathbf{y} - \mathbf{J}\mathbf{y}^*), \\ \mathbf{y}_{ek} = \frac{1}{2}(\mathbf{y}_k + \mathbf{J}\mathbf{y}_k^*), & \mathbf{y}_{ok} = \frac{1}{2}(\mathbf{y}_k - \mathbf{J}\mathbf{y}_k^*), \ k = 2, \dots, K+1, \end{cases} \tag{12.3}$$

which are independent circularly symmetric complex Gaussian vectors with means depending on which hypothesis is in force and covariance matrices given by

$$E\left[\mathbf{y}_e\mathbf{y}_e^\dagger\right] = E\left[\mathbf{y}_o\mathbf{y}_o^\dagger\right] = \mathbf{M}/2, \ E\left[\mathbf{y}_{ek}\mathbf{y}_{ek}^\dagger\right] = E\left[\mathbf{y}_{ok}\mathbf{y}_{ok}^\dagger\right] = \gamma\mathbf{M}/2. \tag{12.4}$$

Thus, we can write

$$H_0 : \begin{cases} \mathbf{y}_e = \mathbf{n}_e, \ \mathbf{y}_o = \mathbf{n}_o, \\ \mathbf{y}_{ek} = \mathbf{n}_{ek}, \ \mathbf{y}_{ok} = \mathbf{n}_{ok}, \qquad k = 2, \dots, K+1, \end{cases} \tag{12.5a}$$

$$H_1 : \begin{cases} \mathbf{y}_e = a_r\mathbf{p} + \mathbf{n}_e, \mathbf{y}_o = a_i\mathbf{p} + \mathbf{n}_o, \\ \mathbf{y}_{ek} = \mathbf{n}_{ek}, \ \mathbf{y}_{ok} = \mathbf{n}_{ok}, \qquad k = 2, \dots, K+1, \end{cases} \tag{12.5b}$$

where

$$\begin{cases} \mathbf{n}_e = \frac{1}{2}(\mathbf{n} + \mathbf{J}\mathbf{n}^*), \ \mathbf{n}_o = \frac{1}{2}(\mathbf{n} - \mathbf{J}\mathbf{n}^*), \\ \mathbf{n}_{ek} = \frac{1}{2}(\mathbf{n}_k + \mathbf{J}\mathbf{n}_k^*), \ \mathbf{n}_{ok} = \frac{1}{2}(\mathbf{n}_k - \mathbf{J}\mathbf{n}_k^*), \ k = 2, \dots, K+1. \end{cases} \tag{12.6}$$

Problem (12.5) can be further recast exploiting the fact that clutter spectrum is real and even. Otherwise stated, the ICM is real and, hence, it is possible to express problem (12.5) in terms of the following quantities:

$$\mathbf{y}_{er} = \mathfrak{Re}(\mathbf{y}_e), \ \mathbf{y}_{ei} = \mathfrak{Im}(\mathbf{y}_e), \mathbf{y}_{or} = \mathfrak{Re}(\mathbf{y}_o), \mathbf{y}_{oi} = \mathfrak{Im}(\mathbf{y}_o),$$
$$\mathbf{y}_{ekr} = \mathfrak{Re}(\mathbf{y}_{ek}), \ \mathbf{y}_{eki} = \mathfrak{Im}(\mathbf{y}_{ek}),$$
$$\mathbf{y}_{okr} = \mathfrak{Re}(\mathbf{y}_{ok}), \mathbf{y}_{oki} = \mathfrak{Im}(\mathbf{y}_{ok}), \ k = 2, \dots, K+1. \tag{12.7}$$

It is easy to know that $\mathbf{y}_{er}$, $\mathbf{y}_{ei}$, $\mathbf{y}_{or}$, $\mathbf{y}_{oi}$, $\mathbf{y}_{ekr}$, $\mathbf{y}_{eki}$, $\mathbf{y}_{okr}$, and $\mathbf{y}_{oki}$ are statistically independent real Gaussian vectors with means depending on which hypothesis is in force and covariance matrices given by

$$E\left[\mathbf{y}_{er}\mathbf{y}_{er}^{\dagger}\right] = E\left[\mathbf{y}_{ei}\mathbf{y}_{ei}^{\dagger}\right] = E\left[\mathbf{y}_{or}\mathbf{y}_{or}^{\dagger}\right] = E\left[\mathbf{y}_{oi}\mathbf{y}_{oi}^{\dagger}\right] = \mathbf{R}, \quad (12.8)$$

$$E\left[\mathbf{y}_{ekr}\mathbf{y}_{ekr}^{\dagger}\right] = E\left[\mathbf{y}_{eki}\mathbf{y}_{eki}^{\dagger}\right] = E\left[\mathbf{y}_{okr}\mathbf{y}_{okr}^{\dagger}\right] = E\left[\mathbf{y}_{oki}\mathbf{y}_{oki}^{\dagger}\right]$$
$$= \gamma\mathbf{R} \quad (12.9)$$

with $\mathbf{R} = \frac{1}{4}\mathbf{M} \in \mathbb{R}^{N \times N}$. Thus, problem (12.5) is equivalent to

$$H_0 : \begin{cases} \mathbf{y}_{er} = \mathbf{n}_{er}, \ \mathbf{y}_{ei} = \mathbf{n}_{ei}, \ \mathbf{y}_{or} = \mathbf{n}_{or}, \ \mathbf{y}_{oi} = \mathbf{n}_{oi}, \\ \mathbf{y}_{ekr} = \mathbf{n}_{ekr}, \ \mathbf{y}_{eki} = \mathbf{n}_{eki}, \\ \mathbf{y}_{okr} = \mathbf{n}_{okr}, \ \mathbf{y}_{oki} = \mathbf{n}_{oki}, \ k = 2, \ldots, K+1, \end{cases} \quad (12.10a)$$

$$H_1 : \begin{cases} \mathbf{y}_{er} = a_r\mathbf{p}_r + \mathbf{n}_{er}, \ \mathbf{y}_{ei} = a_r\mathbf{p}_i + \mathbf{n}_{ei}, \\ \mathbf{y}_{or} = a_i\mathbf{p}_r + \mathbf{n}_{or}, \mathbf{y}_{oi} = a_i\mathbf{p}_i + \mathbf{n}_{oi}, \\ \mathbf{y}_{ekr} = \mathbf{n}_{ekr}, \ \mathbf{y}_{eki} = \mathbf{n}_{eki}, \\ \mathbf{y}_{okr} = \mathbf{n}_{okr}, \ \mathbf{y}_{oki} = \mathbf{n}_{oki}, \ k = 2, \ldots, K+1, \end{cases} \quad (12.10b)$$

where

$$\mathbf{n}_{er} = \Re(\mathbf{n}_e), \ \mathbf{n}_{ei} = \Im(\mathbf{n}_e), \mathbf{n}_{or} = \Re(\mathbf{n}_o), \mathbf{n}_{oi} = \Im(\mathbf{n}_o),$$
$$\mathbf{n}_{ekr} = \Re(\mathbf{n}_{ek}), \ \mathbf{n}_{eki} = \Im(\mathbf{n}_{ek}),$$
$$\mathbf{n}_{okr} = \Re(\mathbf{n}_{ok}), \mathbf{n}_{oki} = \Im(\mathbf{n}_{ok}), \ k = 2, \ldots, K+1. \quad (12.11)$$

Comparison of problem (12.10) with problem (12.1) indicates that such transformation is equivalent to multiplying the number of secondary data by four, and hence, the new receivers obtained by solving problem (12.10) would work under the constraint $4K \geqslant N$ instead of $K \geqslant N$ which is required by the traditional detectors in [4–7].

As a preliminary step, we define the following quantities:

- $\mathbf{Y}_{ps} = [\mathbf{Y}_e, \mathbf{Y}_o] \in \mathbb{R}^{N \times 4}$ denotes the primary data matrix with $\mathbf{Y}_e = [\mathbf{y}_{er}, \mathbf{y}_{ei}] \in \mathbb{R}^{N \times 2}$ and $\mathbf{Y}_o = [\mathbf{y}_{or}, \mathbf{y}_{oi}] \in \mathbb{R}^{N \times 2}$;

- $\mathbf{Y}_{ps1} = [\mathbf{Y}_{e1}, \mathbf{Y}_{o1}] \in \mathbb{R}^{N \times 4K}$ denotes the secondary data matrix with $\mathbf{Y}_{e1} = [\mathbf{y}_{e2r}, \ldots, \mathbf{y}_{e(K+1)r}, \mathbf{y}_{e2i}, \ldots, \mathbf{y}_{e(K+1)i}] \in \mathbb{R}^{N \times 2K}$, and $\mathbf{Y}_{o1} = [\mathbf{y}_{o2r}, \ldots, \mathbf{y}_{o(K+1)r}, \mathbf{y}_{o2i}, \ldots, \mathbf{y}_{o(K+1)i}] \in \mathbb{R}^{N \times 2K}$;

- $\mathbf{P} = [\mathbf{p}_r, \mathbf{p}_i] \in \mathbb{R}^{N \times 2}$ denotes the nominal steering matrix;

- $\boldsymbol{\theta}_r = [a_r, a_i]^T \in \mathbb{R}^{2 \times 1}$ denotes the signal parameter vector;

- $\boldsymbol{\theta}_s = [\gamma, \ \mathbf{g}^T(\mathbf{R})]^T \in \mathbb{R}^{(N-1)N/2+1}$ denotes the nuisance parameter vector, where $\mathbf{g}(\mathbf{R}) \in \mathbb{R}^{(N-1)N/2 \times 1}$ is a vector that contains in univocal way the elements of $\mathbf{R}$. Observe that since $\mathbf{R} \in \mathbf{S}_{++}$, it can be well represented by the $[(N-1)N/2]$-dimensional vector;

- $\boldsymbol{\theta} = \left[\boldsymbol{\theta}_r^T, \boldsymbol{\theta}_s^T\right]^T$ contains all unknown parameters.

Under the above assumptions, the joint PDF of $\mathbf{Y}_{ps}$ and $\mathbf{Y}_{ps1}$ under H_l, $l = 0, 1$, is given by

$$f(\mathbf{Y}_{ps}, \mathbf{Y}_{ps1} | l\boldsymbol{\theta}_r, \boldsymbol{\theta}_s, H_l) = f(\mathbf{Y}_{ps} | l\boldsymbol{\theta}_r, H_l) f(\mathbf{Y}_{ps1} | \boldsymbol{\theta}_s), \qquad (12.12)$$

where

$$\begin{cases} f(\mathbf{Y}_{ps} | l\boldsymbol{\theta}_r, H_l) = \frac{1}{(2\pi)^{2N} \det^2(\mathbf{R})} \exp\left\{ -\frac{1}{2} \operatorname{tr}\left[\mathbf{R}^{-1} \mathbf{T}_l \right] \right\}, \\ f(\mathbf{Y}_{ps1} | \boldsymbol{\theta}_s) = \frac{\gamma^{-2NK}}{(2\pi)^{2NK} \det^{2K}(\mathbf{R})} \exp\left\{ -\frac{1}{2} \operatorname{tr}\left[\mathbf{R}^{-1} \frac{1}{\gamma} \mathbf{Y}_{ps1} \mathbf{Y}_{ps1}^T \right] \right\}, \end{cases} \qquad (12.13)$$

with

$$\begin{cases} \mathbf{T}_0 = \mathbf{Y}_e \mathbf{Y}_e^T + \mathbf{Y}_o \mathbf{Y}_o^T \in \mathbb{R}^{N \times N}, \\ \mathbf{T}_1(\boldsymbol{\theta}_r) = (\mathbf{Y}_e - a_r \mathbf{P})(\mathbf{Y}_e - a_r \mathbf{P})^T + (\mathbf{Y}_o - a_i \mathbf{P})(\mathbf{Y}_o - a_i \mathbf{P})^T \in \mathbb{R}^{N \times N}. \end{cases} \qquad (12.14)$$

12.2 Rao Test

In this section, we devise the detection architecture resorting to the Rao test criterion which is often simpler than and asymptotically equivalent to the GLRT. The Rao test for problem (12.10) can be written as follows:

$$\begin{aligned} \Lambda_0 = &\left[\frac{\partial \ln f(\mathbf{Y}_{ps}, \mathbf{Y}_{ps1} | \boldsymbol{\theta}_r, \boldsymbol{\theta}_s, H_1)}{\partial \boldsymbol{\theta}_r} \right]^T_{\boldsymbol{\theta} = \hat{\boldsymbol{\theta}}_0} \mathbf{J}_{rr}(\boldsymbol{\theta}) \bigg|_{\boldsymbol{\theta} = \hat{\boldsymbol{\theta}}_0} \\ &\times \left[\frac{\partial \ln f(\mathbf{Y}_{ps}, \mathbf{Y}_{ps1} | \boldsymbol{\theta}_r, \boldsymbol{\theta}_s, H_1)}{\partial \boldsymbol{\theta}_r} \right]_{\boldsymbol{\theta} = \hat{\boldsymbol{\theta}}_0} \overset{H_1}{\underset{H_0}{\gtrless}} \eta_0, \end{aligned} \qquad (12.15)$$

where η_0 is the detection threshold, $\mathbf{J}_{rr}(\boldsymbol{\theta})$ is given by (5.16), and $\hat{\boldsymbol{\theta}}_0$ is the MLE of $\boldsymbol{\theta}$ under H_0.

Let us start with the MLEs of the nuisance parameters under H_0. To this end, note that taking the derivative of $\ln f(\mathbf{Y}_{ps}, \mathbf{Y}_{ps1} | \boldsymbol{\theta}_s, H_0)$ with respect to $\mathbf{R}$ and equating it to zero results in

$$\hat{\mathbf{R}}_0 = \frac{1}{4(K+1)} \left(\frac{1}{\gamma} \mathbf{Y}_{ps1} \mathbf{Y}_{ps1}^T + \mathbf{Y}_{ps} \mathbf{Y}_{ps}^T \right). \qquad (12.16)$$

Plugging (12.16) into (12.12), we have

$$\begin{aligned} f(\mathbf{Y}_{ps}, \mathbf{Y}_{ps1} | \hat{\mathbf{R}}_0, \gamma, H_0) &\propto \left[\gamma^{\frac{NK}{K+1}} \det\left(\frac{1}{\gamma} \mathbf{Y}_{ps1} \mathbf{Y}_{ps1}^T + \mathbf{Y}_{ps} \mathbf{Y}_{ps}^T \right) \right]^{-2(K+1)} \\ &= \left[\det\left(\frac{1}{\gamma} \mathbf{I}_4 + \mathbf{Q} \right) \right. \\ &\left. \times \det(\mathbf{Y}_{ps1} \mathbf{Y}_{ps1}^T) \gamma^{4 - \frac{N}{K+1}} \right]^{-2(K+1)}, \end{aligned} \qquad (12.17)$$

where

$$\mathbf{Q} = \mathbf{Y}_{ps}^T \left(\mathbf{Y}_{ps1} \mathbf{Y}_{ps1}^T \right)^{-1} \mathbf{Y}_{ps}, \tag{12.18}$$

and the last equality comes from

$$\det \left(\frac{1}{\gamma} \mathbf{I}_N + \mathbf{AB} \right) = \gamma^{4-N} \det \left(\frac{1}{\gamma} \mathbf{I}_4 + \mathbf{BA} \right) \tag{12.19}$$

with $\mathbf{A} \in \mathbf{R}^{N \times 4}$ and $\mathbf{B} \in \mathbf{R}^{4 \times N}$ rectangular matrices.

Apparently, $\hat{\gamma}_0$ is obtained by minimizing $\gamma^{4-\frac{N}{K+1}} \det \left(\frac{1}{\gamma} \mathbf{I}_4 + \mathbf{Q} \right)$. To this end, we denote by r the rank of matrix $\mathbf{S}$, with

$$\mathbf{S} = \left(\mathbf{Y}_{ps1} \mathbf{Y}_{ps1}^T \right)^{-1/2} \mathbf{Y}_{ps} \mathbf{Y}_{ps}^T \left(\mathbf{Y}_{ps1} \mathbf{Y}_{ps1}^T \right)^{-1/2}, \tag{12.20}$$

note that $r = \min(N, 4) \geqslant 2$. Following the lead of the *Proposition 2* of [8], it is not difficult to show that under the constraint $r > \frac{N}{K+1}$, $\hat{\gamma}_0$ is the unique positive solution of the following equation:

$$\sum_{i=1}^r \frac{\mu_i \gamma}{\mu_i \gamma + 1} = \frac{N}{K + 1} \tag{12.21}$$

with μ_i, $i = 1, \ldots, r$, the non-zero eigenvalues of $\mathbf{S}$. Note that (12.21) can be solved by resorting to the Matlab function "roots" which evaluates the eigenvalues of a companion matrix of order $(r + 1) \times (r + 1)$ at most. Thus the MLE of $\mathbf{R}$ under H_0 can be obtained as

$$\hat{\mathbf{R}}_0 = \frac{1}{4(K+1)} \left(\frac{1}{\hat{\gamma}_0} \mathbf{Y}_{ps1} \mathbf{Y}_{ps1}^T + \mathbf{Y}_{ps} \mathbf{Y}_{ps}^T \right). \tag{12.22}$$

The elements of the first partial derivative of the score function $\ln f(\mathbf{Y}_{ps}, \mathbf{Y}_{ps1} | \boldsymbol{\theta}_r, \boldsymbol{\theta}_s, H_1)$ with respect to $\boldsymbol{\theta}_r$ are

$$\frac{\partial \ln f(\mathbf{Y}_{ps}, \mathbf{Y}_{ps1} | \boldsymbol{\theta}_r, \boldsymbol{\theta}_s, H_1)}{\partial \boldsymbol{\theta}_r} = \left[\begin{array}{c} \frac{\partial \ln f(\mathbf{Y}_{ps}, \mathbf{Y}_{ps1} | \boldsymbol{\theta}_r, \boldsymbol{\theta}_s, H_1)}{\partial a_r} \\ \frac{\partial \ln f(\mathbf{Y}_{ps}, \mathbf{Y}_{ps1} | \boldsymbol{\theta}_r, \boldsymbol{\theta}_s, H_1)}{\partial a_i} \end{array} \right]$$

$$= 2 \left[\begin{array}{c} \mathrm{tr} \left[\mathbf{P}^T \mathbf{R}^{-1} \left(\mathbf{Y}_e - a_r \mathbf{P} \right) \right] \\ \mathrm{tr} \left[\mathbf{P}^T \mathbf{R}^{-1} \left(\mathbf{Y}_o - a_i \mathbf{P} \right) \right] \end{array} \right]. \tag{12.23}$$

Thus, it is easy to show that

$$\frac{\partial \ln f(\mathbf{Y}_{ps}, \mathbf{Y}_{ps1} | \boldsymbol{\theta}_r, \boldsymbol{\theta}_s, H_1)}{\partial \boldsymbol{\theta}_r} \bigg|_{\boldsymbol{\theta} = \hat{\boldsymbol{\theta}}_0} = 2 \left[\begin{array}{c} \mathrm{tr} \left(\mathbf{P}^T \hat{\mathbf{R}}_0^{-1} \mathbf{Y}_e \right) \\ \mathrm{tr} \left(\mathbf{P}^T \hat{\mathbf{R}}_0^{-1} \mathbf{Y}_o \right) \end{array} \right], \tag{12.24}$$

and

$$\mathbf{F}_{rr}(\boldsymbol{\theta}) = 2 \, \mathrm{tr} \left(\mathbf{P}^T \mathbf{R}^{-1} \mathbf{P} \right) \mathbf{I}_2. \tag{12.25}$$

On the other hand, we have $\mathbf{F}_{rs}(\boldsymbol{\theta}) = \mathbf{0}$ since the elements of it are linear functions of $\mathbf{Y}_e - a_r\mathbf{P}$ and $\mathbf{Y}_o - a_i\mathbf{P}$, and $\mathrm{E}\left[\mathbf{Y}_e - a_r\mathbf{P}\right] = \mathrm{E}\left[\mathbf{Y}_o - a_i\mathbf{P}\right] = \mathbf{0}$. As a result,

$$\left.\mathbf{J}_{rr}(\boldsymbol{\theta})\right|_{\boldsymbol{\theta}=\hat{\boldsymbol{\theta}}_0} = \left.\mathbf{F}_{rr}^{-1}(\boldsymbol{\theta})\right|_{\boldsymbol{\theta}=\hat{\boldsymbol{\theta}}_0} = \frac{1}{2\operatorname{tr}\left(\mathbf{P}^T\hat{\mathbf{R}}_0^{-1}\mathbf{P}\right)}\mathbf{I}_2. \tag{12.26}$$

Gathering the above results, the Rao test for PHE can be recast as

$$\frac{\operatorname{tr}^2\left(\mathbf{P}^T\hat{\mathbf{R}}_0^{-1}\mathbf{Y}_e\right) + \operatorname{tr}^2\left(\mathbf{P}^T\hat{\mathbf{R}}_0^{-1}\mathbf{Y}_o\right)}{\operatorname{tr}\left(\mathbf{P}^T\hat{\mathbf{R}}_0^{-1}\mathbf{P}\right)} \underset{H_0}{\overset{H_1}{\gtrless}} \eta. \tag{12.27}$$

Plugging $\hat{\gamma}_0 = 1$ into (12.27) leads to the Rao test for the HE as

$$\begin{aligned}
\Lambda =& \frac{\operatorname{tr}^2\left[\mathbf{P}^T\left(\mathbf{Y}_{ps1}\mathbf{Y}_{ps1}^T + \mathbf{Y}_{ps}\mathbf{Y}_{ps}^T\right)^{-1}\mathbf{Y}_e\right]}{\operatorname{tr}\left[\mathbf{P}^T\left(\mathbf{Y}_{ps1}\mathbf{Y}_{ps1}^T + \mathbf{Y}_{ps}\mathbf{Y}_{ps}^T\right)^{-1}\mathbf{P}\right]} \\
&+ \frac{\operatorname{tr}^2\left[\mathbf{P}^T\left(\mathbf{Y}_{ps1}\mathbf{Y}_{ps1}^T + \mathbf{Y}_{ps}\mathbf{Y}_{ps}^T\right)^{-1}\mathbf{Y}_o\right]}{\operatorname{tr}\left[\mathbf{P}^T\left(\mathbf{Y}_{ps1}\mathbf{Y}_{ps1}^T + \mathbf{Y}_{ps}\mathbf{Y}_{ps}^T\right)^{-1}\mathbf{P}\right]} \underset{H_0}{\overset{H_1}{\gtrless}} \eta.
\end{aligned} \tag{12.28}$$

In the sequel, we refer to detectors (12.27) and (12.28) as the persymmetric and symmetric spectrum Rao test for PHE (PSS-RAO-PHE) and PSS-RAO-HE, respectively.

12.3   Two-Step GLRT and Wald Test

In this section, detection schemes are devised resorting to the two-step GLRT and two-step Wald test, which consist in evaluating the conventional GLRT when some parameters are known and then replacing them with suitable estimates.

12.3.1   Homogeneous Environment

1) Two-Step GLRT for known $\mathbf{R}$ *and unknown* $\boldsymbol{\theta}_r$*:*

Under the assumption that $\mathbf{R}$ is known and $\gamma = 1$, the GLRT based on $\mathbf{Y}_{ps}$ is given by

$$\Lambda_0 = \frac{\max_{\{\boldsymbol{\theta}_r\}} f(\mathbf{Y}_{ps}|\boldsymbol{\theta}_r, \mathbf{R}, H_1)}{f(\mathbf{Y}_{ps}|\mathbf{R}, H_0)} \underset{H_0}{\overset{H_1}{\gtrless}} \eta_0. \tag{12.29}$$

Taking the natural logarithm of the left-hand side of (12.29), we can obtain

$$\Lambda_0 = \text{tr}\left(\mathbf{Y}_e^T \mathbf{R}^{-1} \mathbf{Y}_e + \mathbf{Y}_o^T \mathbf{R}^{-1} \mathbf{Y}_o\right) - \min_{\{\boldsymbol{\theta}_r\}} f(\boldsymbol{\theta}_r) \underset{H_0}{\overset{H_1}{\gtrless}} \eta_0, \qquad (12.30)$$

where

$$f(\boldsymbol{\theta}_r) = \text{tr}[(\mathbf{Y}_e - a_r\mathbf{P})^T \mathbf{R}^{-1} (\mathbf{Y}_e - a_r\mathbf{P}) + (\mathbf{Y}_o - a_i\mathbf{P})^T \mathbf{R}^{-1} (\mathbf{Y}_o - a_i\mathbf{P})]. \qquad (12.31)$$

Setting to zero the gradient of $f(\boldsymbol{\theta}_r)$ with respect to a_r and a_i leads to

$$\begin{cases} \hat{a}_r(\mathbf{R}) = \text{tr}(\mathbf{P}^T\mathbf{R}^{-1}\mathbf{Y}_e)/\text{tr}(\mathbf{P}^T\mathbf{R}^{-1}\mathbf{P}), \\ \hat{a}_i(\mathbf{R}) = \text{tr}(\mathbf{P}^T\mathbf{R}^{-1}\mathbf{Y}_o)/\text{tr}(\mathbf{P}^T\mathbf{R}^{-1}\mathbf{P}). \end{cases} \qquad (12.32)$$

Finally, using the above results and replacing $\mathbf{R}$ with $\mathbf{Y}_{ps1}\mathbf{Y}_{ps1}^T$, the GLRT for HE can be recast as

$$\Lambda = \frac{\text{tr}^2\left(\mathbf{P}^T\left(\mathbf{Y}_{ps1}\mathbf{Y}_{ps1}^T\right)^{-1}\mathbf{Y}_e\right) + \text{tr}^2\left(\mathbf{P}^T\left(\mathbf{Y}_{ps1}\mathbf{Y}_{ps1}^T\right)^{-1}\mathbf{Y}_o\right)}{\text{tr}\left(\mathbf{P}^T\left(\mathbf{Y}_{ps1}\mathbf{Y}_{ps1}^T\right)^{-1}\mathbf{P}\right)} \underset{H_0}{\overset{H_1}{\gtrless}} \eta. \qquad (12.33)$$

2) Two-Step GLRT for known $\boldsymbol{\theta}_r$ and unknown $\mathbf{R}$:

Under the assumption that $\boldsymbol{\theta}_r$ is known and $\gamma = 1$, the GLRT based on $\mathbf{Y}_{ps}$ and $\mathbf{Y}_{ps1}$ is given by

$$\Xi_0 = \frac{\max_{\{\mathbf{R}\}} f(\mathbf{Y}_{ps}, \mathbf{Y}_{ps1}|\boldsymbol{\theta}_r, \mathbf{R}, H_1)}{\max_{\{\mathbf{R}\}} f(\mathbf{Y}_{ps}, \mathbf{Y}_{ps1}|\mathbf{R}, H_0)} \underset{H_0}{\overset{H_1}{\gtrless}} \eta_0. \qquad (12.34)$$

It is easy to show that the maximization over $\mathbf{R}$ yields

$$\Xi_1 = \frac{\det\left(\mathbf{T}_0 + \mathbf{Y}_{ps1}\mathbf{Y}_{ps1}^T\right)}{\det\left(\mathbf{T}_1(\boldsymbol{\theta}_r) + \mathbf{Y}_{ps1}\mathbf{Y}_{ps1}^T\right)} \underset{H_0}{\overset{H_1}{\gtrless}} \eta_1. \qquad (12.35)$$

By replacing $\mathbf{T}(\boldsymbol{\theta}_r)$ with $\mathbf{T}(\hat{\boldsymbol{\theta}}_r)$, where $\hat{\boldsymbol{\theta}}_r = \left[\hat{a}_r(\mathbf{Y}_{ps1}\mathbf{Y}_{ps1}^T), \hat{a}_i(\mathbf{Y}_{ps1}\mathbf{Y}_{ps1}^T)\right]$, (12.35) can be written as

$$\Xi = \frac{\det\left(\mathbf{T}_0 + \mathbf{Y}_{ps1}\mathbf{Y}_{ps1}^T\right)}{\det\left(\mathbf{T}_1(\hat{\boldsymbol{\theta}}_r) + \mathbf{Y}_{ps1}\mathbf{Y}_{ps1}^T\right)} \underset{H_0}{\overset{H_1}{\gtrless}} \eta. \qquad (12.36)$$

In the sequel, we refer to detectors (12.33) and (12.36) as the persymmetry and symmetric spectrum GLRT 1 (PSS-GLRT1) and 2 (PSS-GLRT2), respectively.

3) Two-Step Wald for known $\mathbf{R}$:

Let us first assume that the $\mathbf{R}$ is known. The Wald test for known $\mathbf{R}$ has the following expression [9]:

$$\Lambda_0 = \hat{\boldsymbol{\theta}}_{r,1}^T \left[\mathbf{J}_{rr}(\boldsymbol{\theta})\right]^{-1} \hat{\boldsymbol{\theta}}_{r,1} \underset{H_0}{\overset{H_1}{\gtrless}} \eta_0, \tag{12.37}$$

where $\hat{\boldsymbol{\theta}}_{r,1} = [\hat{a}_r(\mathbf{R}), \hat{a}_i(\mathbf{R})]^T$ is the MLE of $\boldsymbol{\theta}_r$ under H_1. According to Section 12.2,

$$\left[\mathbf{J}_{rr}(\boldsymbol{\theta})\right]^{-1} = \mathbf{F}_{rr}(\boldsymbol{\theta}) = 2\,\mathrm{tr}\left(\mathbf{P}^T\mathbf{R}^{-1}\mathbf{P}\right)\mathbf{I}_2. \tag{12.38}$$

As a result, replacing $\mathbf{R}$ with $\mathbf{Y}_{ps1}\mathbf{Y}_{ps1}^T$, (12.37) can be recast as

$$\Lambda = \frac{\mathrm{tr}^2\left(\mathbf{P}^T\left(\mathbf{Y}_{ps1}\mathbf{Y}_{ps1}^T\right)^{-1}\mathbf{Y}_e\right) + \mathrm{tr}^2\left(\mathbf{P}^T\left(\mathbf{Y}_{ps1}\mathbf{Y}_{ps1}^T\right)^{-1}\mathbf{Y}_o\right)}{\mathrm{tr}\left(\mathbf{P}^T\left(\mathbf{Y}_{ps1}\mathbf{Y}_{ps1}^T\right)^{-1}\mathbf{P}\right)} \underset{H_0}{\overset{H_1}{\gtrless}} \eta. \tag{12.39}$$

It can be observed that the two-step Wald test is equivalent to the PSS-GLRT1.

12.3.2 Partially Homogeneous Environment

Since the mathematical derivation of of the plain GLRT and the Wald test for the problem at hand is a formidable task, we resort to the two-step GLRT to devise the adaptive architecture under PHE. In order to facilitate the derivations of the test, we move the power scaling factor from the secondary data to primary data. Precisely, we define $\boldsymbol{\Sigma} = \gamma\mathbf{R}$, which implies that $\mathbf{R} = \lambda\boldsymbol{\Sigma}$ with $\lambda = \frac{1}{\gamma}$, namely,

$$\mathrm{E}\left[\mathbf{y}_{er}\mathbf{y}_{er}^\dagger\right] = \mathrm{E}\left[\mathbf{y}_{ei}\mathbf{y}_{ei}^\dagger\right] = \mathrm{E}\left[\mathbf{y}_{or}\mathbf{y}_{or}^\dagger\right] = \mathrm{E}\left[\mathbf{y}_{oi}\mathbf{y}_{oi}^\dagger\right] = \lambda\boldsymbol{\Sigma},$$
$$\mathrm{E}\left[\mathbf{y}_{ekr}\mathbf{y}_{ekr}^\dagger\right] = \mathrm{E}\left[\mathbf{y}_{eki}\mathbf{y}_{eki}^\dagger\right] = \mathrm{E}\left[\mathbf{y}_{okr}\mathbf{y}_{okr}^\dagger\right] = \mathrm{E}\left[\mathbf{y}_{oki}\mathbf{y}_{oki}^\dagger\right]$$
$$= \boldsymbol{\Sigma}. \tag{12.40}$$

We first assume that $\boldsymbol{\Sigma}$ is known and derive decision rule according to a specific design criterion. Then, an adaptive detector is obtained by substituting $\boldsymbol{\Sigma}$ with the suitable estimates. Under the assumption that $\boldsymbol{\Sigma}$ is known, the GLRT for PHE can be written as

$$\Xi_0 = \frac{\max_{\{\boldsymbol{\theta}_r,\lambda\}} f(\mathbf{Y}_{ps}|\boldsymbol{\theta}_r,\lambda,\boldsymbol{\Sigma},H_1)}{\max_{\{\lambda\}} f(\mathbf{Y}_{ps}|\lambda,\boldsymbol{\Sigma},H_0)} \underset{H_0}{\overset{H_1}{\gtrless}} \eta_0. \tag{12.41}$$

Previous assumptions imply that the PDF of $\mathbf{Y}_{ps}$ under H_l, $l = 0, 1$ is

$$f(\mathbf{Y}_{ps}|l\boldsymbol{\theta}_r,\boldsymbol{\Sigma},H_l) = \frac{1}{(2\pi)^{2N}\det^2(\lambda\boldsymbol{\Sigma})}\exp\left\{-\frac{1}{2}\,\mathrm{tr}\left[\frac{1}{\lambda}\boldsymbol{\Sigma}^{-1}\mathbf{T}_l\right]\right\}. \tag{12.42}$$

It is easy to show that the MLE of λ under H_l, $l = 0, 1$, is given by

$$\begin{cases} \hat{\lambda}_0 = \frac{1}{4N} \operatorname{tr}\left[\boldsymbol{\Sigma}^{-1}\mathbf{T}_0\right], \\ \hat{\lambda}_1 = \frac{1}{4N} \operatorname{tr}\left[\boldsymbol{\Sigma}^{-1}\mathbf{T}_1(\boldsymbol{\theta}_r)\right]. \end{cases} \tag{12.43}$$

Substituting (12.42) and (12.43) into (12.41), after some algebraic manipulations, results in

$$\Xi_0 = \frac{\operatorname{tr}\left(\mathbf{Y}_e^T\boldsymbol{\Sigma}^{-1}\mathbf{Y}_e + \mathbf{Y}_o^T\boldsymbol{\Sigma}^{-1}\mathbf{Y}_o\right)}{\min_{\{\boldsymbol{\theta}_r\}} f'(\boldsymbol{\theta}_r)} \begin{array}{c} H_1 \\ \gtrless \\ H_0 \end{array} \eta_0, \tag{12.44}$$

where

$$f'(\boldsymbol{\theta}_r) = \operatorname{tr}[(\mathbf{Y}_e - a_r\mathbf{P})^T \boldsymbol{\Sigma}^{-1} (\mathbf{Y}_e - a_r\mathbf{P})$$
$$+ (\mathbf{Y}_o - a_i\mathbf{P})^T \boldsymbol{\Sigma}^{-1} (\mathbf{Y}_o - a_i\mathbf{P})]. \tag{12.45}$$

The estimates of a_r and a_i are given by $\hat{a}_r(\boldsymbol{\Sigma})$ and $\hat{a}_i(\boldsymbol{\Sigma})$, respectively. By replacing $\boldsymbol{\Sigma}$ with $\mathbf{Y}_{ps1}\mathbf{Y}_{ps1}^T$, the GLRT can be recast as

$$\Xi = \frac{\operatorname{tr}^2\left(\mathbf{P}^T \left(\mathbf{Y}_{ps1}\mathbf{Y}_{ps1}^T\right)^{-1} \mathbf{Y}_e\right) + \operatorname{tr}^2\left(\mathbf{P}^T \left(\mathbf{Y}_{ps1}\mathbf{Y}_{ps1}^T\right)^{-1} \mathbf{Y}_o\right)}{\operatorname{tr}\left(\mathbf{P}^T \left(\mathbf{Y}_{ps1}\mathbf{Y}_{ps1}^T\right)^{-1} \mathbf{P}\right) \operatorname{tr}\left(\mathbf{Y}_{ps}^T \left(\mathbf{Y}_{ps1}\mathbf{Y}_{ps1}^T\right)^{-1} \mathbf{Y}_{ps}\right)} \begin{array}{c} H_1 \\ \gtrless \\ H_0 \end{array} \eta. \tag{12.46}$$

In what follows, we refer to detector (12.46) as the persymmetric and symmetric spectrum two-step GLRT for PHE (PSS-2SGLRT-PHE).

12.4   Numerical Examples

This section contains illustrative examples aimed at showing the effectiveness of the proposed approach in terms of probability of detection P_d and in comparison with those decision schemes that did not jointly exploit the persymmetric covariance matrix and symmetric clutter spectrum, including conventional unstructured detectors.

Since in this section, an analytical expression for the PD is not available, we resort to the standard MC counting techniques. Specifically, we evaluate the thresholds necessary to ensure a preassigned value of P_{fa} resorting to $100/P_{fa}$ independent trials, while P_d values are estimated over 10^4 independent trials. All the examples assume $P_{fa} = 0.001$ and $N = 10$. We consider a clutter-dominated environment and the (i, j)th element of the clutter covariance matrix of the primary data is chosen as $\mathbf{R}(i, j) = \sigma^2 0.9^{|i-j|}$, where σ^2 is the clutter power. The SNR is defined as $\text{SNR} = |a|^2 \mathbf{v}_m(\theta_{az})^\dagger \mathbf{R}^{-1}\mathbf{v}_m(\theta_{az})$, where θ_{az} is the target azimuthal angle and

$$
\mathbf{v}_m(\theta_{az}) = \left[\exp\left(\jmath\pi \frac{d}{\lambda}(N-1)\cos\theta_{az} \right), \cdots, 1, \right.
$$
$$
\left. \cdots, \exp\left(-\jmath\pi \frac{d}{\lambda}(N-1)\cos\theta_{az} \right) \right] / \sqrt{N},
\tag{12.47}
$$

with d the element spacing, λ the working wavelength.

12.4.1   Homogeneous Environment

First of all, we consider the HE where $\gamma = 1$ and investigate the detection performance of PSS-GLRT1, PSS-GLRT2, and PSS-RAO-HE in comparison with the unstructured GLRT [10], AMF [11], and Rao [7] detectors, as well as the structured counterparts exploiting either persymmetric covariance matrix (PGLR [3] and PS-AMF [12]) or symmetric clutter spectrum (i.e., the SS-AMF, I-GLRT, and SS-RAO, introduced in [13]).

In Fig. 12.1, we plot the P_d curves versus SNR for all considered detectors assuming $K = 2N$. The figure highlights that the PSS-GLRT1 and the PSS-GLRT2 significantly outperform the traditional unstructured detectors and the state-of-the-art counterparts which exploit either the persymmetry or the symmetric spectrum. Precisely, the PSS-GLRT1, PSS-GLRT2, and PSS-RAO-HE exhibit similar performance and outperform the GLRT and PGLR with a gain of about 3.5 dB and 1.5 dB at $P_d = 0.9$, respectively. Fig. 12.2 contains the P_d with the same configurations with Fig. 12.1, except for $K = 12$. Inspection of the figures implies that when $N < K < 2N$, the gain performed by the PSS-GLRT1 over the GLRT and PGLR increases to around 11 dB and 2.5 dB, respectively. It is also important to highlight that the PSS-GLRT2 exhibits better performance than the PSS-GLRT1, whereas the PSS-RAO-HE outperforms all of the counterparts at low SNRs but worse than the PSS-GLRT1 when SNR > 20 dB. The above-mentioned gain is reduced by increasing K due to the fact that the estimate of the covariance matrix becomes more reliable.

Fig. 12.3 deals with the case where $K < N$ and shows the resulting P_d against SNR for two different configurations of the value of K. Specifically, Fig. 12.3(a) depicts the P_d of the structured detectors when $K = 8$ due to the fact that the conventional unstructured detectors cannot work when $K < N$. It can be seen that the PSS-GLRT1, PSS-GLRT2, and PSS-RAO-HE still outperform their counterparts. Meanwhile, when $K = 8$, the PSS-GLRT2 ensures the best performance and the PSS-RAO-HE cannot achieve $P_d = 1$ even at SNR = 40 dB. Thus, joint exploitation of the clutter and system symmetry properties is a very effective means to improve performance in the presence of a small number of secondary data. Due to the fact that only the detectors introduced in this section can work under $K < N/2$, we plot the P_d of PSS-GLRT1, PSS-GLRT2, and PSS-RAO-HE in Fig. 12.3(b) with $K = 4$. Substantially, the hierarchy observed in Fig. 12.2 and 12.3(a) is confirmed with

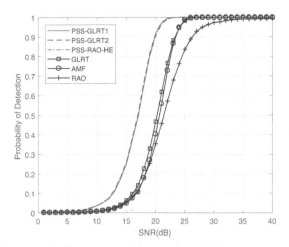

(a) Comparison with unstructured detectors

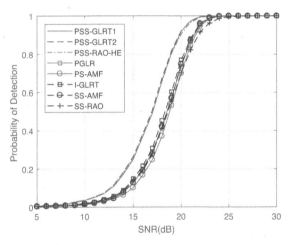

(b) Comparison with structured detectors

FIGURE 12.1
Detection probability versus SNR for $K = 20$, $N = 10$, and $\gamma = 1$.

the difference that the P_d of PSS-RAO-HE does not achieve values greater than 0.35 due to the small number of secondary data.

12.4.2 Partially Homogeneous Environment

In this section, we present some numerical examples to show the performance of the PSS-RAO-PHE and the PSS-2SGLRT-PHE under the PHE. To this

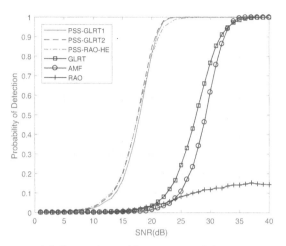

(a) Comparison with unstructured detectors

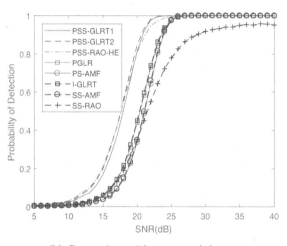

(b) Comparison with structured detectors

FIGURE 12.2
Detection probability versus SNR for $K = 12$, $N = 10$, and $\gamma = 1$.

end, we not only compare our detectors to the symmetric spectrum detectors that neglect the persymmetry, namely, the so-called SS-ACE introduced in [14], but also compare the new receivers with the state-of-the-art persymmetric detectors that ignore the symmetric spectrum, including the persymmetric GLRT for PHE (P-GLRT-PHE) [15], and the persymmetric adaptive coherence estimator (P-ACE) [16]. In the examples, we also include the conventional unstructured ACE [17].

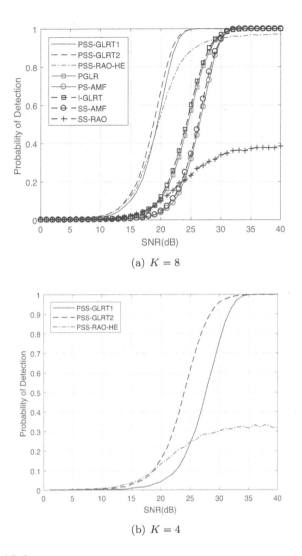

(a) $K = 8$

(b) $K = 4$

FIGURE 12.3
Detection probability versus SNR for $K < N$, $N = 10$, and $\gamma = 1$.

In Fig. 12.4, we plot the P_d versus the SNR for the considered detectors assuming $N = 10$, $K > N$, $\gamma = 5$. The figure shows that the PSS-RAO-PHE and the PSS-2SGLRT-PHE significantly outperform the traditional state-of-the-art counterparts which either ignore the persymmetry or the symmetric spectrum. Precisely, when $K = 20$, the best detection performance is ensured by the PSS-2SGLRT-PHE with a gain of about 0.1 dB at $P_d = 0.9$ over the PSS-RAO-PHE. Such a gain increases to above 2 dB with respect to

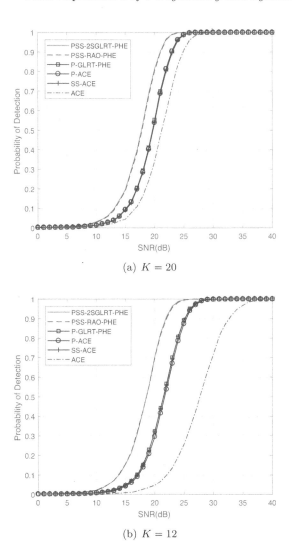

(a) $K = 20$

(b) $K = 12$

FIGURE 12.4
Detection probability versus SNR for $K > N$, $N = 10$, and $\gamma = 5$.

persymmetric detectors and the SS-ACE. Fig. 12.4(b) shows that the gain between the PSS-2SGLRT-PHE and the counterparts increases, which is consistent with what has been observed in the former subsection.

Fig. 12.5 highlights that when $N/2 < K < N$, the PSS-2SGLRT-PHE and PSS-RAO-PHE still outperform their structured counterparts, whereas when $K < N/2$, PSS-2SGLRT-PHE ensures better performance and PSS-RAO-PHE cannot achieve a P_d higher than 0.58 for the considered SNR values.

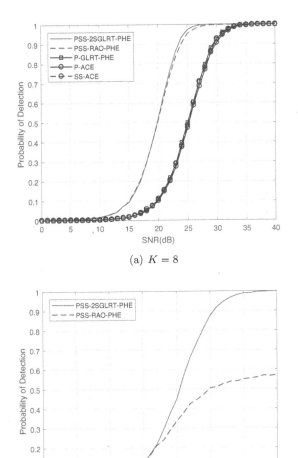

(a) $K = 8$

(b) $K = 4$

FIGURE 12.5
Detection probability versus SNR for $K < N$, $N = 10$, and $\gamma = 5$.

Bibliography

[1] G. Foglia, C. Hao, G. Giunta, and D. Orlando, "Knowledge-aided adaptive detection in partially homogeneous clutter: Joint exploitation of persymmetry and symmetric spectrum," *Digital Signal Processing*, vol. 67, pp. 131–138, 2017.

[2] C. Hao, D. Orlando, G. Foglia, and G. Giunta, "Knowledge-based adaptive detection: Joint exploitation of clutter and system symmetry properties," *IEEE Signal Processing Letters*, vol. 23, no. 10, pp. 1489-1493, October, 2016.

[3] L. Cai and H. Wang, "A persymmetric multiband GLR algorithm," *IEEE Transactions on Aerospace and Electronic Systems*, vol. 28, no. 3, pp. 806–816, July 1992.

[4] E. J. Kelly, "An adaptive detection algorithm," *IEEE Transactions on Aerospace and Electronic Systems*, vol. 22, no. 2, pp. 115–127, March 1986.

[5] F. C. Robey, D. R. Fuhrmann, E. J. Kelly, and R. Nitzberg, "A CFAR adaptive matched filter detector," *IEEE Transactions on Aerospace and Electronic Systems*, vol. 28, no. 1, pp. 208–216, January 1992.

[6] A. De Maio, "A new derivation of the adaptive matched filter," *IEEE Signal Processing Letters*, vol. 11, no. 10, pp. 792–793, October 2004.

[7] ——, "Rao test for adaptive detection in gaussian interference with unknown covariance matrix," *IEEE Transactions on Signal Processing*, vol. 55, no. 7, pp. 3577–3584, 2007.

[8] E. Conte, A. De Maio, and G. Ricci, "GLRT-based adaptive detection algorithms for range-spread targets," *IEEE Transactions on Signal Processing*, vol. 49, no. 7, pp. 1336–1348, July 2001.

[9] S. M. Kay, *Fundamentals of Statistical Signal Processing: Detection Theory*. New Jersey, USA, Prentice Hall, 1998, vol. 2.

[10] E. J. Kelly, "An adaptive detection algorithm," *IEEE Transactions on Aerospace and Electronic Systems*, vol. 22, no. 1, pp. 115–127, March 1986.

[11] F. C. Robey, D. R. Fuhrmann, E. J. Kelly, and R. Nitzberg, "A CFAR adaptive matched filter detector," *IEEE Transactions on Aerospace and Electronic Systems*, vol. 28, no. 1, pp. 208–216, January 1992.

[12] G. Pailloux, P. Forster, J.-P. Ovarlez, and F. Pascal, "Persymmetric adaptive radar detectors," *IEEE Transactions on Aerospace and Electronic Systems*, vol. 47, no. 4, pp. 2376–2390, October 2011.

[13] A. De Maio, D. Orlando, C. Hao, and G. Foglia, "Adaptive detection of point-like targets in spectrally symmetric interference," *IEEE Transactions on Signal Processing*, vol. 64, no. 12, pp. 3207–3220, June 15 2016.

[14] C. Hao, D. Orlando, A. Farina, S. Iommelli, and C. Hou, "Symmetric spectrum detection in the presence of partially homogeneous environment," in *IEEE Radar Conference (RadarConf)*, Philadelphia, PA, USA, May 2016, pp. 1–4.

[15] M. Casillo, A. De Maio, S. Iommelli, and L. Landi, "A persymmetric glrt for adaptive detection in partially-homogeneous environment," *IEEE Signal Processing Letters*, vol. 14, no. 12, pp. 1016–1019, December 2007.

[16] Y. Gao, G. Liao, S. Zhu, X. Zhang, and D. Yang, "Persymmetric adaptive detectors in homogeneous and partially homogeneous environments," *IEEE Transactions on Signal Processing*, vol. 62, no. 2, pp. 331–342, January 15 2014.

[17] E. Conte, M. Lops, and G. Ricci, "Asymptotically optimum radar detection in compound Gaussian noise," *IEEE Transactions on Aerospace and Electronics Systems*, vol. 31, no. 2, pp. 617–625, April 1995.

13

Adaptive Detection after Covariance Matrix Classification

This chapter discusses a radar adaptive detection architecture composed of an ICM structure classifier before a bank of adaptive radar detectors. The former relies on the MOS framework [1,2]. This classifier accounts for six ICM structure classes: Hermitian, centrohermitian, symmetric, centrosymmetric, Hermitian-Toeplitz, and centrotoeplitz ICM structures [3]. This architecture exhibits an improved estimation of covariance matrix, and better detection performance, especially in the presence of a small volume of training samples.

13.1 Problem Formulation

Considering that a radar system receives N echoes from the CUT, then organizes them into an N-dimensional vector after appropriate signal conditioning and sampling operations[1]. More specifically, the echoes with signals can be modeled as

$$\mathbf{z} = \alpha \mathbf{v} + \mathbf{n}, \tag{13.1}$$

where $\alpha \in \mathbb{C}$ is an unknown but deterministic factor accounting for target and channel effects, $\mathbf{v} \in \mathbb{C}^{N \times 1}$ is the target steering vector and $\mathbf{n} \sim \mathcal{CN}_N(\mathbf{0}, \mathbf{M})$ includes both interference and noise effects. As customary, a set of training (secondary) data $\{\mathbf{z}_k\}_{k=1}^{K}$ free of the target echoes is assumed available. Moreover, the training data are supposed IID with $\mathbf{n}$ in the test data.

The target detection problem can be formulated as a binary hypothesis test:

$$\begin{cases} \mathcal{H}_0 : \mathbf{z} = \mathbf{n}, & \mathbf{z}_k = \mathbf{n}_k, \quad k = 1, \dots, K, \\ \mathcal{H}_1 : \mathbf{z} = \alpha \mathbf{v} + \mathbf{n}, & \mathbf{z}_k = \mathbf{n}_k, \quad k = 1, \dots, K, \end{cases} \tag{13.2}$$

where $\mathbf{n}, \mathbf{n}_k \sim \mathcal{CN}_N(\mathbf{0}, \mathbf{M})$ for $k = 1, \dots, K$. Before addressing this problem, we first discuss the structure of ICM.

---

[1]It could represent a spatial snapshot collected by an array of sensors, a sequence of temporal samples from a coherent pulse train, or space-time data.

DOI: 10.1201/9781003340232-13

253

Define $\mathbf{Z} = [\mathbf{z}_1, \ldots, \mathbf{z}_K] \in \mathbb{C}^{N \times K}$ as the entire training dataset. The PDF of $\mathbf{Z}$ is given by

$$f(\mathbf{Z}) = \pi^{-KN} \det(\mathbf{M})^K \exp\left(- \sum_{k=1}^{K} \mathbf{z}_k^\dagger \mathbf{M}^{-1} \mathbf{z}_k \right), \qquad (13.3)$$

where the ICM structure (i.e., $\mathbf{M}$) generally depends on the radar scenario. In particular:

1. ground clutter, observed by a stationary monostatic radar, usually exhibits a symmetrical power spectral density centered around the zero Doppler frequency. The resulting covariance matrix is real and symmetric;

2. symmetrically spaced linear arrays or pulse trains induce a centrohermitian structure in the ICM;

3. systems equipped with uniform spaced linear arrays or pulse trains show, in theory, a Toeplitz structured ICM.

The structure classes taken into account in this chapter are

$$\begin{cases} H_1 : \ \mathbf{M} \in \mathbb{C}^{N \times N} & \text{is Hermitian unstructured,} \\ H_2 : \ \mathbf{M} \in \mathbb{R}^{N \times N} & \text{is symmetric unstructured,} \\ H_3 : \ \mathbf{M} \in \mathbb{C}^{N \times N} & \text{is centrohermitian (persymmetric and Hermitian),} \\ H_4 : \ \mathbf{M} \in \mathbb{R}^{N \times N} & \text{is centrosymmetric (persymmetric and symmetric),} \\ H_5 : \ \mathbf{M} \in \mathbb{C}^{N \times N} & \text{is Toeplitz and Hermitian,} \\ H_6 : \ \mathbf{M} \in \mathbb{R}^{N \times N} & \text{is Toeplitz and symmetric.} \end{cases}$$

$$(13.4)$$

Note that it is possible to identify nested hypotheses among those listed in (13.4) (e.g. $H_2 \subset H_1, H_3 \subset H_1, H_4 \subset H_2, H_5 \subset H_3, H_6 \subset H_4$, etc.). Next, an architecture of adaptive detection after covariance matrix classification is provided. Specifically, the classification of the ICM structure among the six classes in (13.4) is first performed, and then the most suitable detector exploiting the corresponding structure is chosen to make decision whether the target is present or not.

13.2 Architecture Design

In this section, we discuss the detection scheme, which includes the proper MOS classifier and the detector corresponding to the classification result. The architecture's details are described below. The first part deals with the

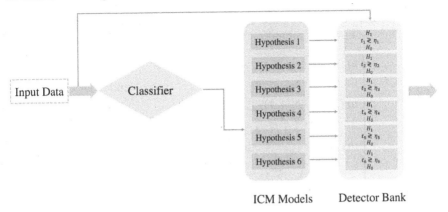

FIGURE 13.1
Overall architecture scheme, including both the classification and the detection stages.

classification stage: in order to select among the hypotheses of (13.4), the classifier exploits the asymptotic BIC. Once a hypothesis on the ICM is selected, the corresponding decision statistics and thresholds (t_i and η_i, respectively) are exploited. In respect of the detection stage, instead, the GLRT and its modifications are included in the detector bank. The overall processing scheme is shown in Figure 13.1.

13.2.1 Classification Stage

In [1] several MOS rules have been devised to perform the covariance structure selection. Among them, the BIC has been singled out as the classifier in order to guarantee a good tradeoff between performance, complexity, and stability. Its goal is achieved by using only training (secondary) data.

The asymptotic BIC can be expressed as:

$$\hat{H} = \arg \min_{H_1,\ldots,H_6} \{-2\ln f(\mathbf{Z}; \hat{\mathbf{M}}_i, H_i) + m_i \ln(K)\}, \qquad (13.5)$$

where m_i is the number of unknown parameters under the hypothesis H_i, and $\hat{\mathbf{M}}_i$ is the MLE of $\mathbf{M}$ under the hypothesis H_i ($i = 1, \ldots, 6$), and

$$\ln f(\mathbf{Z}; \hat{\mathbf{M}}_i, H_i) = -K[N\ln\pi + \ln\det(\hat{\mathbf{M}}_i)] - \mathrm{tr}(\hat{\mathbf{M}}_i^{-1}\mathbf{S}), \qquad (13.6)$$

with $\mathbf{S} = \mathbf{ZZ}^\dagger$ being the SCM up to a scalar. It is easy to check that the number of unknown parameters under each hypothesis H_i is given by

$$
\begin{cases}
m_1 = N^2 & \text{under } H_1, \\
m_2 = N(N+1)/2 & \text{under } H_2, \\
m_3 = N(N+1)/2 & \text{under } H_3, \\
m_4 = \begin{cases} \dfrac{N}{2}\left(\dfrac{N}{2}+1\right) & \text{if } N \text{ is even} \\ \left(\dfrac{N+1}{2}\right)^2 & \text{if } N \text{ is odd} \end{cases} & \text{under } H_4, \\
m_5 = 2N-1 & \text{under } H_5, \\
m_6 = N & \text{under } H_6.
\end{cases}
\tag{13.7}
$$

In [1] the ICM estimates were provided for the hypothesises $H_i, i = 1, 2, 3, 4$. Under the hypothesis H_5, we adopt an EM algorithm given in [4] to obtain the ICM estimate $\hat{\mathbf{M}}_5$. Further, the ICM estimate $\hat{\mathbf{M}}_6$ can be obtained by selecting the real part of $\hat{\mathbf{M}}_5$. In summary, the suitable estimate of the ICM under each hypothesis can be written as

$$
\begin{cases}
\hat{\mathbf{M}}_1 = \frac{1}{K}\mathbf{S} & \text{under } H_1, \\
\hat{\mathbf{M}}_2 = \frac{1}{K}\mathfrak{Re}[\mathbf{S}] & \text{under } H_2, \\
\hat{\mathbf{M}}_3 = \frac{1}{2K}(\mathbf{S}+\mathbf{JS}^*\mathbf{J}) & \text{under } H_3, \\
\hat{\mathbf{M}}_4 = \frac{1}{2K}\mathfrak{Re}[\mathbf{S}+\mathbf{JS}^*\mathbf{J}] & \text{under } H_4, \\
\hat{\mathbf{M}}_5 \text{ obtained by the EM algorithm} & \text{under } H_5, \\
\hat{\mathbf{M}}_6 = \mathfrak{Re}[\hat{\mathbf{M}}_5] & \text{under } H_6,
\end{cases}
\tag{13.8}
$$

where a detailed description of the EM algorithm can be found in [4].

13.2.2 Detection Stage

According to the classifier's choice, the corresponding decision scheme for the detection problem (13.2) can be selected as follows.

13.2.2.1 Detector under H_1

For the hypothesis H_1, a well-known detector is the GLRT [5], whose expression is

$$
t_1 = \frac{|\mathbf{v}^\dagger \hat{\mathbf{M}}_1^{-1} \mathbf{z}|^2}{(\mathbf{v}^\dagger \hat{\mathbf{M}}_1^{-1} \mathbf{v})(1 + \mathbf{z}^\dagger \hat{\mathbf{M}}_1^{-1} \mathbf{z})} \underset{\mathcal{H}_0}{\overset{\mathcal{H}_1}{\gtrless}} \eta_1.
\tag{13.9}
$$

13.2.2.2 Detector under H_2

When the hypothesis H_2 is selected, the symmetric spectrum adaptive matched filter is used [6]:

$$t_2 = \frac{T_1(\hat{\mathbf{M}}_2) + T_2(\hat{\mathbf{M}}_2)}{\mathbf{v}^\dagger \hat{\mathbf{M}}_2^{-1} \mathbf{v}} \underset{\mathcal{H}_0}{\overset{\mathcal{H}_1}{\gtrless}} \eta_2, \tag{13.10}$$

where:

$$T_1(\hat{\mathbf{M}}_2) = (\mathfrak{Re}[\mathbf{v}]^T \hat{\mathbf{M}}_2^{-1} \mathfrak{Re}[\mathbf{z}] + \mathfrak{Im}[\mathbf{v}]^T \hat{\mathbf{M}}_2^{-1} \mathfrak{Im}[\mathbf{z}])^2, \tag{13.11}$$

$$T_2(\hat{\mathbf{M}}_2) = (\mathfrak{Re}[\mathbf{v}]^T \hat{\mathbf{M}}_2^{-1} \mathfrak{Im}[\mathbf{z}] + \mathfrak{Im}[\mathbf{v}]^T \hat{\mathbf{M}}_2^{-1} \mathfrak{Re}[\mathbf{z}])^2. \tag{13.12}$$

13.2.2.3 Detector under H_3

If $\hat{H} = H_3$, the ICM is centrohermitian, and the corresponding detector can be written as [7]

$$t_3 = \frac{\mathbf{v}^\dagger \hat{\mathbf{M}}_3^{-1} \mathbf{X}_p [\mathbf{I} + \mathbf{X}_p^\dagger \hat{\mathbf{M}}_3^{-1} \mathbf{X}_p]^{-1} \mathbf{X}_p^\dagger \hat{\mathbf{M}}_3^{-1} \mathbf{v}}{\mathbf{v}^\dagger \hat{\mathbf{M}}_3^{-1} \mathbf{v}} \underset{\mathcal{H}_0}{\overset{\mathcal{H}_1}{\gtrless}} \eta_3, \tag{13.13}$$

where $\mathbf{X}_p = [\mathbf{z}_e \ \mathbf{z}_o]$ with

$$\begin{cases} \mathbf{z}_e = \dfrac{\mathbf{z} + \mathbf{J}\mathbf{z}^*}{2}, \\ \mathbf{z}_o = \dfrac{\mathbf{z} - \mathbf{J}\mathbf{z}^*}{2}. \end{cases} \tag{13.14}$$

13.2.2.4 Detector under H_4

The GLRT for centrosymmetric ICM can be written as [8]

$$t_4 = \frac{\text{tr}^2(\mathbf{V}^T \hat{\mathbf{M}}_4^{-1} \mathbf{Z}_e) + \text{tr}^2(\mathbf{V}^T \hat{\mathbf{M}}_4^{-1} \mathbf{Z}_o)}{\text{tr}(\mathbf{V}^T \hat{\mathbf{M}}_4^{-1} \mathbf{V})} \underset{\mathcal{H}_0}{\overset{\mathcal{H}_1}{\gtrless}} \eta_4, \tag{13.15}$$

where

$$\begin{cases} \mathbf{V} = [\mathfrak{Re}\{\mathbf{v}\} \ \mathfrak{Im}\{\mathbf{v}\}], \\ \mathbf{Z}_e = [\mathfrak{Re}\{\mathbf{z}_e\} \ \mathfrak{Im}\{\mathbf{z}_e\}], \\ \mathbf{Z}_o = [\mathfrak{Re}\{\mathbf{z}_o\} \ \mathfrak{Im}\{\mathbf{z}_o\}]. \end{cases} \tag{13.16}$$

13.2.2.5 Detector under H_5

Under the hypothesis H_5, the GLRT for Toeplitz and Hermitian ICM can be written as

$$t_5 = \frac{|\mathbf{v}^\dagger \hat{\mathbf{M}}_5^{-1} \mathbf{z}|^2}{\mathbf{v}^\dagger \hat{\mathbf{M}}_5^{-1} \mathbf{v}} \underset{\mathcal{H}_0}{\overset{\mathcal{H}_1}{\gtrless}} \eta_5, \tag{13.17}$$

where we resort to the EM algorithm derived in [4] to come up with the MLE $\hat{\mathbf{M}}_5$.

13.2.2.6   Detector under H_6

When $\hat{H} = H_6$, the GLRT can be given by

$$t_6 = \frac{|\mathbf{v}^\dagger \hat{\mathbf{M}}_6^{-1} \mathbf{z}|^2}{\mathbf{v}^\dagger \hat{\mathbf{M}}_6^{-1} \mathbf{v}} \underset{\mathcal{H}_0}{\overset{\mathcal{H}_1}{\gtrless}} \eta_6, \tag{13.18}$$

where $\hat{\mathbf{M}}_6 = \mathfrak{Re}[\hat{\mathbf{M}}_5]$.

13.2.3   Threshold Setting

The various detection thresholds are obtained by exploiting the first of the two approaches presented in [2]. Precisely, this method is based on an approximation regarding the false alarm probability function when a sufficient number of training samples guarantees reliable classification performance.

For the system of first classification and then detection proposed in this chapter, the problem of setting the detection threshold $\eta_1, ..., \eta_6$ of each detection scheme needs to be solved. For optional H_i, we have

$$P_{fa} = \sum_{i=1}^{6} P_{fa|H_i} P(H_i), \tag{13.19}$$

where $P_{fa|H_i}$ represents the PFA for given H_i and $P(H_i)$ refers to probability of occurrence of H_i. In order to omit discussion of $P(H_i)$ assumptions, we can fix P_{fa} to a specific value γ by setting $P_{fa|H_i} = \gamma$, with $i = 1, ..., 6$. Below we discuss choosing appropriate thresholds such that each detector has $P_{fa|H_i} = \gamma$.

When $K \geqslant \frac{3}{2}m$, the asymptotic BIC $P_{cc|H_i}$ is approximately 1. So the following approximate equation can be derived:

$$\begin{aligned}
P_{fa|H_i} &= \sum_{k=1}^{6} P\left(\mathcal{H}_1 \mid \mathcal{H}_0, H_i, \hat{H} = H_k\right) P\left(\hat{H} = H_k \mid H_i\right) \\
&= \sum_{k=1}^{6} P\left(t_k > \eta_k \mid \mathcal{H}_0, H_i, \hat{H} = H_k\right) P\left(\hat{H} = H_k \mid H_i\right) \\
&\approx P\left(t_i > \eta_i \mid \mathcal{H}_0, H_i, \hat{H} = H_i\right) P\left(\hat{H} = H_i \mid H_i\right) \\
&= P\left(t_i > \eta_i, \hat{H} = H_i \mid \mathcal{H}_0, H_i\right) \\
&\approx P\left(t_i > \eta_i \mid \mathcal{H}_0, H_i\right).
\end{aligned} \tag{13.20}$$

Consequently, the various thresholds η_i can be set by closed-form expressions (when available) or, otherwise, by MC simulations for each detector such that $P(t_i > \eta_i | \mathcal{H}_0, H_i) = \gamma$, with $i = 1, ..., 6$ and γ being a preassigned PFA.

13.3 Numerical Results

This section is dedicated to analyzing the architecture described in the previous section. Probability of correct classification (P_{cc}) under each hypothesis is obtained from 1000 independent trials. The detection thresholds (or the actual probability of false alarm (P_{fa})) and the probability of detection (P_d) are obtained from $100/(\text{nominal } P_{fa})$ and 5000 independent trials, respectively.

We generate the interference as a circular complex normal random vector with covariance matrix

$$\mathbf{M}_i = \mathbf{A}_i \mathbf{R}_i \mathbf{A}_i^\dagger + \sigma_n^2 \mathbf{I}, \quad i = 1, \ldots, 6, \tag{13.21}$$

where σ_n^2 denotes the white noise power. In order to distinguish between different hypotheses, we use $\mathbf{R}_i, i = 1, 2, 5, 6$, to represent the clutter contributions and incorporates the clutter power, and $\mathbf{A}_i, i = 1, 2, 5, 6$, accounts for matrix factor modeling possible array channel errors (amplification and/or delay errors, calibration residuals, and mutual coupling); $\mathbf{R}_i, i = 3, 4$, represents the power matrix for different interference signal, and $\mathbf{A}_i, i = 3, 4$, accounts for interference signals with different Doppler frequencies. The specific forms of $\mathbf{R}_i$ and $\mathbf{A}_i$ are given as below.

We generate different interference sources by using the matrix $\mathbf{R}_i$ with the following expressions [2]

$$\begin{cases} \mathbf{R}_i(h, k) = \sigma_c^2 \rho^{|h-k|} e^{j2\pi(h-k)f}, & i = 1, 2, 5, 6, \\ \mathbf{R}_i = \sigma_c^2 \mathbf{I}_s, & i = 3, 4, \end{cases} \tag{13.22}$$

where $\sigma_c^2 > 0$ is the clutter power, ρ is the one-lag correlation coefficient, and f is the normalized Doppler frequency. We define the CNR as $\text{CNR} = \sigma_c^2/\sigma_n^2$. We choose different parameters to deal with different hypotheses:

- under H_1: $\mathbf{A}_1 = \mathbf{I} + \sigma_d \mathbf{W}_1$, $f \neq 0$, where $\sigma_d = 0.15$, $\mathbf{W}_1(h, k) \sim \mathcal{CN}_1(0, 1)$ and are IID;

- under H_2: $\mathbf{A}_2 = \mathbf{I} + \sigma_d \mathbf{W}_2$, $f = 0$, where $\sigma_d = 0.15$, $\mathbf{W}_2(h, k) \sim \mathcal{N}_1(0, 1)$ and are IID;

- under H_3: $\mathbf{A}_3 = [\mathbf{a}(f_1), \ \mathbf{a}(f_2), \ \mathbf{a}(f_3), \ \mathbf{a}(f_4)]$, where $[f_1, f_2, f_3, f_4] = [0.285, 0.005, 0.125, 0.25]$, and

$$\mathbf{a}(f_t) = \frac{1}{\sqrt{N}}[e^{-j2\pi f_t \frac{N-1}{2}} \ \cdots \ e^{-j2\pi f_t} \ 1 \ e^{j2\pi f_t} \ \cdots \ e^{j2\pi f_t \frac{N-1}{2}}], \tag{13.23}$$

  for $t = 1, \ldots, 4$;

- under H_4: $\mathbf{M}_4 = \Re \left[\mathbf{A}_3 \mathbf{R}_3 \mathbf{A}_3^\dagger \right] + \sigma_n^2 \mathbf{I}$;

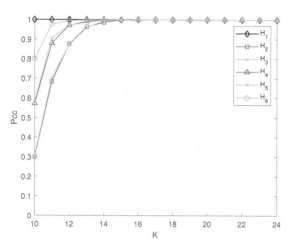

FIGURE 13.2
P_{cc} versus K assuming $N = 10$ for $H_i, i = 1, \ldots, 6$.

TABLE 13.1
Classification probability of the asymptotic BIC for different K.

	H_1	H_2	H_3	H_4	H_5	H_6
$K = 10$	100.0%	30.00%	30.70%	57.20%	58.20%	80.00%
$K = 11$	100.0%	68.30%	70.20%	87.90%	90.50%	90.70%
$K = 12$	100.0%	87.60%	87.30%	97.10%	97.10%	99.40%

- under H_5: $\mathbf{A}_5 = \mathbf{I}$, $f \neq 0$;

- under H_6: $\mathbf{A}_6 = \mathbf{I}$, $f = 0$.

As to the target parameters, we define the SINR as

$$\text{SINR} = |\alpha|^2 \mathbf{v}^\dagger \mathbf{M}_i^{-1} \mathbf{v}, \quad i = 1, \ldots, 6, \quad\quad (13.24)$$

where the signal steering vector $\mathbf{v}$ is chosen as $\mathbf{v} = \mathbf{a}(0.01)$ by assuming N odd.

First, we study the classification performance of asymptotic BIC. In Fig. 13.2, we plot P_{cc} versus K for $N = 10$ under $H_i, i = 1, \ldots, 6, f = 0.285$ (whenever $f \neq 0$), CNR $= 30$ dB, and $\rho = 0.95$. As shown, the application of the asymptotic BIC in the classification stage can distinguish between the Toeplitz structure and other structures. When K increases, the performance of classifier improves. For $H_i, i = 1, \ldots, 6$, we can see that $P_{cc} \approx 1$ when $K \geqslant 14$. More specifically, the classification probability statistics for different K are shown in Table 13.1. It has to be emphasized that for $H_i, i = 1, \ldots, 6$,

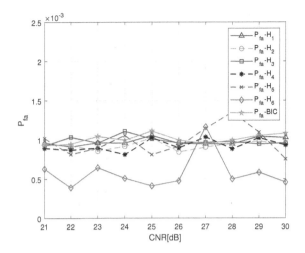

FIGURE 13.3
P_{fa} versus CNR assuming $N = 10$ for $H_i, i = 1, \ldots, 6$, and nominal $P_{fa} = 10^{-3}$.

the values of K are required to be 9, 13, 13, 12, 11, 11, respectively, to satisfy $P_{cc} > 90\%$.

Fig. 13.3 shows the actual P_{fa} as a function of CNR, where $N = 10, K = 18$, and nominal $P_{fa} = 10^{-3}$. The number of MC runs is $1000/(\text{nominal } P_{fa})$. It can be observed that the proposed detection architecture exhibits a quasi-CFAR property, namely, the actual P_{fa} is very close to the nominal one when the CNR varies. Note that it does not possess the CFAR property in H_5 and H_6 because of the exploitation of Toeplitz in (13.17) and (13.18).

Fig. 13.4 refers to the classification results of asymptotic BIC for $K = 10$. Other parameters are the same as those in Fig. 13.2. From Fig. 13.4, we can clearly observe what misclassification occurs. We take Fig. 13.4(e) as an example. The probability of correct classification in this case is 99.4%, the probability of misclassifying H_5 as H_1 is 0.3%, and the probability of misclassifying H_5 as H_3 is 0.3%. In addition, H_5 is never classified to be H_2 or H_4. Inspection of Fig. 13.4 reveals that the Toeplitz structure can be classified well in these parameter settings.

In Fig. 13.5, we plot P_d versus SINR under H_5 and H_6 for $N = 10, f = 0.285$, CNR $= 30$ dB, and $\rho = 0.95$. The normal P_{fa} is set to be 10^{-3}. For comparison purposes, we consider the Kelly's GLRT [5] and the AMF [9,10]. We can observe from Fig. 13.5 that our proposed architecture outperforms its counterparts, especially in the case of limited training data.

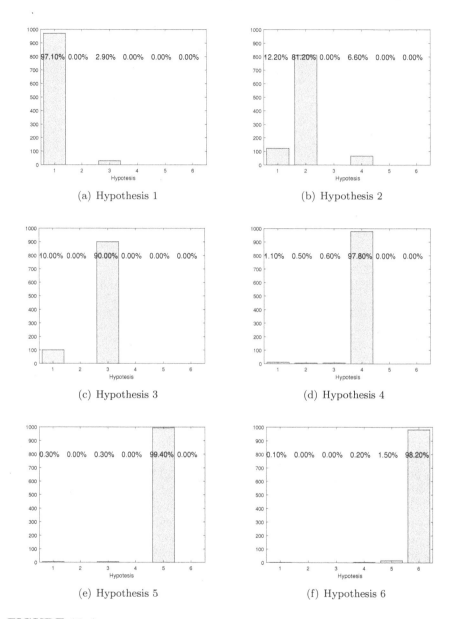

(a) Hypothesis 1

(b) Hypothesis 2

(c) Hypothesis 3

(d) Hypothesis 4

(e) Hypothesis 5

(f) Hypothesis 6

FIGURE 13.4
Classification probability of the asymptotic BIC, for $N = 10$ and $K = 10$.

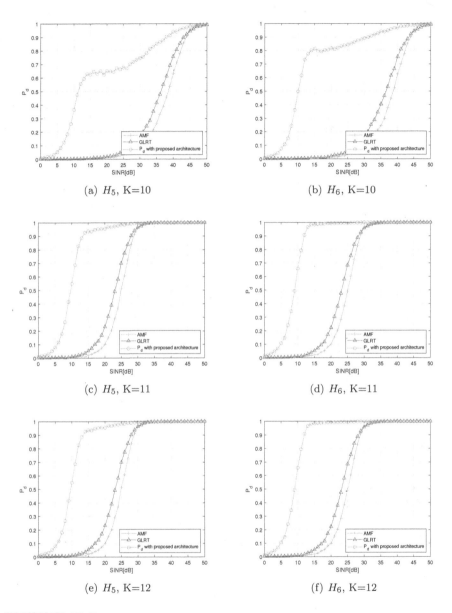

(a) H_5, K=10

(b) H_6, K=10

(c) H_5, K=11

(d) H_6, K=11

(e) H_5, K=12

(f) H_6, K=12

FIGURE 13.5
Performance comparisons for $N = 10, K = 10, 11, 12$.

Bibliography

[1] V. Carotenuto, A. De Maio, D. Orlando, and P. Stoica, "Model order selection rules for covariance structure classification in radar," *IEEE*

Transactions on Signal Processing, vol. 65, no. 20, pp. 5305–5317, October 2017.

[2] V. Carotenuto, A. De Maio, D. Orlando, and P. Stoica, "Radar detection architecture based on interference covariance structure classification," *IEEE Transactions on Aerospace and Electronic Systems*, vol. 55, no. 2, pp. 607–618, April 2019.

[3] J. Liu, Z. Gao, Z. Sun, T. Jian, and W. Liu, "Detection architecture with improved classification capabilities for covariance structures," *Digital Signal Processing*, vol. 123, 103404 2022.

[4] M. J. Turmon and M. I. Miller, "Maximum-likelihood estimation of complex sinusoids and Toeplitz covariances," *IEEE Transactions on Signal Processing*, vol. 42, no. 5, pp. 1074–1086, May 1994.

[5] E. J. Kelly, "An adaptive detection algorithm," *IEEE Transactions on Aerospace and Electronic Systems*, vol. 22, no. 1, pp. 115–127, March 1986.

[6] A. De Maio, D. Orlando, C. Hao, and G. Foglia, "Adaptive detection of point-like targets in spectrally symmetric interference," *IEEE Transactions on Signal Processing*, vol. 64, no. 12, pp. 3207–3220, December 2016.

[7] L. Cai and H. Wang, "A persymmetric multiband GLR algorithm," *IEEE Transactions on Aerospace and Electronic Systems*, vol. 28, no. 3, pp. 806–816, July 1992.

[8] C. Hao, D. Orlando, G. Foglia, and G. Giunta, "Knowledge-based adaptive detection: Joint exploitation of clutter and system symmetry properties," *IEEE Signal Processing Letters*, vol. 23, no. 10, pp. 1489–1493, Oct 2016.

[9] F. C. Robey, D. R. Fuhrmann, E. J. Kelly, and R. Nitzberg, "A CFAR adaptive matched filter detector," *IEEE Transactions on Aerospace and Electronic Systems*, vol. 28, no. 1, pp. 208–216, January 1992.

[10] A. De Maio, "A new derivation of the adaptive matched filter," *IEEE Signal Processing Letters*, vol. 11, no. 10, pp. 792–793, October 2004.

14

MIMO Radar Target Detection

MIMO concept, which stems from communications, is widely used in radar [1–5]. MIMO radar falls into two classes: colocated radar and distributed radar. The former has colocated antennas [6–10], which uses waveform diversity to design flexible beampattern [11,12], enhance detection performance [13], and improve the capability of rejecting interference [14,15], and so on. The latter has widely distributed antennas [16–19], which produces spatial diversity to reduce the scintillations of RCSs of targets, and hence achieves good performance in several aspects (e.g., detection probability [20,21] and estimation accuracy [22]).

It has been shown in [23] that colocated MIMO radar can directly use adaptive techniques. In other words, target detection and parameter estimation can be done without secondary data or even range compression. In [14], the parameter estimation problem was addressed for the case of a point-like target in MIMO radar with unknown disturbance covariance matrix. In order to suppress strong jammer, a GLRT detector was derived without resorting to training data. One of the attractive properties of this GLRT detector is CFAR with respect to the covariance matrix. When multiple targets exist, an iterative GLRT detector was developed in MIMO radar [24]. Different from the GLRT detector, Rao and Wald tests were derived in [25] without requiring training data. In [26], a tunable detector is proposed in MIMO radar by introducing a tunable parameter, which includes the Rao and Wald tests as special cases. Particularly, the robustness of the tunable detector can be flexibly adjusted by tuning the parameter.

In this chapter, we discuss the target detection problems for MIMO radar:

- We consider the problem of adaptive detection by exploiting the persymmetric structure of disturbance covariance matrix in colocated MIMO radar. A persymmetric adaptive detector is designed according to the GLRT principle [10,29], and expressions for the PFA and detection probability are derived.

- We also consider the adaptive detection problem in colored Gaussian noise with unknown persymmetric covariance matrix in MIMO radar with spatially dispersed antennas. To this end, a set of secondary data for each transmit-receive pair is assumed to be available. Distributed MIMO versions of the persymmetric GLRT detector and the persymmetric SMI detector are designed [19]. Compared to the first detector, the second

DOI: 10.1201/9781003340232-14

detector has a simple form and is computationally more efficient. Numerical examples are provided to demonstrate that the designed two detection algorithms can significantly alleviate the requirement of the amount of secondary data and allow for a noticeable improvement in detection performance.

14.1 Persymmetric Detection in Colocated MIMO Radar

14.1.1 Problem Formulation

Consider a MIMO radar system consisting of M colocated transmit antennas and $\tilde{N}$ colocated receive antennas. The baseband signal sent by the mth transmit antenna is represented by $s_m(k)$, $k = 1, 2, \ldots K$, where K is the sample number. Define

$$\mathbf{s}_k \triangleq [s_1(k), s_2(k), \ldots, s_M(k)]^T \in \mathbb{C}^M. \tag{14.1}$$

Assuming that the signal is narrowband, we can write the signal at a target location as

$$\sum_{m=1}^{M} \exp(-\jmath 2\pi \bar{f}_0 \xi_m(\theta)) s_m(k) = \mathbf{a}_t^\dagger \mathbf{s}_k, k = 1, 2, \ldots, K, \tag{14.2}$$

where $\xi_m(\theta)$ is the time required by the signal omitted by the mth transmit antenna to arrive at the target, $\bar{f}_0$ is the carrier frequency, θ denotes target location parameters (e.g., its DOA), and

$$\mathbf{a}_t \triangleq [\exp(\jmath 2\pi \bar{f}_0 \xi_1(\theta)), \ldots, \exp(\jmath 2\pi \bar{f}_0 \xi_M(\theta))]^T \in \mathbb{C}^M \tag{14.3}$$

is called the transmit steering vector.

Let $x_k(n, p)$ be the signal received by the nth receive antenna at the pth pulse, and define

$$\mathbf{x}_k(p) \triangleq [x_k(1, p), x_k(2, p), \ldots, x_k(\tilde{N}, p)]^T \in \mathbb{C}^{\tilde{N}}, \tag{14.4}$$

where $k = 1, 2, \ldots, K$, and $p = 1, 2, \ldots, P$ with P being the total number of pulses in a CPI. In the point-like target cases, $\mathbf{x}_k(p)$ is given by

$$\mathbf{x}_k(p) = a \exp(\jmath 2\pi f_p) \tilde{\mathbf{a}}_1 \mathbf{a}_t^\dagger \mathbf{s}_k + \mathbf{n}_k(p), \tag{14.5}$$

where a is an unknown complex scalar accounting for the target reflectivity and the channel propagation effects, f_p is the Doppler frequency of the target in the pth pulse, $\mathbf{n}_k(p) \in \mathbb{C}^{\tilde{N}}$ denotes the disturbance or noise term, and

$$\tilde{\mathbf{a}}_1 \triangleq [\exp(\jmath 2\pi \bar{f}_0 \bar{\xi}_1(\theta)), \ldots, \exp(\jmath 2\pi \bar{f}_0 \bar{\xi}_{\tilde{N}}(\theta))]^T \in \mathbb{C}^{\tilde{N}}, \tag{14.6}$$

with $\bar{\xi}_n(\theta)$ representing the time required by the signal reflected by the target located at θ to arrive at the nth receive antenna.

Further, the received data within the CPI can be written as

$$
\begin{aligned}
\mathbf{x}_k &= [\mathbf{x}_k^T(1), \mathbf{x}_k^T(2), \ldots, \mathbf{x}_k^T(P)]^T \in \mathbb{C}^{\tilde{N}P} \\
&= a\mathbf{a}_r \mathbf{a}_t^\dagger \mathbf{s}_k + \mathbf{n}_k,
\end{aligned}
\tag{14.7}
$$

where $\mathbf{a}_r = \tilde{\mathbf{a}}_2 \otimes \tilde{\mathbf{a}}_1 \in \mathbb{C}^{\tilde{N}P}$ is the receive steering vector with

$$
\tilde{\mathbf{a}}_2 \triangleq [\exp(\jmath 2\pi f_1), \ldots, \exp(\jmath 2\pi f_P)]^T \in \mathbb{C}^P,
\tag{14.8}
$$

and

$$
\mathbf{n}_k = [\mathbf{n}_k^T(1), \mathbf{n}_k^T(2), \ldots, \mathbf{n}_k^T(P)]^T \in \mathbb{C}^{\tilde{N}P}.
\tag{14.9}
$$

As in [13, 14, 23], we assume that the disturbance $\mathbf{n}_k$ is IID and has a circularly symmetric, complex Gaussian distribution with zero mean and covariance matrix $\mathbf{R}$, i.e., $\mathbf{n}_k \sim \mathcal{CN}_{\tilde{N}P}(\mathbf{0}, \mathbf{R})$ for $k = 1, 2, \ldots, K$. The problem considered herein involves structured $\mathbf{R}$ and $\mathbf{a}_r$. More precisely, $\mathbf{R}$ is persymmetric, i.e.,

$$
\mathbf{R} = \mathbf{J}\mathbf{R}^*\mathbf{J},
\tag{14.10}
$$

where $\mathbf{J}$ is defined in (1.2). In addition, the receive steering vector is also persymmetric, satisfying $\mathbf{a}_r = \mathbf{J}\mathbf{a}_r^*$. The above structures of $\mathbf{R}$ and $\mathbf{a}_r$ are valid when the MIMO radar receiver uses a symmetrically spaced linear array and/or symmetrically spaced pulse trains. As a specific example, the steering vector $\mathbf{a}_r$ in the case of uniformly linear array with a single pulse has the form

$$
\mathbf{a}_r = \left[e^{-\jmath 2\pi \bar{f} \frac{(\tilde{N}-1)}{2}}, \ldots, e^{-\jmath 2\pi \bar{f} \frac{1}{2}}, \; e^{\jmath 2\pi \bar{f} \frac{1}{2}}, \ldots, e^{\jmath 2\pi \bar{f} \frac{(\tilde{N}-1)}{2}} \right]^T,
\tag{14.11}
$$

for even $\tilde{N}$, and

$$
\mathbf{a}_r = \left[e^{-\jmath 2\pi \bar{f} \frac{(\tilde{N}-1)}{2}}, \ldots, e^{-\jmath 2\pi \bar{f}}, \; 1, \; e^{\jmath 2\pi \bar{f}}, \ldots, e^{\jmath 2\pi \bar{f} \frac{(\tilde{N}-1)}{2}} \right]^T,
\tag{14.12}
$$

for odd $\tilde{N}$, where $\bar{f}$ is a quantity relative to the location parameters of the target.

Arranging the received data $\mathbf{x}_k$'s in matrix form, we have

$$
\mathbf{X} = a\mathbf{a}_r\mathbf{a}_t^\dagger \mathbf{S} + \mathbf{N},
\tag{14.13}
$$

where

$$
\mathbf{X} = [\mathbf{x}_1, \mathbf{x}_2, \ldots, \mathbf{x}_K] \in \mathbb{C}^{N \times K},
\tag{14.14}
$$

$$
\mathbf{S} = [\mathbf{s}_1, \mathbf{s}_2, \ldots, \mathbf{s}_K] \in \mathbb{C}^{M \times K},
\tag{14.15}
$$

and

$$
\mathbf{N} = [\mathbf{n}_1, \mathbf{n}_2, \ldots, \mathbf{n}_K] \in \mathbb{C}^{N \times K},
\tag{14.16}
$$

with

$$N = \tilde{N}P. \tag{14.17}$$

Note that the data model in (14.13) is widely adopted in MIMO radar [11–15, 23–25, 27]. We aim to decide the presence or absence of a target based on the observed data $\mathbf{X}$.

We can formulate the detection problem as a binary hypothesis test of H_1 against H_0:

$$\begin{cases} H_0 : \mathbf{X} = \mathbf{N}, \\ H_1 : \mathbf{X} = a\mathbf{a}_r\mathbf{a}_t^\dagger\mathbf{S} + \mathbf{N}, \end{cases} \tag{14.18}$$

where a and $\mathbf{R}$ are both unknown. For mathematical tractability, we assume $K \geqslant \lceil \frac{N}{2} + 1 \rceil$. This condition is less restrictive than the one (i.e., $K \geqslant N + 1$) required in [14, 25, 26]. At the end of the next section, we will give the reasons why this condition has to be satisfied for the detection problem considered here.

It should be emphasized that $\mathbf{R}$ and $\mathbf{a}_r$ have persymmetric structures in (14.18), i.e., $\mathbf{R} = \mathbf{J}\mathbf{R}^*\mathbf{J}$ and $\mathbf{a}_r = \mathbf{J}\mathbf{a}_r^*$. The persymmetric structures imposed on $\mathbf{R}$ and $\mathbf{a}_r$ make the detection problem considered here different from the one in [14, 25, 26]. In the following, we design an adaptive detector for the detection problem (14.18) by exploiting the persymmetric structures.

14.1.2 Adaptive Detector

In this section, we design an adaptive detector according to GLRT criterion for the detection problem (14.18) by exploiting the persymmetric structures.

Define

$$\mathbf{b} \triangleq [b_1, b_2, \dots, b_K]^T = \mathbf{S}^\dagger\mathbf{a}_t \in \mathbb{C}^K, \tag{14.19}$$

and

$$\mathbf{z} = \mathbf{X}\mathbf{b}/(\mathbf{b}^\dagger\mathbf{b})^{\frac{1}{2}} \in \mathbb{C}^N. \tag{14.20}$$

Then, we have

$$\mathbf{z} \sim \begin{cases} \mathcal{CN}_N(\mathbf{0}, \mathbf{R}), & \text{under } H_0, \\ \mathcal{CN}_N(a(\mathbf{b}^\dagger\mathbf{b})^{1/2}\mathbf{a}_r, \mathbf{R}), & \text{under } H_1. \end{cases} \tag{14.21}$$

Define

$$\bar{\mathbf{b}} = (\mathbf{b}^\dagger\mathbf{b})^{-\frac{1}{2}}\mathbf{b} \in \mathbb{C}^K. \tag{14.22}$$

Obviously, $\bar{\mathbf{b}}$ is a complex-valued unit vector of $K \times 1$ dimension. We can use it to construct a complete set of orthogonal basis vectors (denoted by $\{\bar{\mathbf{b}}, \mathbf{t}_1, \mathbf{t}_2, \dots, \mathbf{t}_{K-1}\}$) for the K-dimension complex-valued vector space. This is to say, there exists a matrix

$$\mathbf{T} \triangleq [\mathbf{t}_1, \mathbf{t}_2, \dots, \mathbf{t}_{K-1}] \in \mathbb{C}^{K \times (K-1)} \tag{14.23}$$

such that

$$\mathbf{T}^\dagger\bar{\mathbf{b}} = \mathbf{0}_{(K-1) \times 1}, \text{ and } \mathbf{T}^\dagger\mathbf{T} = \mathbf{I}_{K-1}. \tag{14.24}$$

Define

$$\mathbf{Z} \triangleq [\mathbf{z}_1, \mathbf{z}_2, \ldots, \mathbf{z}_{K-1}] = \mathbf{XT} \in \mathbb{C}^{N \times (K-1)}. \quad (14.25)$$

It is obvious that

$$\mathbf{z}_k = \sum_{n=1}^{K} t_{k,n} \mathbf{x}_n \in \mathbb{C}^N, \quad \text{for} \quad k = 1, 2, \ldots, K-1, \quad (14.26)$$

where $t_{k,n}$ is the nth element of the column vector $\mathbf{t}_k$. As a result,

$$
\begin{aligned}
\mathrm{E}(\mathbf{z}_k) &= \sum_{n=1}^{K} t_{k,n} \mathrm{E}(\mathbf{x}_n) = \sum_{n=1}^{K} t_{k,n} a \mathbf{a}_r b_n^* \\
&= a \mathbf{a}_r \sum_{n=1}^{K} t_{k,n} b_n^* = \mathbf{0}_{N \times 1},
\end{aligned}
\quad (14.27)
$$

$$\mathrm{E}(\mathbf{z}_k \mathbf{z}_k^\dagger) = \sum_{n=1}^{K} |t_{k,n}|^2 \mathrm{E}(\mathbf{x}_n \mathbf{x}_n^\dagger) = \mathbf{R}, \quad \text{for} \quad k = 1, 2, \ldots, K-1, \quad (14.28)$$

and

$$\mathrm{E}(\mathbf{z}_k \mathbf{z}_m^\dagger) = \sum_{n=1}^{K} t_{k,n} t_{m,n}^* \mathrm{E}(\mathbf{x}_n \mathbf{x}_n^\dagger) = \mathbf{0}, \quad \text{for} \quad k \neq m. \quad (14.29)$$

This is to say,

$$\mathbf{z}_k \sim \mathcal{CN}_N(\mathbf{0}, \mathbf{R}), \quad \text{for} \quad k = 1, 2, \ldots, K-1, \quad (14.30)$$

and they are independent of each other. In addition, the independence between $\mathbf{z}$ and $\mathbf{z}_k$ can be easily checked by the following equality:

$$
\begin{aligned}
\mathrm{E}(\mathbf{z} \mathbf{z}_k^\dagger) &= (\mathbf{b}^\dagger \mathbf{b})^{-1/2} \mathrm{E}\left(\sum_{m=1}^{K} b_m \mathbf{x}_m \sum_{n=1}^{K} t_{k,n}^* \mathbf{x}_n^\dagger \right) \\
&= (\mathbf{b}^\dagger \mathbf{b})^{-1/2} \sum_{m=1}^{K} b_m t_{k,m}^* \mathrm{E}\left(\mathbf{x}_m \mathbf{x}_m^\dagger \right) \\
&= \mathbf{0}_{N \times N}.
\end{aligned}
\quad (14.31)
$$

Due to the unknown parameters a and $\mathbf{R}$, the Neyman-Pearson criterion cannot be employed. According to the GLRT, a practical detector can be obtained by replacing all the unknown parameters with their MLEs, i.e., the detector is obtained by

$$\frac{\max_{\{a, \mathbf{R}\}} f(\mathbf{z}, \mathbf{Z}|H_1)}{\max_{\{\mathbf{R}\}} f(\mathbf{z}, \mathbf{Z}|H_0)} \underset{H_0}{\overset{H_1}{\gtrless}} \lambda_0, \quad (14.32)$$

where λ_0 is the detection threshold, $f(\cdot)$ denotes PDF. Due to the independence, the joint PDF of $\mathbf{z}$ and $\mathbf{Z}$ under H_q ($q = 0, 1$) can be represented as

$$f(\mathbf{z}, \mathbf{Z}|H_q) = \frac{1}{\pi^{NK} \det(\mathbf{R})^K} \exp\left[-\mathrm{tr}(\mathbf{R}^{-1} \mathbf{F}_q) \right], \quad q = 0, 1, \quad (14.33)$$

where

$$\mathbf{F}_q = \hat{\mathbf{R}} + \left[\mathbf{z} - qa(\mathbf{b}^\dagger \mathbf{b})^{1/2}\mathbf{a}_r\right]\left[\mathbf{z} - qa(\mathbf{b}^\dagger \mathbf{b})^{1/2}\mathbf{a}_r\right]^\dagger, \qquad (14.34)$$

with

$$\hat{\mathbf{R}} = \mathbf{ZZ}^\dagger \in \mathbb{C}^{N \times N}. \qquad (14.35)$$

Using (14.10), we have

$$\begin{aligned}\text{tr}(\mathbf{R}^{-1}\mathbf{F}_q) &= \text{tr}[\mathbf{J}(\mathbf{R}^*)^{-1}\mathbf{J}\mathbf{F}_q] \\ &= \text{tr}[(\mathbf{R}^*)^{-1}\mathbf{J}\mathbf{F}_q\mathbf{J}] \qquad (14.36) \\ &= \text{tr}[\mathbf{R}^{-1}\mathbf{J}\mathbf{F}_q^*\mathbf{J}], \qquad q = 0,1.\end{aligned}$$

As a result, (14.33) can be rewritten as

$$f(\mathbf{z}, \mathbf{Z}|H_q) = \frac{1}{\pi^{NK}\det(\mathbf{R})^K}\exp\left[-\text{tr}(\mathbf{R}^{-1}\mathbf{H}_q)\right], \quad q = 0,1, \qquad (14.37)$$

where

$$\mathbf{H}_q = \frac{1}{2}(\mathbf{F}_q + \mathbf{J}\mathbf{F}_q^*\mathbf{J}). \qquad (14.38)$$

Applying (14.34) into (14.38), after some algebra, yields

$$\mathbf{H}_q = \hat{\mathbf{R}}_p + \left[\mathbf{Y}_p - q(\mathbf{b}^\dagger \mathbf{b})^{1/2}\mathbf{a}_r\mathbf{a}_p^\dagger\right]\left[\mathbf{Y}_p - q(\mathbf{b}^\dagger \mathbf{b})^{1/2}\mathbf{a}_r\mathbf{a}_p^\dagger\right]^\dagger, \qquad (14.39)$$

where

$$\hat{\mathbf{R}}_p = \frac{1}{2}\left(\hat{\mathbf{R}} + \mathbf{J}\hat{\mathbf{R}}^*\mathbf{J}\right) \in \mathbb{C}^{N \times N}, \qquad (14.40)$$

$$\mathbf{Y}_p = [\mathbf{z}_e, \mathbf{z}_o] \in \mathbb{C}^{N \times 2}, \qquad (14.41)$$

and

$$\mathbf{a}_p = [a_e, a_o]^T \in \mathbb{C}^2, \qquad (14.42)$$

with

$$\mathbf{z}_e = \frac{1}{2}\left(\mathbf{z} + \mathbf{J}\mathbf{z}^*\right), \quad \mathbf{z}_o = \frac{1}{2}\left(\mathbf{z} - \mathbf{J}\mathbf{z}^*\right), \qquad (14.43)$$

and

$$a_e = \frac{1}{2}(a + a^*) = \mathfrak{Re}(a), \ a_o = \frac{1}{2}(a - a^*) = j\mathfrak{Im}(a). \qquad (14.44)$$

Similar to [28], we can obtain an adaptive detector as

$$\Lambda = \frac{\mathbf{a}_r^\dagger \hat{\mathbf{R}}_p^{-1}\mathbf{Y}_p\left(\mathbf{I}_2 + \mathbf{Y}_p^\dagger \hat{\mathbf{R}}_p^{-1}\mathbf{Y}_p\right)^{-1}\mathbf{Y}_p^\dagger \hat{\mathbf{R}}_p^{-1}\mathbf{a}_r}{\mathbf{a}_r^\dagger \hat{\mathbf{R}}_p^{-1}\mathbf{a}_r} \underset{H_0}{\overset{H_1}{\gtrless}} \lambda, \qquad (14.45)$$

where λ is the detection threshold. This detector is referred to as persymmetric GLRT in colocated MIMO radar (PGLRT-CMIMO). According to (14.40), the condition $K \geqslant \lceil\frac{N}{2} + 1\rceil$ is required to ensure with probability one the nonsingularity of the estimated disturbance covariance matrix.

14.1.3 Analytical Performance

In this section, we give a simple way to calculate the detection threshold λ of the PGLRT-CMIMO detector (14.45) for any given PFA. To this end, an analytical expression is derived for the PFA. Moreover, we also derive an analytical expression for the detection probability in order to facilitate the performance evaluation.

14.1.3.1 Transformation from Complex Domain to Real Domain

Define a unitary matrix

$$\mathbf{D} = \frac{1}{2}[(\mathbf{I}_N + \mathbf{J}) + \jmath\,(\mathbf{I}_N - \mathbf{J})] \in \mathbb{C}^{N \times N}. \tag{14.46}$$

Using (14.41), (14.43), and (14.46), we have

$$\mathbf{D}\mathbf{Y}_p = [\mathbf{y}_e, \jmath\mathbf{y}_o], \tag{14.47}$$

where

$$\mathbf{y}_e = \mathbf{D}\mathbf{z}_e = \frac{1}{2}\left[(\mathbf{I}_N + \mathbf{J})\Re(\mathbf{z}) - (\mathbf{I}_N - \mathbf{J})\Im(\mathbf{z})\right], \tag{14.48}$$

and

$$\mathbf{y}_o = -\jmath\mathbf{D}\mathbf{z}_o = \frac{1}{2}\left[(\mathbf{I}_N - \mathbf{J})\Re(\mathbf{z}) + (\mathbf{I}_N + \mathbf{J})\Im(\mathbf{z})\right]. \tag{14.49}$$

Note that $\mathbf{y}_e$ and $\mathbf{y}_o$ are N-dimensional real vectors.

Define another unitary matrix

$$\mathbf{V} = \begin{bmatrix} 1 & 0 \\ 0 & -\jmath \end{bmatrix} \in \mathbb{C}^{2 \times 2}. \tag{14.50}$$

We use the two unitary matrices $\mathbf{D}$ and $\mathbf{V}$ to transform all complex quantities in (14.45) into real ones, i.e.,

$$\hat{\mathbf{R}}_r = \mathbf{D}\hat{\mathbf{R}}_p\mathbf{D}^\dagger = \Re(\hat{\mathbf{R}}_p) + \mathbf{J}\Im(\hat{\mathbf{R}}_p) \in \mathbb{R}^{N \times N}, \tag{14.51}$$

$$\mathbf{p} = \mathbf{D}\mathbf{a}_r = \Re(\mathbf{a}_r) - \Im(\mathbf{a}_r) \in \mathbb{R}^N, \tag{14.52}$$

and

$$\mathbf{Y} = \mathbf{D}\mathbf{Y}_p\mathbf{V} = [\mathbf{y}_e, \mathbf{y}_o] \in \mathbb{R}^{N \times 2}. \tag{14.53}$$

Then, the detector (14.45) can be rewritten as

$$\Lambda = \frac{\mathbf{p}^\dagger\hat{\mathbf{R}}_r^{-1}\mathbf{Y}\left(\mathbf{I}_2 + \mathbf{Y}^\dagger\hat{\mathbf{R}}_r^{-1}\mathbf{Y}\right)^{-1}\mathbf{Y}^\dagger\hat{\mathbf{R}}_r^{-1}\mathbf{p}}{\mathbf{p}^\dagger\hat{\mathbf{R}}_r^{-1}\mathbf{p}} \underset{H_0}{\overset{H_1}{\gtrless}} \lambda. \tag{14.54}$$

It is worth pointing out that all quantities in (14.54) are real.

14.1.3.2 Statistical Properties

Similar to the derivations in [28], the detector in (14.54) can be equivalently expressed as

$$\frac{\nu}{\tau} \underset{H_0}{\overset{H_1}{\gtrless}} \eta, \tag{14.55}$$

where

$$\eta = \frac{\lambda}{1 - \lambda}, \tag{14.56}$$

$$\tau \sim \chi_{2K-N-1}^2, \tag{14.57}$$

and

$$\nu \sim \begin{cases} \chi_2^2, & \text{under } H_0, \\ \chi_2'^2(2\rho\delta), & \text{under } H_1, \end{cases} \tag{14.58}$$

with

$$\delta = |a|^2 a_r^\dagger R^{-1} a_r a_t^\dagger S S^\dagger a_t, \tag{14.59}$$

and

$$\rho \sim \beta\left(\frac{2K - N + 1}{2}, \frac{N - 1}{2}\right). \tag{14.60}$$

In addition, the random quantities ν and τ are independent of each other.

From (14.57) and (14.60), we obtain that the PDFs of τ and ρ are

$$f_\tau(\tau) = \frac{1}{2^{\frac{2K-N-1}{2}}\Gamma\left(\frac{2K-N-1}{2}\right)} \tau^{\frac{2K-N-3}{2}} \exp\left(-\frac{\tau}{2}\right), \quad \tau > 0, \tag{14.61}$$

and

$$f_\rho(\rho) = \frac{\Gamma(K)\rho^{\frac{2K-N-1}{2}}(1-\rho)^{\frac{N-3}{2}}}{\Gamma\left(\frac{2K-N+1}{2}\right)\Gamma\left(\frac{N-1}{2}\right)}, \quad 0 < \rho < 1, \tag{14.62}$$

respectively.

14.1.3.3 Detection Probability

The random variable ρ is fixed temporarily. We first derive the detection probability conditioned on ρ and then obtain the unconditional detection probability by averaging over the random variable ρ. For fixed ρ, the CCDF of ν under H_1 is [30, eq. (29.2)]

$$\begin{aligned} G(\nu) &= 1 - \exp(-\rho\delta) \sum_{j=0}^\infty \frac{(\rho\delta)^j}{2^{1+j}\Gamma(1+j)j!} \times \int_0^\nu y^j \exp\left(-\frac{y}{2}\right) dy \\ &= 1 - \exp(-\rho\delta) \sum_{j=0}^\infty \frac{(\rho\delta)^j}{\Gamma(j+1)j!} \gamma\left(1+j, \frac{\nu}{2}\right), \end{aligned} \tag{14.63}$$

where the second equality is obtained from [31, eq. (3.381.1)], and γ is the incomplete Gamma function defined in [31, eq. (8.350.1)]. So, the detection probability conditioned on ρ can be expressed as

$$
\begin{aligned}
P_{D|\rho} &= \int_0^{+\infty} G(\eta\tau) f_\tau(\tau) \mathrm{d}\tau \\
&= 1 - \sum_{j=0}^{\infty} \frac{(\rho\delta)^j \exp(-\rho\delta)}{2^{\frac{2K-N-1}{2}} \Gamma(1+j)\Gamma\left(\frac{2K-N-1}{2}\right) j!} \qquad (14.64) \\
&\quad \times \underbrace{\int_0^{\infty} \gamma\left(1+j,\frac{\eta\tau}{2}\right) \tau^{\frac{2K-N-3}{2}} \exp\left(-\frac{\tau}{2}\right) \mathrm{d}\tau}_{\triangleq W_1}.
\end{aligned}
$$

According to [31, eq. (6.455.2)], we have

$$
W_1 = \frac{\left(\frac{\eta}{2}\right)^{j+1} \Gamma\left(j + \frac{2K-N+1}{2}\right)}{(j+1)\left(\frac{1+\eta}{2}\right)^{j+\frac{2K-N+1}{2}}} {}_2F_1\left(1, j + \frac{2K-N+1}{2}; j+2; \frac{\eta}{1+\eta}\right).
$$

$$(14.65)$$

Further, the detection probability of the PGLRT-CMIMO detector is obtained by averaging over ρ, i.e.,

$$
\begin{aligned}
P_D &= \int_0^1 P_{D|\rho} f_\rho(\rho)\,\mathrm{d}\rho, \\
&= 1 - \sum_{j=0}^{\infty} \frac{\delta^j W_1 \Gamma(K)}{2^{\frac{2K-N-1}{2}} \Gamma(1+j)\Gamma\left(\frac{2K-N-1}{2}\right) j!\,\Gamma\left(\frac{2K-N+1}{2}\right)\Gamma\left(\frac{N-1}{2}\right)} \quad (14.66) \\
&\quad \times \underbrace{\int_0^1 \exp(-\rho\delta)\rho^{\frac{2K+2j-N-1}{2}}(1-\rho)^{\frac{N-3}{2}}\,\mathrm{d}\rho}_{\triangleq W_2}.
\end{aligned}
$$

Using [31, eq. (3.383.1)], we have

$$
W_2 = \frac{\Gamma\left(\frac{N-1}{2}\right)\Gamma\left(\frac{2K+2j-N+1}{2}\right)}{\Gamma(K+j)} {}_1F_1\left(\frac{2K+2j-N+1}{2}; K+j; -\delta\right).
$$

$$(14.67)$$

Substituting (14.65) and (14.67) into (14.66), we derive that the detection probability for the case of deterministic a is given by

$$
\begin{aligned}
P_{\mathrm{D}} =1 - & \frac{\Gamma\left(K\right)}{2^{\frac{2K-N-1}{2}}\Gamma\left(\frac{2K-N-1}{2}\right)\Gamma\left(\frac{2K-N+1}{2}\right)} \\
& \times \sum_{j=0}^{\infty} \frac{\delta^{j}\left(\frac{\eta}{2}\right)^{j+1}\Gamma\left(j+\frac{2K-N+1}{2}\right)\Gamma\left(j+\frac{2K-N+1}{2}\right)}{j!\,(j+1)!\left(\frac{1+\eta}{2}\right)^{j+\frac{2K-N+1}{2}}\Gamma\left(K+j\right)} \\
& \times {}_2F_1\left(1, \frac{2K+2j-N+1}{2}; j+2; \frac{\eta}{1+\eta}\right) \\
& \times {}_1F_1\left(\frac{2K+2j-N+1}{2}; K+j; -\delta\right),
\end{aligned}
\tag{14.68}
$$

where $\eta = \frac{\lambda}{1-\lambda}$, and δ is defined in (14.59).

When a random model is adopted for the target amplitude, i.e., $a \sim \mathcal{CN}_1(0, \sigma_a^2)$ where σ_a^2 is the power of target echo, we define

$$
\zeta = \sigma_a^2 \mathbf{a}_r^\dagger \mathbf{R}^{-1}\mathbf{a}_r \mathbf{a}_t^\dagger \mathbf{S}\mathbf{S}^\dagger \mathbf{a}_t.
\tag{14.69}
$$

The PDF of δ can be given by

$$
f_\delta(\delta) = \zeta^{-1}\exp(-\zeta^{-1}\delta), \quad \delta \geqslant 0.
\tag{14.70}
$$

The detection probability for the random case can be expressed as

$$
\begin{aligned}
\tilde{P}_{\mathrm{D}} = & \int_0^\infty P_{\mathrm{D}} f_\delta(\delta)\mathrm{d}\delta \\
=1 - & \frac{\Gamma\left(K\right)}{2^{\frac{2K-N-1}{2}}\Gamma\left(\frac{2K-N-1}{2}\right)\Gamma\left(\frac{2K-N+1}{2}\right)} \\
& \times \sum_{j=0}^{\infty} \frac{\zeta^{j}\left(\frac{\eta}{2}\right)^{j+1}\Gamma\left(j+\frac{2K-N+1}{2}\right)\Gamma\left(j+\frac{2K-N+1}{2}\right)}{(j+1)!\left(\frac{1+\eta}{2}\right)^{j+\frac{2K-N+1}{2}}\Gamma\left(K+j\right)} \\
& \times {}_2F_1\left(1, \frac{2K+2j-N+1}{2}; j+2; \frac{\eta}{1+\eta}\right) \\
& \times {}_2F_1\left(\frac{2K+2j-N+1}{2}, j+1; K+j; -\zeta\right),
\end{aligned}
\tag{14.71}
$$

where the second equality is obtained from [32].

It has to be emphasized that the analytical expressions (14.68) and (14.71) for the detection probability hold true for any positive integer N, whereas the result in [28] is exact only for odd positive integer N. It means that the analytical result for the detection probability obtained here is different from that in [28].

14.1.3.4 Probability of False Alarm

Under H_0, the PDF of ν is

$$f_{\nu|H_0}(\nu) = \frac{1}{2}\exp\left(-\frac{\nu}{2}\right), \quad \nu > 0. \tag{14.72}$$

According to (14.55), we can obtain the PFA as

$$\begin{aligned} P_{\text{FA}} &= \int_0^\infty \int_{\eta\tau}^\infty f_{\nu|H_0}(\nu)\mathrm{d}\nu f_\tau(\tau)\mathrm{d}\tau \\ &= (1+\eta)^{-\frac{2K-N-1}{2}} \\ &= (1-\lambda)^{\frac{2K-N-1}{2}}. \end{aligned} \tag{14.73}$$

It follows that the PGLRT-CMIMO detector exhibits the desirable CFAR property against the disturbance covariance matrix, since the PFA in (14.73) is irrelevant to $\mathbf{R}$. For a given PFA, the detection threshold can be calculated as

$$\lambda = 1 - P_{\text{FA}}^{\frac{2}{2K-N-1}}. \tag{14.74}$$

14.1.4 Numerical Examples

In this section, numerical simulations are provided to verify the above theoretical results. Assume that the MIMO radar receiver (or transmitter) uses a uniformly spaced linear array antennas consisting of 10 elements (i.e., $M = \tilde{N} = 10$). Without loss of generality, we consider the case where $P = 1$. In this case, we have $N = \tilde{N}P = 10$. Assume that the DOA of a target is $5°$. The (i,j)th element of the noise covariance matrix is chosen to be $\mathbf{R}(i,j) = \sigma^2 0.9^{|i-j|}$, where σ^2 is the disturbance power. Unless specifically stated, the PFA is set to be 10^{-4}, and the numbers of independent trials used for simulating the probabilities of false alarm and detection are $100/P_{\text{FA}}$ and 10^4, respectively.

The SNR in decibel is defined by

$$\text{SNR} = 10\log_{10}\frac{|a|^2}{\sigma^2}. \tag{14.75}$$

For comparison purposes, the Rao and Wald tests in [25] and the GLRT detector in [14] are considered, i.e.,

$$t_{\text{Rao}} = \frac{\left|\mathbf{a}_r^\dagger(\mathbf{XX}^\dagger)^{-1}\mathbf{XS}^\dagger\mathbf{a}_t\right|^2}{\mathbf{a}_r^\dagger(\mathbf{XX}^\dagger)^{-1}\mathbf{a}_r\mathbf{a}_t^\dagger\mathbf{SS}^\dagger\mathbf{a}_t}, \tag{14.76}$$

$$t_{\text{Wald}} = \frac{\left|\mathbf{a}_r^\dagger\left(\mathbf{XP}_{\mathbf{S}^\dagger\mathbf{a}_t}^\perp\mathbf{X}^\dagger\right)^{-1}\mathbf{XS}^\dagger\mathbf{a}_t\right|^2}{\mathbf{a}_r^\dagger\left(\mathbf{XP}_{\mathbf{S}^\dagger\mathbf{a}_t}^\perp\mathbf{X}^\dagger\right)^{-1}\mathbf{a}_r\mathbf{a}_t^\dagger\mathbf{SS}^\dagger\mathbf{a}_t}, \tag{14.77}$$

and

$$t_{\text{GLRT}} = 1 - \frac{\mathbf{a}_r^\dagger (\mathbf{X}\mathbf{X}^\dagger)^{-1}\mathbf{a}_r}{\mathbf{a}_r^\dagger \mathbf{Q}^{-1}\mathbf{a}_r}, \tag{14.78}$$

where

$$\mathbf{Q} = \mathbf{X}\mathbf{X}^\dagger - \frac{\mathbf{X}\mathbf{S}^\dagger \mathbf{a}_t \mathbf{a}_t^\dagger \mathbf{S}\mathbf{X}^\dagger}{\mathbf{a}_t^\dagger \mathbf{S}\mathbf{S}^\dagger \mathbf{a}_t}. \tag{14.79}$$

Note that (14.68) is an infinite-sum expression which cannot be directly used for calculation of the detection probability. In practice, we use the first finite terms to approximate the detection probability. Denote by $\tilde{P}_{\text{D}}(L)$ the detection probability including the first L terms in the sum, i.e.,

$$
\begin{aligned}
\tilde{P}_{\text{D}}(L) = 1 &- \frac{\Gamma(K)}{2^{\frac{2K-N-1}{2}}\Gamma\left(\frac{2K-N-1}{2}\right)\Gamma\left(\frac{2K-N+1}{2}\right)} \\
&\times \sum_{j=0}^{L} \frac{\delta^j \left(\frac{\eta}{2}\right)^{j+1}\Gamma\left(j+\frac{2K-N+1}{2}\right)\Gamma\left(j+\frac{2K-N+1}{2}\right)}{j!\,(j+1)!\left(\frac{1+\eta}{2}\right)^{j+\frac{2K-N+1}{2}}\Gamma(K+j)} \\
&\times {}_2F_1\left(1, \frac{2K+2j-N+1}{2}; j+2; \frac{\eta}{1+\eta}\right) \\
&\times {}_1F_1\left(\frac{2K+2j-N+1}{2}; K+j; -\delta\right).
\end{aligned}
\tag{14.80}
$$

The RE is defined as

$$\text{RE} = \frac{\left|\tilde{P}_{\text{D}}(L+1) - \tilde{P}_{\text{D}}(L)\right|}{\tilde{P}_{\text{D}}(L+1)}. \tag{14.81}$$

Fig. 14.1(a) depicts the RE as the increase of L, where $K = 2N$. It can be observed that the RE is much small when L is moderately large (e.g., more than 20 in this example). Fig. 14.1(b) plots the curve of $\tilde{P}_{\text{D}}(L)$ as a function of L. For comparison purposes, we also provide the detection probability obtained by MC simulations. We can observe that $\tilde{P}_{\text{D}}(L)$ approaches the MC result when the number of terms used is no less than 15 in the chosen parameter setting.

As observed in Fig. 14.1, $\tilde{P}_{\text{D}}(L)$ is very close to the true detection probability when the RE is less than 10^{-4}. As a rule of thumb, L is selected in practice as the one, for which the RE is less than 10^{-4}. Such a rule is adopted for the detection probability calculation in the following simulations. It should be pointed out that similar results can be obtained for the calculation of detection probability in the random target case. Hence, it is not repeated for brevity.

The curve of detection probability as a function of SNR is plotted for different K in Fig. 14.2. The dotted line denotes the results obtained by using (14.68), and the symbol "$\diamond$" denotes MC results. It can be seen that they

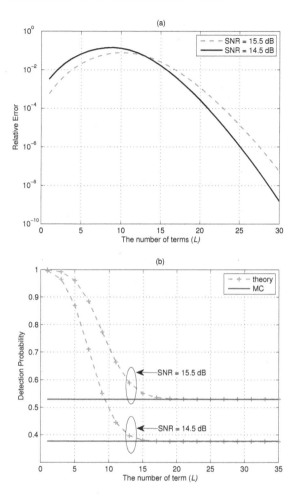

FIGURE 14.1
Approximation accuracy by using finite terms in the sum of (14.68) for the calculation of detection probability.

are in good agreement. As shown in Fig. 14.2, the PGLRT-CMIMO detector has better detection performance than the counterparts. This is because the PGLRT-CMIMO detector exploits the persymmetric structure of the covariance matrix. Comparing Fig. 14.2(a) with Fig. 14.2(b), we find out that the detection gain in the case of $K = 1.5N$ is higher than that in the case of $K = 2N$. It can be explained as follows. The data matrix $\mathbf{Z}$ can be seen as training data, and the column vectors in $\mathbf{Z}$ serve as the training data. The covariance matrix estimation accuracy is low for small K, and hence the GLRT, Rao, and Wald detectors exhibit poor performance. The improvement in the covariance matrix estimation accuracy is obvious when the persymmtric structure is exploited, especially in the case of small K. As a result, the

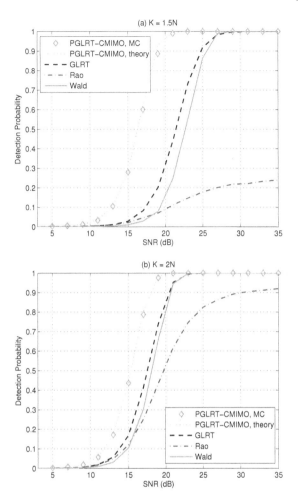

FIGURE 14.2
Detection probability versus SNR for the deterministic target model. (a) $K = 1.5N$; (b) $K = 2N$.

PGLRT-CMIMO detector has larger performance gain than the counterparts for small K.

In Fig. 14.3, we compare the detection performance in the random case where $a \sim \mathcal{CN}_1(0, \sigma_a^2)$. In such a case, the PGLRT-CMIMO detector still performs the best. The performance gain of the PGLRT-CMIMO detector with respect to the counterparts is higher when the parameter K becomes small. It means that our PGLRT-CMIMO detector is more suitable for the case of small K.

In the above simulations, we assume $K \geqslant N + 1$. Under this condition, the classic adaptive detectors (i.e., the GLRT in [14], Rao and Wald tests in [25]) can work. Now we examine the case where $N + 1 > K \geqslant \lceil \frac{N}{2} + 1 \rceil$.

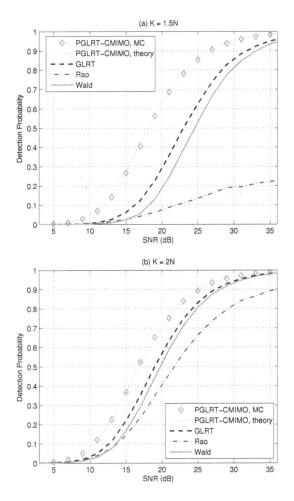

FIGURE 14.3
Detection probability versus SNR for the random target model. (a) $K = 1.5N$;
(b) $K = 2N$.

Fig. 14.4 shows the detection probability of the PGLRT-CMIMO detector as a function of SNR. Here, we choose $K = 10$ and $M = N = 10$. Notice that for this small K, the classic adaptive detectors no longer work. Therefore, we do not provide their performance for comparisons. We can observe from Fig. 14.4 that for high SNRs, the detection probability in the deterministic target model is higher than that in the random target model. However, for low SNRs the detection performance in the deterministic case is slightly worse than that in the random case. This is because the amplitude fluctuation can result in a gain in the detection performance for a low SNR, but lead to a loss in the detection performance for a high SNR. The similar phenomenon can be seen in [33].

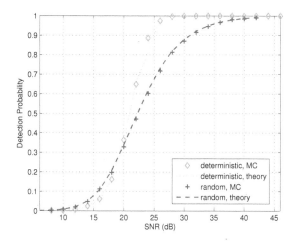

FIGURE 14.4
Detection probability versus SNR in the case of small $K = 10$.

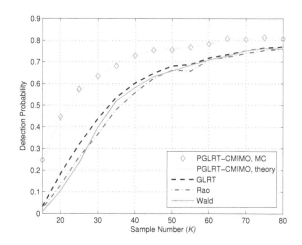

FIGURE 14.5
Detection probability versus K with SNR $= 15$ dB.

14.2 Persymmetric Detection in Distributed MIMO Radar

14.2.1 Signal Model

In this section, a signal model for MIMO radar with widely distributed antennas is presented. Suppose that a MIMO radar consists of M transmit antennas and N receive antennas which are geographically dispersed. The total number of transmit-receive paths available is $V = MN$. Assume that the tth transmitter sends Q_t pulses and a target to be detected does not leave the CUT during these pulses. We further impose the standard assumption that all transmit waveforms are orthogonal to each other, and each receiver uses a bank of M matched filters corresponding to the M orthogonal waveforms.

Sampled at the pulse rate via slow-time sampling, the signal received by the rth receive antenna due to the transmission from the tth transmit antenna, which is usually called test data (primary data), can be expressed as a $Q_t \times 1$ vector, i.e.,

$$\mathbf{x}_{r,t} = a_{r,t}\, \mathbf{s}_{r,t} + \mathbf{n}_{r,t}, \tag{14.82}$$

where $\mathbf{s}_{r,t} \in \mathbb{C}^{Q_t \times 1}$ denotes a known $Q_t \times 1$ steering vector for the target relative to the tth transmitter and rth receiver pair [20, 34]; $a_{r,t} \in \mathbb{C}$ is a deterministic but unknown complex scalar accounting for the target reflectivity and the channel propagation effects in the tth transmitter and rth receiver pair; the noise $\mathbf{n}_{r,t} \sim \mathcal{CN}_{Q_t}(\mathbf{0}, \mathbf{R}_{r,t})$, where $\mathbf{R}_{r,t}$ is a positive definite covariance matrix of dimension $Q_t \times Q_t$.

These primary data vectors $\{\mathbf{x}_{r,t}\}$ can be assumed independent of each other due to the widely distributed antennas in the MIMO radar. Notice that in the above model (14.82), these steering vectors $\mathbf{s}_{r,t}$'s are not necessarily identical even though they describe the same target, since the relative position and velocity of the target with respect to different widely dispersed radars may be distinct. In addition, the covariance matrices $\mathbf{R}_{r,t}$'s are also not constrained to be the same, because the statistical properties of the noise may be unique for each transmit-receive perspective. We further assume that $Q_t > 1, t = 1, 2, \ldots, M$, such that coherent processing for each test data is possible. Note that $Q_t, t = 1, 2, \ldots, M$ are not constrained to be identical, which means that the numbers of the pulses transmitted by different transmit antennas may be distinct. Another standard assumption we impose is that for each test data vector $\mathbf{x}_{r,t}$, there exists a set of training data (secondary data) free of target signal components, i.e., $\{\mathbf{y}_{r,t}(k), k = 1, 2, \ldots, K_{r,t} | \mathbf{y}_{r,t}(k) \sim \mathcal{CN}_{Q_t}(\mathbf{0}, \mathbf{R}_{r,t})\}$. Here, we require $K_{r,t} \geqslant Q_t/2$ to guarantee a nonsingular covariance matrix estimate with unit probability [28]. Note that the numbers of secondary data vectors $K_{r,t}$ are not constrained to be the same. Suppose further that these secondary data vectors are independent of each other and of the primary data vectors.

The detection problem considered herein involves structured $\mathbf{R}_{r,t}$ and $\mathbf{s}_{r,t}$. Specifically, it is supposed that each of the $\mathbf{R}_{r,t}$'s has the persymmetric property, i.e., $\mathbf{R}_{r,t} = \mathbf{J}\mathbf{R}_{r,t}^*\mathbf{J}$ where $\mathbf{J}$ is a permutation matrix defined in (1.2). In addition, the steering vector is also assumed to be a persymmetric one satisfying $\mathbf{s}_{r,t} = \mathbf{J}\mathbf{s}_{r,t}^*$. The above assumption on the structures of $\mathbf{R}_{r,t}$ and $\mathbf{s}_{r,t}$ is valid when each antenna in the MIMO radar uses a pulse train symmetrically spaced with respect to its mid time delay for temporal domain processing. In the common case of pulse trains with uniform spacing, the steering vector $\mathbf{s}_{r,t}$

has the form:

$$
\mathbf{s}_{r,t} = \left[e^{-\jmath \frac{(Q_t-1)2\pi \bar{f}_{r,t}}{2}}, \ldots, e^{-\jmath 2\pi \bar{f}_{r,t}}, 1, e^{\jmath 2\pi \bar{f}_{r,t}}, \ldots, e^{\jmath \frac{(Q_t-1)2\pi \bar{f}_{r,t}}{2}} \right]^T, \quad (14.83)
$$

where $\bar{f}_{r,t}$ defined similarly as [34, eq. (2)] is the normalized target Doppler shift corresponding to the tth transmitter and rth receiver pair.

The detection problem is to decide between the null hypothesis and the alternative one:

$$
H_0 : \begin{cases} \mathbf{x}_{r,t} \sim \mathcal{CN}_{Q_t}(\mathbf{0}, \mathbf{R}_{r,t}) \\ \mathbf{y}_{r,t}(k) \sim \mathcal{CN}_{Q_t}(\mathbf{0}, \mathbf{R}_{r,t}) \end{cases} \quad (14.84a)
$$

and

$$
H_1 : \begin{cases} \mathbf{x}_{r,t} \sim \mathcal{CN}_{Q_t}(a_{r,t}\,\mathbf{s}_{r,t}, \mathbf{R}_{r,t}) \\ \mathbf{y}_{r,t}(k) \sim \mathcal{CN}_{Q_t}(\mathbf{0}, \mathbf{R}_{r,t}) \end{cases} \quad (14.84b)
$$

for $t = 1, 2, \ldots, M$, $r = 1, 2, \ldots, N$, and $k = 1, 2, \ldots, K_{r,t}$. In [20, 35, 36], MIMO detection algorithms were developed without using any prior knowledge about the special structure of the noise covariance matrix. In the sequel, two adaptive detectors are designed by exploiting the persymmetric structures of $\mathbf{R}_{r,t}$ and $\mathbf{s}_{r,t}$. It will be seen that the exploitation of persymmetry can bring in a noticeable detection gain.

14.2.2   Persymmetric GLRT Detector

In this section, we derive two adaptive detectors exploiting the persymmetric structures for the detection problem (14.84).

14.2.2.1   Detector Design

Due to the unknown parameters $a_{r,t}$ and $\mathbf{R}_{r,t}$, the Neyman-Pearson criterion cannot be employed. According to the GLRT, a practical detector can be obtained by replacing all the unknown parameters with their MLEs, i.e., the detector is obtained by

$$
\frac{\max_{\{a_{r,t}, \mathbf{R}_{r,t} \mid t=1,2,\ldots,M,\ r=1,2,\ldots,N\}} f\left(\mathbf{X}|H_1\right)}{\max_{\{\mathbf{R}_{r,t} \mid t=1,2,\ldots,M,\ r=1,2,\ldots,N\}} f\left(\mathbf{X}|H_0\right)} \begin{array}{c} H_1 \\ \gtrless \\ H_0 \end{array} \lambda_0, \quad (14.85)
$$

where λ_0 is the detection threshold, $f(\cdot)$ denotes PDF, and $\mathbf{X} = \{\mathbf{X}_{1,1}, \ldots, \mathbf{X}_{N,M}\}$ with

$$
\mathbf{X}_{r,t} = [\mathbf{x}_{r,t}, \mathbf{y}_{r,t}(1), \mathbf{y}_{r,t}(2), \ldots, \mathbf{y}_{r,t}(K_{r,t})]. \quad (14.86)
$$

Due to the independent assumption on these test data vectors, the PDF of $\mathbf{X}$ under H_q ($q = 0, 1$) can be represented as

$$
f(\mathbf{X}|H_q) = \prod_{r=1}^{N} \prod_{t=1}^{M} \underbrace{f_{\mathbf{X}_{r,t}}(\mathbf{X}_{r,t}|H_q)}_{\triangleq f_{r,t,q}}, \quad q = 0, 1, \quad (14.87)
$$

where

$$f_{r,t,q} = \left\{ \frac{1}{\pi^{Q_t} \det(\mathbf{R}_{r,t})} \exp\left[-\operatorname{tr}(\mathbf{R}_{r,t}^{-1} \mathbf{T}_{r,t,q}) \right] \right\}^{K_{r,t}+1} \tag{14.88}$$

with

$$\mathbf{T}_{r,t,q} = \frac{1}{K_{r,t}+1} \left[\sum_{k=1}^{K_{r,t}} \mathbf{y}_{r,t}(k) \mathbf{y}_{r,t}^{\dagger}(k) \right. \\ \left. + (\mathbf{x}_{r,t} - q a_{r,t}\, \mathbf{s}_{r,t})(\mathbf{x}_{r,t} - q a_{r,t}\, \mathbf{s}_{r,t})^{\dagger} \right]. \tag{14.89}$$

Define

$$\hat{\mathbf{R}}_{r,t} = \frac{1}{2} \sum_{k=1}^{K_{r,t}} \left\{ \mathbf{y}_{r,t}(k) \mathbf{y}_{r,t}^{\dagger}(k) + \mathbf{J}[\mathbf{y}_{r,t}(k) \mathbf{y}_{r,t}^{\dagger}(k)]^{*} \mathbf{J} \right\}, \tag{14.90}$$

$$\mathbf{x}_{r,t}^{e} = \frac{1}{2} \left[(\mathbf{I} + \mathbf{J})\Re(\mathbf{x}_{r,t}) - (\mathbf{I} - \mathbf{J})\Im(\mathbf{x}_{r,t}) \right], \tag{14.91}$$

$$\mathbf{x}_{r,t}^{o} = \frac{1}{2} \left[(\mathbf{I} - \mathbf{J})\Re(\mathbf{x}_{r,t}) + (\mathbf{I} + \mathbf{J})\Im(\mathbf{x}_{r,t}) \right], \tag{14.92}$$

and

$$\tilde{\mathbf{X}}_{r,t} = [\mathbf{x}_{r,t}^{e}, \mathbf{x}_{r,t}^{o}]. \tag{14.93}$$

As derived in Appendix 14.A, the detector is given by

$$\prod_{r=1}^{N} \prod_{t=1}^{M} \left(\frac{1}{1 - \Phi_{r,t}} \right)^{K_{r,t}+1} \underset{H_0}{\overset{H_1}{\gtrless}} \lambda_0, \tag{14.94}$$

where

$$\Phi_{r,t} = \frac{\tilde{\mathbf{s}}_{r,t}^{\dagger} \tilde{\mathbf{R}}_{r,t}^{-1} \tilde{\mathbf{X}}_{r,t} (\mathbf{I}_2 + \tilde{\mathbf{X}}_{r,t}^{\dagger} \tilde{\mathbf{R}}_{r,t}^{-1} \tilde{\mathbf{X}}_{r,t})^{-1} \tilde{\mathbf{X}}_{r,t}^{\dagger} \tilde{\mathbf{R}}_{r,t}^{-1} \tilde{\mathbf{s}}_{r,t}}{\tilde{\mathbf{s}}_{r,t}^{\dagger} \tilde{\mathbf{R}}_{r,t}^{-1} \tilde{\mathbf{s}}_{r,t}} \tag{14.95}$$

with

$$\tilde{\mathbf{s}}_{r,t} = \Re(\mathbf{s}_{r,t}) - \Im(\mathbf{s}_{r,t}), \tag{14.96}$$

and

$$\tilde{\mathbf{R}}_{r,t} = \Re(\hat{\mathbf{R}}_{r,t}) + \mathbf{J}\Im(\hat{\mathbf{R}}_{r,t}). \tag{14.97}$$

Here, (14.94) is referred to as persymmetric GLRT in distributed MIMO radar (PGLRT-DMIMO).

14.2.2.2   Performance Analysis

In order to complete the construction of the test in (14.94), we should provide an approach to set the detection threshold. In this section, we derive a closed-form expression for the PFA of the PGLRT-DMIMO detector, which can be employed to compute the detection threshold for any given PFA. In doing so, we take the logarithm of (14.94), namely,

$$\Lambda = \sum_{r=1}^{N} \sum_{t=1}^{M} (K_{r,t} + 1) \ln \left(\frac{1}{1 - \Phi_{r,t}} \right) \underset{H_0}{\overset{H_1}{\gtrless}} \lambda, \tag{14.98}$$

where $\lambda = \ln \lambda_0$.

Define

$$\alpha_{r,t} = \frac{2K_{r,t} + 2}{2K_{r,t} - Q_t + 1}. \tag{14.99}$$

We relabel $\alpha_{1,1}, \ldots, \alpha_{1,M}, \ldots, \alpha_{N,1}, \ldots, \alpha_{N,M}$ as $\alpha_1, \alpha_2, \ldots, \alpha_V$, respectively, where $V = MN$. As shown in Appendix 14.B, (14.98) has a statistically equivalent form as follows:

$$\Lambda = \sum_{i=1}^{V} \alpha_i \Omega_i \underset{H_0}{\overset{H_1}{\gtrless}} \lambda, \tag{14.100}$$

where the PDF of Ω_i under hypothesis H_0 is the standard exponential distribution as in (14.B.6), and the random variables Ω_i are independent of one another. It is worth noting that the test statistic Λ under H_0 is exactly a sum of weighted exponential variables Ω_i.

Recall that some of Q_t's may be identical, and so are some of $K_{r,t}$'s. Thus, some of α_i's may be equal. We denote these coefficients with the same value by a new symbol, i.e., $\alpha_1, \alpha_2, \ldots, \alpha_V$ are denoted by $e_1, e_2, \ldots, e_s$ ($e_n \neq e_m$ for $n \neq m$), where s is the total number of the coefficients with different values, and e_i corresponds to some coefficient whose value is repeated $d_i + 1$ times among $\alpha_1, \alpha_2, \ldots, \alpha_V$. In addition, $d_i \geqslant 0, i = 1, 2, \ldots, s$ and $s + \sum_{i=1}^{s} d_i = V$.

According to Theorem 3 of [37], the PFA of the PGLRT-DMIMO detector can be expressed as

$$P_{\text{FA}} = \left(\prod_{i=1}^{s} d_i! \right)^{-1} \frac{\partial^{d_1 + d_2 + \ldots + d_s}}{\partial e_1^{d_1} \partial e_2^{d_2} \ldots \partial e_s^{d_s}} \left[\sum_{n=1}^{s} \frac{J_V(e_n, \lambda)}{E_n} \right], \tag{14.101}$$

where the operator $\frac{\partial^{d_1 + d_2 + \ldots + d_s}}{\partial e_1^{d_1} \partial e_2^{d_2} \ldots \partial e_s^{d_s}} (\cdot)$ denotes the mixed $(d_1 + d_2 + \ldots + d_s)$th order partial derivatives of a function with respect to $e_1, e_2, \ldots, e_s$,

$$J_L(x, T) = x^L \exp(-Tx^{-1}), \tag{14.102}$$

and

$$E_n = e_n \prod_{j=1, \, j \neq n}^{s} (e_n - e_j). \tag{14.103}$$

Note that for $n = 1$, we have $E_1 = e_1$.

In particular, the expression (14.101) bears a simple form for the following two cases.

Case I: $\alpha_1, \alpha_2, \ldots, \alpha_V$ are all the same. At this moment, $s = 1$, $E_1 = e_1 = \alpha_1$, and $d_1 = V - 1$. Then, we have

$$\sum_{n=1}^{s} \frac{J_V(e_n, \lambda)}{E_n} = J_{V-1}(e_1, \lambda). \tag{14.104}$$

Thus, the PFA for this case is

$$
\begin{aligned}
P_{\text{FA}} &= \frac{1}{(V-1)!} \frac{\partial^{V-1}}{\partial e_1^{V-1}} [J_{V-1}(e_1, \lambda)] \\
&= \frac{\exp(-\lambda e_1^{-1})}{(V-1)!} \sum_{n=0}^{V-1} C_{V-1}^n (V-1-n)! \lambda^n e_1^{-n} \\
&= \exp(-\lambda e_1^{-1}) \sum_{n=0}^{V-1} \frac{1}{n!} (\lambda e_1^{-1})^n,
\end{aligned}
\tag{14.105}
$$

where the second equality is obtained with the Lemma in [38].

Case II: $\alpha_1, \alpha_2, \ldots, \alpha_V$ are all distinct. In this case, $s = V$ and $d_k = 0$, $e_k = \alpha_k$ for $k = 1, 2, \ldots, V$. Therefore, the PFA for this case can be simplified as

$$P_{\text{FA}} = \sum_{n=1}^{V} \frac{J_V(e_n, \lambda)}{E_n} = \sum_{n=1}^{V} \frac{e_n^V \exp(-\lambda e_n^{-1})}{E_n}. \tag{14.106}$$

For the general case where parts of α_i's are identical, we can use (14.101) to obtain the false alarm rate of the PGLRT-DMIMO detector. As an example, we consider the case in which $V = 4$, $\alpha_2 = \alpha_3$, but α_1, α_2, and α_4 are all different. In this case, the expression for the PFA of the PGLRT-DMIMO detector can be represented by

$$
\begin{aligned}
P_{\text{FA}} &= \frac{1}{0!1!0!} \frac{\partial}{\partial e_2} \left[\sum_{j=1}^{3} \frac{J_4(e_j, \lambda)}{E_j} \right] \\
&= \frac{e_1^3 \exp(-\lambda e_1^{-1})}{(e_1 - e_2)^2 (e_1 - e_3)} + \frac{e_2(3e_2 + \lambda) \exp(-\lambda e_2^{-1})}{(e_2 - e_1)(e_2 - e_3)} \\
&\quad - \frac{e_2^3(2e_2 - e_1 - e_3) \exp(-\lambda e_2^{-1})}{(e_2 - e_1)^2 (e_2 - e_3)^2} \\
&\quad + \frac{e_3^3 \exp(-\lambda e_3^{-1})}{(e_3 - e_1)(e_3 - e_2)^2}.
\end{aligned}
\tag{14.107}
$$

It is obvious that the PFA of the PGLRT-DMIMO detector does not depend on the noise covariance matrix. Therefore, the PGLRT-DMIMO detector exhibits the desirable CFAR property against the noise covariance matrices.

It can be seen from the above derivation that the closed-form expression for the PFA of the PGLRT-DMIMO detector is obtained with the fact that under H_0, the individual test statistic $\Phi_{r,t}$ in the PGLRT-DMIMO detector has a simple right-tail probability as (14.B.2), and thus has a simple PDF as (14.B.4). However, the right-tail probability of $\Phi_{r,t}$ under H_1 contains a one-dimensional integral (see [28, eq. (16)]). Therefore, the detection probability of the PGLRT-DMIMO detector is analytically intractable.

14.2.3   Persymmetric SMI Detector

In this section, an alternative solution to the detection problem (14.84) is designed, which has a lower computational burden than the PGLRT-DMIMO detector. To this end, an approach similar to that in [39] is employed. More specifically, for each transmit-receive pair we use the following test statistic:

$$\Xi_{r,t} = |\mathbf{w}_{r,t}^\dagger \mathbf{x}_{r,t}^e|^2 + |\mathbf{w}_{r,t}^\dagger \mathbf{x}_{r,t}^o|^2, \tag{14.108}$$

where $\mathbf{x}_{r,t}^e$ and $\mathbf{x}_{r,t}^o$ are defined in (14.91) and (14.92), respectively, and the weight vector $\mathbf{w}_{r,t}$ is given by

$$\mathbf{w}_{r,t} = \frac{\tilde{\mathbf{R}}_{r,t}^{-1}\tilde{\mathbf{s}}_{r,t}}{(\tilde{\mathbf{s}}_{r,t}^\dagger \tilde{\mathbf{R}}_{r,t}^{-1}\tilde{\mathbf{s}}_{r,t})^{1/2}}. \tag{14.109}$$

Jointly processing the independent data received by all the transmit-receive pairs, we have the following decision rule:

$$\Xi = \sum_{r=1}^N \sum_{t=1}^M \Xi_{r,t} = \sum_{r=1}^N \sum_{t=1}^M \frac{\tilde{\mathbf{s}}_{r,t}^\dagger \tilde{\mathbf{R}}_{r,t}^{-1}\tilde{\mathbf{X}}_{r,t}\tilde{\mathbf{X}}_{r,t}^\dagger \tilde{\mathbf{R}}_{r,t}^{-1}\tilde{\mathbf{s}}_{r,t}}{\tilde{\mathbf{s}}_{r,t}^\dagger \tilde{\mathbf{R}}_{r,t}^{-1}\tilde{\mathbf{s}}_{r,t}} \underset{H_0}{\overset{H_1}{\gtrless}} \xi, \tag{14.110}$$

where ξ is the detection threshold. Here, (14.110) is referred to as the persymmetric SMI in distributed MIMO radar (PSMI-DMIMO).

Note that Ξ in (14.110) is different from the test statistic in [39, eq. (13)], since the weight vectors $\mathbf{w}_{r,t}$ in (14.109) are distinct for different transmit-receive pairs, whereas the weight vectors in [39, eq. (13)] are all the same. Compared with the PGLRT-DMIMO detector in (14.94), the PSMI-DMIMO detector in (14.110) has a simpler structure. More specifically, (14.110) does not need to compute the term $(\mathbf{I}_2 + \tilde{\mathbf{X}}_{r,t}^\dagger \tilde{\mathbf{R}}_{r,t}^{-1}\tilde{\mathbf{X}}_{r,t})^{-1}$, and thus is computationally more efficient.

As to the theoretical performance of the PSMI-DMIMO detector, unfortunately, closed-form expressions for the probabilities of false alarm and detection are both intractable, since the right-tail probability of its constituent test statistic $\Xi_{r,t}$ under H_0 or H_1 includes one-dimensional integral (see [39, eqs. (14) and (15)]). Nevertheless, we can observe from [39, eq. (14)] that the statistical property of the test statistic $\Xi_{r,t}$ under H_0 is irrelevant to the noise covariance matrix. As expected, the PSMI-DMIMO detector consisting of $\Xi_{r,t}$'s possesses the CFAR property with respect to all the noise covariance matrices.

14.2.4   Simulations Results

In this section, numerical simulations are conducted to validate the above theoretical analysis and illustrate the performance of the two designed detectors. For simplicity, we consider a MIMO radar comprised of two transmit antennas and two receive antennas (i.e., $M = N = 2$), and each transmit antenna sends nine coherent pulses with equal spacing (i.e., $Q_1 = Q_2 = 9$). The steering vector $\mathbf{s}_{r,t}$ has the form as (14.83). Suppose further that the normalized Doppler shifts of the target are 0.1, 0.2, 0.3, and 0.4 for the (1,1), (1,2), (2,1), and (2,2) transmit-receive pairs, respectively. The (i,j)th element of the noise covariance matrix is chosen to be $\mathbf{R}(i,j) = \sigma^2 0.95^{|i-j|}$, where σ^2 represents the noise power. Note that the covariance matrices for all transmit-receive pairs are set to be identical for simplicity. Nevertheless, we estimate the covariance matrix for a specific transmit-receive pair by using only the training data collected in that corresponding transmit-receive pair, instead of using all training data from all transmit-receive pairs. Without loss of generality, $a_{r,t}$'s are supposed to be the same and then can be uniformly denoted by a. The SNR is defined by

$$\text{SNR} = 10 \log_{10} \frac{|a|^2}{\sigma^2}. \tag{14.111}$$

We select $K_{1,1} = K_{1,2} = K_{2,1} = K_{2,2} = 10$, i.e., α_i's defined in (14.99) are all identical.

The detection probability curves versus SNR are plotted with MC simulations in Fig. 14.6, where the PFA is set to be 10^{-4}. For comparison purposes, the GLRT-DMIMO detector in [36] and the AMF-DMIMO detector (19) in [20], both of which do not utilize a-priori knowledge about the persymmetric structure of the noise covariance matrix, are considered. The number of independent trials used to obtain each value of the detection probability is 5 000.

Comparing the detection performance of the same detector, we can see that the more the number of the secondary data, the better the performance. This is because the use of more secondary data samples can improve the accuracy in estimating the noise covariance matrix, and hence lead to a gain in detection performance.

It can also be observed in Fig. 14.6 that the PGLRT-DMIMO detector performs the best, the PSMI-DMIMO detector has slightly inferior performance, and the GLRT-DMIMO or AMF-DMIMO detector achieves the worst performance. Obviously, the PSMI-DMIMO and PGLRT-DMIMO detectors outperform the GLRT-DMIMO and AMF-DMIMO detectors due to the exploitation of *a priori* knowledge about the persymmetric structures in the received signals. In this parameter setting, the PSMI-DMIMO detector performs worse than the PGLRT-DMIMO detector at the benefit of lower computational burden.

Nevertheless, these four detectors considered here perform almost similarly when the number of secondary data is sufficient. This can be seen in

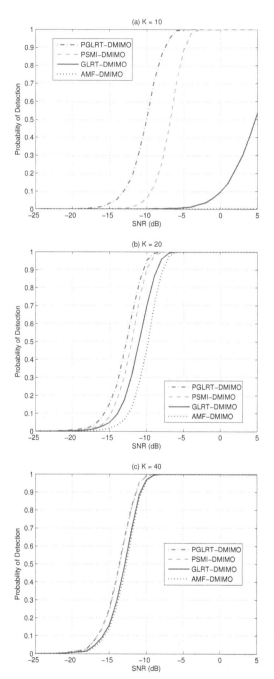

FIGURE 14.6

Performance comparisons with $P_{\text{FA}} = 10^{-4}$.

Fig. 14.6(c). This is due to the fact that using sufficient secondary data, one can obtain a high accuracy in the noise covariance matrix estimate, even without exploiting the persymmetric structures. Hence, when sufficient secondary data are available, the PSMI-DMIMO detector is highly recommended because of its relatively low computational burden and negligible performance loss.

14.A Derivation of (14.94)

Since the random variables $\mathbf{X}_{r,t}$ are independent, the maximization of the left-hand side of (14.85) can be performed term by term. Using (14.87), we can rewrite (14.85) as

$$\prod_{r=1}^{N}\prod_{t=1}^{M}\underbrace{\frac{\max_{\{a_{r,t},\mathbf{R}_{r,t}\}}f_{r,t,1}}{\max_{\{\mathbf{R}_{r,t}\}}f_{r,t,0}}}_{\triangleq\Upsilon_{r,t}} \underset{H_0}{\overset{H_1}{\gtrless}} \lambda_0. \tag{14.A.1}$$

In the following, we simplify $\Upsilon_{r,t}$ by using the approach in the case of a single band in [28]. Exploiting the persymmetric structure of $\mathbf{R}_{r,t}$, we have

$$\mathrm{tr}(\mathbf{R}_{r,t}^{-1}\mathbf{T}_{r,t,q}) = \mathrm{tr}(\mathbf{R}_{r,t}^{-1}\mathbf{J}\mathbf{T}_{r,t,q}^{*}\mathbf{J}). \tag{14.A.2}$$

Then,

$$\mathrm{tr}(\mathbf{R}_{r,t}^{-1}\mathbf{T}_{r,t,q}) = \mathrm{tr}(\mathbf{R}_{r,t}^{-1}\hat{\mathbf{T}}_{r,t,q}), \tag{14.A.3}$$

where

$$\hat{\mathbf{T}}_{r,t,q} = (\mathbf{T}_{r,t,q} + \mathbf{J}\mathbf{T}_{r,t,q}^{*}\mathbf{J})/2. \tag{14.A.4}$$

Using (14.89), $\hat{\mathbf{T}}_{r,t,q}$ can be rewritten as

$$\hat{\mathbf{T}}_{r,t,q} = \frac{\hat{\mathbf{R}}_{r,t} + (\hat{\mathbf{X}}_{r,t} - q\,\mathbf{s}_{r,t}\hat{\mathbf{a}}_{r,t})(\hat{\mathbf{X}}_{r,t} - q\,\mathbf{s}_{r,t}\hat{\mathbf{a}}_{r,t})^{\dagger}}{K_{r,t}+1}, \tag{14.A.5}$$

where

$$\hat{\mathbf{a}}_{r,t} = [\mathfrak{Re}(a_{r,t}),\, \jmath\,\mathfrak{Im}(a_{r,t})], \tag{14.A.6}$$

$$\hat{\mathbf{R}}_{r,t} = \frac{1}{2}\sum_{k=1}^{K_{r,t}}\left\{\mathbf{y}_{r,t}(k)\mathbf{y}_{r,t}^{\dagger}(k) + \mathbf{J}[\mathbf{y}_{r,t}(k)\mathbf{y}_{r,t}^{\dagger}(k)]^{*}\mathbf{J}\right\}, \tag{14.A.7}$$

and

$$\hat{\mathbf{X}}_{r,t} = [\hat{\mathbf{x}}_{r,t}^{\mathrm{e}}, \hat{\mathbf{x}}_{r,t}^{\mathrm{o}}] \tag{14.A.8}$$

with

$$\hat{\mathbf{x}}_{r,t}^{\mathrm{e}} = (\mathbf{x}_{r,t} + \mathbf{J}\mathbf{x}_{r,t}^{*})/2 \tag{14.A.9}$$

and

$$\hat{\mathbf{x}}_{r,t}^{\text{o}} = (\mathbf{x}_{r,t} - \mathbf{J}\mathbf{x}_{r,t}^{*})/2. \tag{14.A.10}$$

According to [40], the MLEs of $\mathbf{R}_{r,t}$ under H_0 and H_1 are $\hat{\mathbf{T}}_{r,t,0}$ and $\hat{\mathbf{T}}_{r,t,1}$, respectively. Using these MLEs, we can write $\Upsilon_{r,t}$ defined in (14.A.1) as

$$\Upsilon_{r,t} = \left[\frac{\det(\hat{\mathbf{T}}_{r,t,0})}{\min_{\{\hat{\mathbf{a}}_{r,t}\}} \det(\hat{\mathbf{T}}_{r,t,1})} \right]^{K_{r,t}+1}. \tag{14.A.11}$$

It follows from (14.A.5) that

$$\det(\hat{\mathbf{T}}_{r,t,1}) = \frac{1}{(K_{r,t}+1)^{Q_t}} \det(\hat{\mathbf{R}}_{r,t})$$
$$\times \det[\mathbf{I}_2 + (\hat{\mathbf{X}}_{r,t} - \mathbf{s}_{r,t}\hat{\mathbf{a}}_{r,t})^{\dagger}\hat{\mathbf{R}}_{r,t}^{-1}(\hat{\mathbf{X}}_{r,t} - \mathbf{s}_{r,t}\hat{\mathbf{a}}_{r,t})].$$

It is straightforward to show that the value of $\hat{\mathbf{a}}_{r,t}$ minimizing the denominator of (14.A.11) is $(\mathbf{s}_{r,t}^{\dagger}\hat{\mathbf{R}}_{r,t}^{-1}\mathbf{s}_{r,t})^{-1}\mathbf{s}_{r,t}^{\dagger}\hat{\mathbf{R}}_{r,t}^{-1}\hat{\mathbf{X}}_{r,t}$. Substituting this MLE into (14.A.11) and after some algebraic manipulations, we can obtain

$$\Upsilon_{r,t} = \left(\frac{1}{1 - \Phi_{r,t}} \right)^{K_{r,t}+1}, \tag{14.A.12}$$

where

$$\Phi_{r,t} = \frac{\mathbf{s}_{r,t}^{\dagger}\hat{\mathbf{R}}_{r,t}^{-1}\hat{\mathbf{X}}_{r,t}(\mathbf{I}_2 + \hat{\mathbf{X}}_{r,t}^{\dagger}\hat{\mathbf{R}}_{r,t}^{-1}\hat{\mathbf{X}}_{r,t})^{-1}\hat{\mathbf{X}}_{r,t}^{\dagger}\hat{\mathbf{R}}_{r,t}^{-1}\mathbf{s}_{r,t}}{\mathbf{s}_{r,t}^{\dagger}\hat{\mathbf{R}}_{r,t}^{-1}\mathbf{s}_{r,t}}. \tag{14.A.13}$$

Define two unitary matrices

$$\mathbf{D} = \frac{1}{2}[(\mathbf{I} + \mathbf{J}) + \jmath\,(\mathbf{I} - \mathbf{J})], \tag{14.A.14}$$

and

$$\mathbf{V} = \begin{bmatrix} 1 & 0 \\ 0 & -\jmath \end{bmatrix}. \tag{14.A.15}$$

Then, $\Phi_{r,t}$ in (14.A.13) can be rewritten in the real domain as

$$\Phi_{r,t} = \frac{\tilde{\mathbf{s}}_{r,t}^{\dagger}\tilde{\mathbf{R}}_{r,t}^{-1}\tilde{\mathbf{X}}_{r,t}(\mathbf{I}_2 + \tilde{\mathbf{X}}_{r,t}^{\dagger}\tilde{\mathbf{R}}_{r,t}^{-1}\tilde{\mathbf{X}}_{r,t})^{-1}\tilde{\mathbf{X}}_{r,t}^{\dagger}\tilde{\mathbf{R}}_{r,t}^{-1}\tilde{\mathbf{s}}_{r,t}}{\tilde{\mathbf{s}}_{r,t}^{\dagger}\tilde{\mathbf{R}}_{r,t}^{-1}\tilde{\mathbf{s}}_{r,t}}, \tag{14.A.16}$$

where

$$\tilde{\mathbf{X}}_{r,t} = \mathbf{D}\hat{\mathbf{X}}_{r,t}\mathbf{V}^{\dagger}, \tag{14.A.17}$$

$$\tilde{\mathbf{R}}_{r,t} = \mathbf{D}\hat{\mathbf{R}}_{r,t}\mathbf{D}^{\dagger} = \mathfrak{Re}(\hat{\mathbf{R}}_{r,t}) + \mathbf{J}\mathfrak{Im}(\hat{\mathbf{R}}_{r,t}), \tag{14.A.18}$$

and

$$\tilde{\mathbf{s}}_{r,t} = \mathbf{D}\mathbf{s}_{r,t} = \mathfrak{Re}(\mathbf{s}_{r,t}) - \mathfrak{Im}(\mathbf{s}_{r,t}). \tag{14.A.19}$$

Furthermore, $\tilde{\mathbf{X}}_{r,t}$ can be expressed as (14.93).

14.B Equivalent Transformation of Λ

First, we examine the possible range where the test statistic $\Phi_{r,t}$ can take values. As derived in [28, eq. (B29)], the test statistic $\Phi_{r,t}$ can be transformed into

$$\Phi_{r,t} = 1 - [1 + \psi_{AA}\mathbf{W}(\mathbf{I}+\Sigma_B)^{-1}\mathbf{W}^{\dagger}]^{-1}, \qquad (14.\text{B}.1)$$

where the quantity $\psi_{AA}\mathbf{W}(\mathbf{I}+\Sigma_B)^{-1}\mathbf{W}^{\dagger}$ defined in [28] is a positive number. It is obvious that $0 < \Phi_{r,t} < 1$. Notice that the right-tail probability of the test statistic $\Phi_{r,t}$ is equal to its PFA. According to [28, eq. (15)], the right-tail probability of $\Phi_{r,t}$ under hypothesis H_0 is

$$Pr(\Phi_{r,t} > \phi_{r,t}|H_0) = (1-\phi_{r,t})^{\frac{2K_{r,t}-Q_t+1}{2}}, \quad 0 < \phi_{r,t} < 1. \qquad (14.\text{B}.2)$$

Then, the CDF of $\Phi_{r,t}$ under hypothesis H_0 is

$$Pr(\Phi_{r,t} < \phi_{r,t}|H_0) = 1 - (1-\phi_{r,t})^{\frac{2K_{r,t}-Q_t+1}{2}}. \qquad (14.\text{B}.3)$$

Therefore, the PDF of $\Phi_{r,t}$ under hypothesis H_0 can be obtained by taking the derivative of the CDF with respect to $\phi_{r,t}$, namely,

$$f_{\Phi_{r,t}}(\phi_{r,t}) = \frac{2K_{r,t}-Q_t+1}{2}(1-\phi_{r,t})^{\frac{2K_{r,t}-Q_t-1}{2}}. \qquad (14.\text{B}.4)$$

Define a monotone transform

$$\Omega_{r,t} = \frac{2K_{r,t}-Q_t+1}{2}\ln\frac{1}{1-\Phi_{r,t}}, \qquad (14.\text{B}.5)$$

we can then obtain the PDF of $\Omega_{r,t}$ under hypothesis H_0

$$f_{\Omega_{r,t}}(w_{r,t}) = \exp(-w_{r,t}), \quad w_{r,t} > 0. \qquad (14.\text{B}.6)$$

Using (14.B.5), (14.98) can be written as

$$\Lambda = \sum_{r=1}^{N}\sum_{t=1}^{M}\alpha_{r,t}\Omega_{r,t} \underset{H_0}{\overset{H_1}{\gtrless}} \lambda, \qquad (14.\text{B}.7)$$

where

$$\alpha_{r,t} = (2K_{r,t}+2)/(2K_{r,t}-Q_t+1). \qquad (14.\text{B}.8)$$

For ease of notation, $\alpha_{1,1},\ldots,\alpha_{1,M},\ldots\alpha_{N,1},\ldots,\alpha_{N,M}$ are relabeled as $\alpha_1,\alpha_2,\ldots,\alpha_V$, respectively, where $V = MN$. Similarly, $\Omega_{1,1},\ldots,\Omega_{N,M}$ are relabeled as $\Omega_1,\Omega_2,\ldots,\Omega_V$, respectively. Then, (14.B.7) can be rewritten as

$$\Lambda = \sum_{i=1}^{V}\alpha_i\Omega_i \underset{H_0}{\overset{H_1}{\gtrless}} \lambda, \qquad (14.\text{B}.9)$$

where the random variable Ω_i under H_0 has the standard exponential distribution as in (14.B.6).

Bibliography

[1] J. Li and P. Stoica, *MIMO Radar Signal Processing*. Hoboken, NJ: John Wiley & Sons, 2009.

[2] B. Tang, M. M. Naghsh, and J. Tang, "Relative entropy-based waveform design for MIMO radar detection in the presence of clutter and interference," *IEEE Transactions on Signal Processing*, vol. 63, no. 14, pp. 3783–3796, July 2015.

[3] X. Li, D. Wang, X. Ma, and W.-Q. Wang, "FDS-MIMO radar low-altitude beam coverage performance analysis and optimization," *IEEE Transactions on Signal Processing*, vol. 66, no. 9, pp. 2494–2506, May 2018.

[4] B. Tang, Y. Zhang, and J. Tang, "An efficient minorization maximization approach for MIMO radar waveform optimization via relative entropy," *IEEE Transactions on Signal Processing*, vol. 66, no. 2, pp. 400–411, January 15 2018.

[5] K. Gao, W.-Q. Wang, and J. Cai, "Frequency diverse array and MIMO hybrid radar transmitter design via Cramer-Rao lower bound minimisation," *IET Radar, Sonar & Navigation*, vol. 10, no. 9, pp. 1660–1670, 2016.

[6] J. Li and P. Stoica, "MIMO radar with colocated antennas," *IEEE Signal Processing Magazine*, vol. 24, no. 5, pp. 106–114, September 2007.

[7] A. Aubry, A. D. Maio, and Y. Huang, "MIMO radar beampattern design via PSL/ISL optimization," *IEEE Transactions on Signal Processing*, vol. 65, no. 14, pp. 3955–3967, August 1 2016.

[8] G. Cui, X. Yu, V. Carotenuto, and L. Kong, "Space-time transmit code and receive filter design for colocated MIMO radar," *IEEE Transactions on Signal Processing*, vol. 65, no. 5, pp. 1116–1129, March 2017.

[9] J. Xu, G. Liao, Y. Zhang, H. Ji, and L. Huang, "An adaptive range-angle-Doppler processing approach for FDA-MIMO radar using three-dimensional localization," *IEEE Journal of Selected Topics in Signal Processing*, vol. 11, no. 2, pp. 309–320, March 2017.

[10] J. Liu, W. Liu, J. Han, B. Tang, Y. Zhao, and H. Yang, "Persymmetric GLRT detection in MIMO radar," *IEEE Transactions on Vehicular Technology*, vol. 67, no. 12, pp. 11 913–11 923, December 2018.

[11] H. Li and B. Himed, "Transmit subaperturing for MIMO radars with colocated antennas," *IEEE Journal of Selected Topics in Signal Processing*, vol. 4, no. 1, pp. 55–65, February 2010.

[12] G. Cui, H. Li, and M. Rangaswamy, "MIMO radar waveform design with constant modulus and similarity constraints," *IEEE Transactions on Signal Processing*, vol. 62, no. 2, pp. 343–353, January 2014.

[13] I. Bekkerman and J. Tabrikian, "Target detection and localization using MIMO radars and sonars," *IEEE Transactions on Signal Processing*, vol. 54, no. 10, pp. 3873–3883, October 2006.

[14] L. Xu, J. Li, and P. Stoica, "Target detection and parameter estimation for MIMO radar systems," *IEEE Transactions on Aerospace and Electronic Systems*, vol. 44, no. 3, pp. 927–939, July 2008.

[15] B. Tang and J. Tang, "Joint design of transmit waveforms and receive filters for MIMO radar space-time adaptive processing," *IEEE Transactions on Signal Processing*, vol. 64, no. 18, pp. 4707– 4722, September 2016.

[16] Q. He and R. S. Blum, "Diversity gain for MIMO Neyman-Pearson signal detection," *IEEE Transactions on Signal Processing*, vol. 59, no. 3, pp. 869–881, March 2011.

[17] N. Li, G. Cui, H. Yang, L. Kong, and Q. H. Liu, "Adaptive detection of moving target with MIMO radar in heterogeneous environments based on Rao and Wald tests," *Signal Processing*, vol. 114, 2015, pp. 198–208.

[18] S. Zhou, H. Liu, Y. Zhao, and L. Hu, "Target spatial and frequency scattering diversity property for diversity MIMO radar," *Signal Processing*, vol. 91, no. 2, pp. 269–276, February 2011.

[19] J. Liu, H. Li, and B. Himed, "Persymmetric adaptive target detection with distributed MIMO radar," *IEEE Transactions on Aerospace and Electronic Systems*, vol. 51, no. 1, pp. 372–382, January 2015.

[20] Q. He, N. H. Lehmann, R. S. Blum, and A. M. Haimovich, "MIMO radar moving target detection in homogeneous clutter," *IEEE Transactions on Aerospace and Electronic Systems*, vol. 46, no. 3, pp. 1290–1301, July 2010.

[21] P. Wang, H. Li, and B. Himed, "A parametric moving target detector for distributed MIMO radar in non-homogeneous environment," *IEEE Transactions on Signal Processing*, vol. 61, no. 9, pp. 1351–1356, April 2013.

[22] Q. He, R. S. Blum, and A. M. Haimovich, "Noncoherent MIMO radar for location and velocity estimation: More antennas means better performance," *IEEE Transactions on Signal Processing*, vol. 58, no. 7, pp. 3661–3680, July 2010.

[23] J. Li, L. Xu, P. Stoica, K. W. Forsythe, and D. W. Bliss, "Range compression and waveform optimization for MIMO radar: A Cramer-Rao bound based study," *IEEE Transactions on Signal Processing*, vol. 56, no. 1, pp. 218–232, January 2008.

[24] L. Xu and J. Li, "Iterative generalized-likelihood ratio test for MIMO radar," *IEEE Transactions on Signal Processing*, vol. 55, no. 6, pp. 2375–2385, June 2007.

[25] W. Liu, Y. Wang, J. Liu, and W. Xie, "Adaptive detection without training data in colocated MIMO radar," *IEEE Transactions on Aerospace and Electronic Systems*, vol. 51, no. 3, pp. 2469–2479, July 2015.

[26] J. Liu, S. Zhou, W. Liu, J. Zheng, H. Liu, and J. Li, "Tunable adaptive detection in colocated MIMO radar," *IEEE Transactions on Signal Processing*, vol. 66, no. 4, pp. 1080–1092, February 2018.

[27] J. Li, P. Stoica, L. Xu, and W. Roberts, "On parameter identifiability of MIMO radar," *IEEE Signal Processing Letters*, vol. 14, no. 12, pp. 968–971, December 2007.

[28] L. Cai and H. Wang, "A persymmetric multiband GLR algorithm," *IEEE Transactions on Aerospace and Electronic Systems*, vol. 28, no. 3, pp. 806–816, July 1992.

[29] H. Yang, Y. Wang, W. Xie, and Y. Di, "Persymmetric adaptive target detection without training data in collocated MIMO radar," in *CIE International Conference on Radar (RADAR)*, Guangzhou, China 2016, pp. 1–4.

[30] N. L. Johnson, S. Kotz, and N. Balakrishnan, *Continuous Univariate Distributions (Volumn 2)*, 2nd ed. Hoboken, New Jersey, John Wiley & Sons, Inc., 1995.

[31] I. S. Gradshteyn and I. M. Ryzhik, *Table of Integrals, Series, and Products*, 7th ed. San Diego: Academic Press, 2007.

[32] "The Wolfram function site." [Online]. Available: http://functions.wolfram.com/07.20.21.0012.01

[33] G. Cui, A. De Maio, and M. Piezzo, "Performance prediction of the incoherent radar detector for correlated generalized Swerling-Chi fluctuating targets," *IEEE Transactions on Aerospace and Electronic Systems*, vol. 49, no. 1, pp. 356–368, January 2013.

[34] P. Wang, H. Li, and B. Himed, "Knowledge-aided parametric tests for multichannel adaptive signal detection," *IEEE Transactions on Signal Processing*, vol. 59, no. 12, pp. 5970–5982, December 2011.

[35] A. Sheikhi and A. Zamani, "Temporal coherent adaptive target detection for multi-input multi-output radars in clutter," *IET Radar, Sonar & Navigation*, vol. 2, no. 2, pp. 86–96, April 2008.

[36] J. Liu, Z.-J. Zhang, Y. Cao, and S. Yang, "A closed-form expression for false alarm rate of adaptive MIMO-GLRT detector with distributed MIMO radar," *Signal Processing*, vol. 93, no. 9, pp. 2771–2776, September 2013.

[37] M. M. Ali and M. Obaidullah, "Distribution of linear combination of exponential variates," *Communications in Statistics-Theory and Methods*, vol. 11, no. 13, pp. 1453–1463, 1982.

[38] J. Liu, Z.-J. Zhang, P.-L. Shui, and H. Liu, "Exact performance analysis of an adaptive subspace detector," *IEEE Transactions on Signal Processing*, vol. 60, no. 9, pp. 4945–4950, September 2012.

[39] L. Cai and H. Wang, "A persymmetric modified-SMI algorithm," *Signal Processing*, vol. 23, no. 1, pp. 27–34, January 1991.

[40] R. Nitzberg, "Application of maximum likelihood estimation of persymmetric covariance matrices to adaptive processing," *IEEE Transactions on Aerospace and Electronic Systems*, vol. AES-16, no. 1, pp. 124–127, January 1980.